This series publishes the work of the Ernst Strüngmann Forum, an independent academic initiative dedicated to advancing knowledge in basic science.

The Ernst Strüngmann Forum provides a collaborative environment for invited experts to address contemporary research challenges. Guided by an international advisory board, topics chosen are of high-priority interest to various areas of research, requiring multidisciplinary input to expand understanding of the issues. Past topics include digital ethology, emergent brain dynamics, deliberate ignorance, psychiatric genetics, agrobiodiversity, and computational psychiatry. The insights and future research strategies that emerge from a topic are disseminated through the Strüngmann Forum Report series.

Established in 2006, the Ernst Strüngmann Forum has gained a reputation as a place where intellectual "dead ends" are overcome, where novel ways of conceptualizing issues are articulated, and where new collaborations are established. For further information, see https://esforum.de

This is an Open Access book series.

Strüngmann Forum Reports

Series Editor

Julia R. Lupp (iD), Ernst Strüngmann Forum, Frankfurt Institute for Advanced Studies, Frankfurt, Germany

L. Zachary DuBois • Anelis Kaiser Trujillo
Margaret M. McCarthy
Editors

Sex and Gender

Toward Transforming Scientific Practice

 Springer

Editors
L. Zachary DuBois
Department of Anthropology
University of Oregon
Eugene, OR, USA

Anelis Kaiser Trujillo
Center for Gender Studies
University of Basel
Basel, Switzerland

Margaret M. McCarthy
Department of Pharmacology
University of Maryland, Baltimore
Baltimore, MD, USA

ISSN 3059-3786 ISSN 3059-3794 (electronic)
Strüngmann Forum Reports
ISBN 978-3-031-91370-9 ISBN 978-3-031-91371-6 (eBook)
https://doi.org/10.1007/978-3-031-91371-6

This work was supported by the Ernst Strüngmann Foundation.

This Springer imprint is published by the registered company Springer Nature Switzerland AG
The registered company address is: Gewerbestrasse 11, 6330 Cham, Switzerland

If disposing of this product, please recycle the paper.

This book is dedicated to the many people who have contributed to our understanding of sex and gender, as well as to those currently persevering amidst incomprehensible hurdles to continue their invaluable work.

Preface

Science is a highly specialized enterprise—one that enables areas of enquiry to be minutely pursued, establishes working paradigms and normative standards, and supports rigor in experimental research. Yet all too often problems are encountered that fall outside the scope of any one discipline, and to make progress, new perspectives are needed to expand conceptualization, increase understanding, and chart new research directions.

The Ernst Strüngmann Forum was established in 2006 to address these types of topics. Founded on the tenets of scientific independence and the inquisitive nature of the human mind, the Forum provides a platform that fosters rigorous cross-disciplinary analysis. Our meetings take the form of intensive intellectual retreats: existing perspectives are questioned, knowledge gaps are identified, and strategies are collectively sought to bridge these gaps. The resulting insights are disseminated through this publication series.

This volume presents the outcomes of an extended discourse that examined *sex* and *gender*—concepts that are understood, measured, and applied in diverse ways across scientific disciplines and within society—and explored the complex ways in which they are entangled. The need for this project was brought to our attention by a member of our scientific advisory board, Amber Wutich, who recommended that L. Zachary DuBois take the lead on developing a proposal. After an extensive developmental process (for an overview, see Chapter 1), the proposal was accepted. From October 27–29, 2022, the Program Advisory Committee met to transform the proposal into a framework that would support an extended, multidisciplinary discussion. Committee members (L. Zachary DuBois, Anelis Kaiser Trujillo, Julia R. Lupp, Margaret M. McCarthy, Stacey A. Ritz, Rebecca M. Shansky and Paula-Irene Villa) worked together to delineate discussion topics, identify potential participants, and establish the overarching goal for this project: to advance conceptualizations of gender and sex, to align dialogue across disciplines, and to promote sound application in research, policy, medicine, and public health.

Central to the philosophy of the Ernst Strüngmann Forum is the principle that consensus is not the goal. Understanding where perspectives diverge and exploring the underlying reasons are critical components in the Forum's process. Accordingly, the committee sought to engage open-minded individuals who would not only challenge perspectives from others areas of expertise but also critically reflect on their own.

The committee put together the following areas for examination:

1. Entanglement of gender/sex dynamics in basic and developmental systems biology
   - How might the concept of "entanglement" of gender and sex transform research?
   - What might be lost if gender and sex are viewed as entangled?
   - How do gender and sex vary across the life course?
   - What do we know about gender and sex during critical periods of development (pre- and perinatal, juvenile/childhood, adolescence, adulthood, and senescence)?

- How can we better operationalize sex/gender entanglement across the life course?

2. Entanglement of gender/sex dynamics and issues of operationalization and measurement

- What do the categories of binary sex (male/female) and gender (man/woman) enable us to achieve? What harms and biases can result?
- What does the gender/sex entanglement mean for nonhuman animal and *in vitro* studies?
- Can a framework be devised to guide how entanglement can be effectively incorporated into research design?
- How can we guide and make transparent decisions regarding when to use proxy categorical variables of gender/sex or continuous variables, targeting instead pathways and mechanisms?
- How can we get beyond biological essentialism (e.g., centering sex assigned at birth) and individualized gender fixation (e.g., centering self-reported gender identity only) in research?
- How do analytic methods produce differences between groups?

3. Entanglement of gender/sex dynamics in human biomedical and clinical research

- What do the categories of binary sex (male/female) and gender (man/woman) enable us to achieve? What harms and biases can result?
- How can gender, gender identity, and gender experience be better integrated with systems biology-based understandings of physiology, epigenetics, clinical research, and public health practice?
- How can gender identity (as distinct from "gender"), including the dynamism of gender identity within individuals as well as across historical and cultural contexts, be better integrated into health research?

4. Entanglement of gender/sex dynamics in policy and practice

- What happens when biological definitions of sex that exclude entanglement with gender are used in social policies?
- Where is scientific clarification of sex-linked biologies, gendered behaviors, and health outcomes most needed?
- What alternative concepts/frameworks might more effectively account for human variation and diversity?
- How can the complexities of gender/sex entanglement be leveraged to foster positive change?

Then, from October 8–13, 2023, researchers from anthropology, behavioral neuroendocrinology, cellular and molecular neuroscience, clinical psychology, epidemiology, feminist studies, genetics, psychiatry, sex, gender and transgender studies gathered in Frankfurt to engage in a most lively debate. This volume synthesizes the ideas and perspectives that emerged throughout this project.

An endeavor of this kind, especially one that brings together disciplines that do not typically interact, creates unique group dynamics and places demands on everyone. I wish to thank each person who participated in this Forum for their time, efforts, and constructive engagement. A special word of gratitude goes to the Program

Advisory Committee as well as to the authors and reviewers of the background papers. In addition, I wish to recognize the efforts of the moderators of the discussion groups—Catherine Woolley, Stacey Ritz, Robert-Paul Juster, and Amber Wutich—and the rapporteurs—Colin Saldanha, Donna Maney, Tonia Poteat, and Alexandra Brewis. To facilitate lively debate and transform these discussions into a coherent document is no simple task, and their work toward this end was invaluable. Finally, I would like to extend my appreciation to Zachary DuBois, Anelis Kaiser Trujillo, and Peg McCarthy, whose leadership and editorial efforts were instrumental in ensuring the successful completion of this volume.

The Ernst Strüngmann Forum carries out its work in the service of science and society thanks to the generous support of the Ernst Strüngmann Foundation, established by Dr. Andreas and Dr. Thomas Strüngmann in honor of their father. I also wish to acknowledge the contributions of our scientific advisory board, as well as our valued partnership with the Ernst Strüngmann Institute, which shared its vibrant setting with us during the Forum.

The expansion of knowledge is an ongoing process of critical scrutiny and continual reassessment. Central to this process must be a willingness to question and, when necessary, to reconsider or set aside long-held viewpoints. While this can be challenging, once the first step is taken, the act of moving forward can be both intellectually stimulating and deeply rewarding. In this spirit, we hope this book will inspire further discourse and provide a foundation for future enquiry and discovery.

Julia R. Lupp, Director, Ernst Strüngmann Forum
Frankfurt Institute for Advanced Studies
Ruth-Moufang-Str. 1, 60438 Frankfurt am Main, Germany
https://esforum.de/

Contents

List of Contributors

AuBuchon, Katarina E. Lombardi Comprehensive Cancer Center, Georgetown University, Washington DC 20057, USA

Bauer, Greta Eli Coleman Institute for Sexual and Gender Health, University of Minnesota Medical School, Minneapolis, MN 55455, USA

Bentley, Gillian R. Dept. of Anthropology, Durham University, Durham DH1 3LE, UK

Bowleg, Lisa Dept. of Psychological and Brain Sciences, The George Washington University, Washington, DC 20052, USA, and The Intersectionality Training Institute, Philadelphia, PA 19119, USA

Brewis, Alexandra School of Human Evolution and Social Change, Arizona State University, Tempe, AZ 85287-2402, USA

Christiansen, Dorte M. The National Center of Psychotraumatology, Dept. of Psychology, University of Southern Denmark, 5230 Odense M, Denmark

Ciccia, Lu Gender in Science, Technology, and Innovation, Center for Research and Gender Studies, Universidad Nacional Autónoma de México, Ciudad de México, Ciudad Universitaria 04510, México

Cornil, Charlotte A. Laboratory of Neuroendocrinology, GIGA Neurosciences, University of Liège, 4000 Liège, Belgium, Brussels

Currah, Paisley Dept. of Political Science, Brooklyn College, City University of New York, Brooklyn, NY 11210, USA

de Vries, Geert J. Dept. of Biology, Georgia State University, Atlanta, GA 30303, USA

DuBois, L. Zachary Dept. of Anthropology, University of Oregon, Eugene, OR 97401, USA

Duchesne, Annie Dept. of Psychology, University of Northern British Columbia, Prince George, BC VN2 4Z9, Canada, and Dept. of Psychology, University du Québec in Trois Rivières, Trois-Rivières, QC G9A 5H7, Canada

Dunsworth, Holly Dept. of Sociology and Anthropology, University of Rhode Island, Kingston, RI 02881, USA

Greaves, Lorraine Centre of Excellence for Women's Health, BC Women's Hospital + Health Centre, Vancouver, BC V6H 3N1, Canada, and School of Population and Public Health, Faculty of Medicine, University of British Columbia, Vancouver, BC V6T 1Z3, Canada

Hoppe, Katharina Institute of Sociology, Goethe-University Frankfurt, 60323 Frankfurt am Main, Germany

Juster, Robert-Paul Dept. of Psychiatry and Addiction, University of Montreal, Montreal, QC H3T 1J4, Canada

Kaiser Trujillo, Anelis   Center for Gender Studies, University of Basel, 4051 Basel, Switzerland

Karkazis, Katrina   SWAGS (Sexuality, Women, and Gender Studies), Amherst College, Amherst, MA 01060, USA

Maney, Donna L.   Dept. of Psychology, Emory University, Atlanta, GA 30322, USA

Malekzadeh, Arianne N.   Dept. of Psychological and Brain Sciences, The George Washington University, Washington, DC 20052, USA

McCarthy, Margaret M.   Dept. of Pharmacology, University of Maryland School of Medicine, Baltimore, MD 21230, USA

Pape, Madeleine   Institute of Social Sciences, University of Lausanne, 1015 Lausanne, Switzerland

Poteat, Tonia   School of Nursing and Co-Director of the Duke SGM Health Program, Duke University, Durham, NC 27708, USA

Ritz, Stacey A.   Dept. of Pathology and Molecular Medicine, McMaster University, Hamilton, ON L8S 4L8, Canada

Rubin, Joshua B.   Depts. of Pediatrics and Neuroscience, Washington University School of Medicine, St. Louis, MO 63110, USA

Saldanha, Colin J.   Dept. of Neuroscience, American University, Washington, DC 20016, USA

Sanchis-Segura, Carla   Dept. Psicologia Bàsica, Clínica i Psicobiología, Universitat Jaume I, 12071 AP Castellón de la Plana, Spain

Schweiger, Susann   Institute of Human Genetics, University Medical Center Mainz, 55131 Mainz, Germany

Shansky, Rebecca M.   Dept. of Psychology, Northeastern University, Boston, MA 02115, USA

Sievert, Lynnette Leidy   Dept. of Anthropology, University of Massachusetts Amherst, Amherst, MA 01003, USA

Vilain, Eric   Institute for Clinical and Translational Science, University of California, Irvine, Irvine, CA 92617, USA

Villa, Paula-Irene   Dept. of Sociology, Ludwig Maximilian University of Munich, 80801 Munich, Germany

Ware, Libby   Dept. of Anthropology, The George Washington University, Washington, DC 20052, USA

Woolley, Catherine S.   Dept. of Neurobiology, Northwestern University, Evanston, IL 60208, USA

Wong, Jason   Dept. of Pediatrics, Washington University School of Medicine, St. Louis, MO 63110, USA

Wutich, Amber   School of Human Evolution and Social Change, Arizona State University, Tempe, AZ 85281, USA

Yang, Wei   Dept. of Genetics, Washington University School of Medicine, St. Louis, MO 63110, USA

1

Sex and Gender

Toward Transforming Scientific Practice

L. Zachary DuBois, Stacey A. Ritz, Margaret M. McCarthy, and Anelis Kaiser Trujillo

Abstract   From October 8–13, 2023, a highly diverse group of scholars gathered at the 36th Ernst Strüngmann Forum in Frankfurt am Main, Germany, to advance conceptualizations of gender and sex, to align dialogue across disciplines, and to promote sound application in research, policy, medicine, and public health. The ensuing interdisciplinary discussions clearly revealed that no single approach is capable of conceptualizing or operationalizing sex/gender entanglement, yet they also revealed a high degree of convergence in perspectives. This chapter provides background to the discourse, introduces key topics discussed in this volume, and suggests a path forward for future research to consider.

Keywords   Gender, sex, entanglement, binary notions of sex, intersectionality, transgender health equity, neuroendocrinology, policy, public health

L. Zachary DuBois (✉)
Dept. of Anthropology, University of Oregon, Eugene, OR 97401, USA
Email: zdubois@uoregon.edu

Stacey A. Ritz
Dept. of Pathology and Molecular Medicine, McMaster University,
Hamilton, ON L8S 4L8, Canada

Margaret M. McCarthy
Dept. of Pharmacology, University of Maryland School of Medicine,
Baltimore, MD 21230, USA

Anelis Kaiser Trujillo
Center for Gender Studies, University of Basel,
4051 Basel, Switzerland

© Frankfurt Institute for Advanced Studies (FIAS) 2025
L. Z. DuBois et al. (eds.), *Sex and Gender*, Strüngmann Forum Reports,
https://doi.org/10.1007/978-3-031-91371-6_1

1 Introduction

Across most human societies, concepts of sex and gender are significant elements that shape social structures; indeed, the terms *sex* and *gender* are two of the most contested and scrutinized today, both in science and in society more broadly. Despite their ubiquitous use, the terms are often understood, applied, and measured differently in different social and scientific contexts, and there are no universally agreed upon definitions of either sex or gender that would fit all conditions for all time. For researchers, the issue goes beyond needing to define a universally applicable set of terms, concepts, and measures or to resolve whether sex or gender are themselves binary. Instead, we face challenges inherent to the use of measures and categories themselves—the challenge of addressing what these represent as well as how and when to use them with as much rigor, precision, and transparency as possible.

When people use the binary categories of female/male or woman/man, they reference something far more complex than might initially be suggested; simplifying complexity is, after all, one reason to categorize (Scott 2010). Categorization itself is a decision-making process that varies widely, determined by how terms are defined and operationalized. Consider what is actually being referenced when the term sex is deployed, whether this is chromosomes, hormones, secondary sex characteristics, gendered behaviors, gender identity, or a combination of traits. Not all of these individual factors are static, discrete, isolatable, or binary under all circumstances, but they are often treated as such.

In daily life, most people categorize sex and gender in a relatively unconscious way that reflects split-second decisions based on initial assumptions and impressions of another's body or gender expression. In everyday life as well as in science, there is significant complexity underlying the categorization of sex and gender. Usage of these terms can be found in some of the most divisive debates today, and these are neither simply theoretical nor narrow academic arguments. The concept of what is often referred to as *biological sex*, for example, is currently being invoked in several global jurisdictions to justify discriminatory laws and policies intended to enforce strict binaries based on biologically essentialist assumptions. These discriminatory laws and policies mandate, for instance, infringements on bodily autonomy, reduced access to gender-specific facilities (e.g., restrooms, toilets), eligibility to participate in competitive sports, the ability to modify legal identification documents, and access to health care, including reproductive and gender-affirming care. Other infringements are more insidious and may be less likely to make news headlines, such as the social and cultural reinforcement of rigid gender norms and roles that simultaneously drive and limit self-expression of individuals, including those from majority populations like cisgender men. Conflicting and synergizing forces are converging to impact new scientific research mandates and agendas and evolving societal views around sex, gender, and gender identity.

Science—replete with all its disciplines, approaches, and practices—provides an enormously powerful system for generating knowledge, addressing pressing issues (including those that pertain to individual and public health), and transforming society in sometimes unforeseeable ways. The discovery of antibiotics, for instance, revolutionized how infections are treated, just as noninvasive diagnostics now enable life-saving measures in critical care. In many areas of science, research produces insights that can be broadly applied to benefit our physical and mental well-being. Science exists *within* society, and thus, there is a direct, bidirectional relationship between society and science. Sociocultural contexts impact our individual and collective perspectives and shape scientific practice, just as scientific knowledge and its application inform societal understanding on multiple levels, including normative expectations. These relationships drive the types of issues addressed by science and determine what and how data are collected, analyzed, interpreted, and disseminated. Acknowledging this enables us to advance scientific practice in ways that will best serve and advance society as a whole.

Institutional mandates regarding sex and/or gender in research have increased over recent decades. The European Union (EU), the Canadian Institutes of Health Research (CIHR), and the U.S. National Institutes of Health (NIH) now require scientific and medical research to consider sex and/or gender, though specific requirements vary between institutes. The CIHR model requires all applicants across all disciplines to explain how sex and gender are "taken into account in the research design, methods, analysis, and interpretation, and/or dissemination of findings" or to explain why they are not taken into account. However, it does not mandate anything about the means through which sex and gender are considered (Government of Canada 2018). Similarly, the EU requires applicants to integrate "sex and/or gender analysis in the design and delivery of research and innovation" or to explain why this analysis will not be considered (European Commission 2021; European Research Executive Agency n.d.). In contrast, the NIH policy on *Consideration of Sex as a Biological Variable in NIH-Funded Research* (NIH 2015) for preclinical research names specific requirements, including an expectation for "researchers to study both male and female vertebrate animals and humans, where applicable" and the "disaggregation of data by sex [to allow] for sex-based comparisons" (ORWH 2015). Moreover, the NIH Policy on *Inclusion of Women and Minorities as Participants in Research Involving Human Subjects* requires the inclusion of women in all clinical research (NIH 2001). Such policies are intended to correct a long history of biomedical research that was overly reliant on male subjects or models for the generation of knowledge, which often excluded females from scientific studies on the grounds of the potential risks of pregnancy, or which presumed that variability arising from estrus or menstrual cycles would be detrimental to the analysis (an assumption that has subsequently been debunked) (Becker et al. 2016; Levy et al. 2023; Prendergast et al. 2014; Shansky 2019). This widespread lack of attention to sex- and gender-related considerations resulted in significant gaps in knowledge that have certainly contributed to the perpetuation of avoidable health disparities (Criado-Perez 2019).

Science must address these societal discussions and polarizing debates, as the meanings of sex and gender are anything but straightforward, yet they wield an enormous impact on people's daily lives, research practice, and policy (Figure 1.1). To date, in research, implementation has focused predominantly on sex and gender as separate but interacting domains, with a heavy reliance on the invocation of a female–male binary comparison. Such ways of thinking about sex and gender have generated useful knowledge and insights. However, when all comparisons are lumped under the binary of average female versus average male, there are, often unintended, consequences.

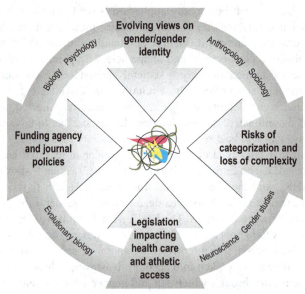

Figure 1.1   Convergence of competing agendas. Simultaneous activity from multiple scientific disciplines, government agencies, and policy makers creates a nexus of fission and fusion around the conceptualization and categorization of gender, sex, and their entanglement.

To develop concepts and approaches required to interrogate issues about sex and gender in scientific and clinical research, we need to revisit and unpack the assumptions tied to the concepts, categories, and scientific measures of sex and gender. Importantly, we need to understand how sex and gender are *entangled*. Here, the term *entanglement* does not refer simply to an additive model (*sex + gender* or *gender + sex*), whereby variables reflecting aspects of gender and variables reflecting aspects of sex are included in a model. Instead, entanglement recognizes that sex and gender are literally inseparable and co-constituted. An example that exemplifies this idea of co-constitution and inseparability (detailed below and throughout the volume), is testosterone,[1] a factor often included in the category of sex but which is also responsive to gendered aspects of behavior and experience related, for example, to nurturance, competition, aggression, and sexual activity (Bateup et al. 2002; Berg and Wynne-Edwards 2001; Bernhardt et al. 1998; Casto and Edwards 2016; Czarna et al. 2022; Gettler et al. 2011; Oliveira et al. 2009; van Anders et al. 2012; van Anders and Watson 2007; Wingfield et al. 1990).

Recognition of this specific entanglement is often signaled through the use of a combined term: sex/gender (Kaiser et al. 2007) or gender/sex (van Anders 2009) (see also Table 1.1). These combined terms should not be understood to signal an "either/or" orientation, but rather to unsettle the assumption of separateness of the two constructs and to signal recognition of their complex inseparability. We agree with the use of either construction but have elected to use sex/gender, except when we are deliberately noting the use of separate constructions. Although these terms are useful, integration of entanglement presents a conceptual, analytical, and

---

[1]   This is true of all people, not just those assigned male at birth. Testosterone and other androgens are crucially important mediators of many functions in all human bodies, including those assigned female at birth.

Table 1.1   General explanation of terms and their usage throughout this volume.

Terms and Concepts	Usage
Sex and gender Gender and sex	Both formulations distinguish two separate constructs. In general, sex refers to biological and gender to sociocultural aspects.
Sex and gender entanglement Sex/gender entanglement Sex/gender	These formulations signal recognition of entangled co-constitution of sex and gender. The order of terms reflects common usage (sex preceding gender) at a given time. Both terms are equal in importance (Kaiser 2012; Kaiser et al. 2007, 2009).
Gender and sex entanglement Gender/sex entanglement Gender/sex	These formulations signal recognition of entangled co-constitution of gender and sex (van Anders and Dunn 2009). Term order aims to center gender and avoids "leaving sex uncritiqued" (van Anders 2015:1182 and 2024:9).

empirical challenge that goes well beyond grammatical or terminological solutions. The challenge is not simply to seek or construct a universally applicable set of terms nor to debate whether the concepts of sex and gender are rigidly binary. As researchers, we must carefully consider the implications of entanglement and address how and when to use these concepts, employing as much rigor, critical reflection, transparency, and interdisciplinarity as possible when we do.

2 Interdisciplinary Dialogue: The Way Forward

Despite the steady increase in publications, seminars, and media coverage of sex/gender-related issues from a range of perspectives, progress to meet these challenges has been limited. One likely reason is the impact of disciplinary silos in science. Often, scholars have only limited opportunities to engage with scholars outside of their immediate field. Most scholars attend conferences and workshops that are intended to advance their specific field of study. Even when interdisciplinary interactions take place, the lack of a shared vocabulary or concepts stands as a real barrier to group-level discussions among scholars with diverse backgrounds and research interests. Yet to address the challenges inherent to the entanglement of sex and gender, a dynamic, ongoing dialogue is needed across scientific and scholarly disciplines and experiences.

It was this conviction, this hope, and this intent that led to the convening of the 36th Ernst Strüngmann Forum on *Sex and Gender: Transforming Scientific Practice*, held in Frankfurt, Germany, from October 8–13, 2023.

The development of this Forum began in August 2020, when Amber Wutich (a member of the advisory board to the Ernst Strüngmann Forum) initiated the idea for a Forum around this topic at a board meeting and introduced L. Zachary DuBois to Julia Lupp (director of the Forum). Together, they discussed ways to approach preparing a proposal focused around his ideas and understanding of the problems being confronted in science and society. From the summer of 2020 through winter of 2021, DuBois then invited 14 scholars and scientists (active in researching gender and sex conceptualization, operationalization, and measurement) to participate in exploratory interviews and share their perspectives on (a) cutting-edge questions and pressing issues, (b) crucial disciplines to involve, and (c) what they considered to be the most important issues and societal implications to be considered by an interdisciplinary Forum. These conversations identified the following key areas and were instrumental in the development of the initial proposal:

- Dialogue should involve social, lab-based, clinical, and other scientists who work with animal models as well as with humans.
- There is a crucial need to address the lack of clarity and consensus on terms and their definitions as well as methods of operationalization and measurement.
- One goal should focus on advancing recognition of sex and gender co-constitution and entanglement across disciplines.
- Another concerns the recognition of sex and gender entanglement and its importance in biomedical and clinical work, social policy, and practice.

After drafting an initial proposal, because of their substantial intellectual contributions to thinking about the application of sex and gender in science and on sex and gender entanglement, DuBois invited Sarah Richardson and Anelis Kaiser Trujillo in August 2021 to join in the revisions and preparation of a final proposal. Once approved, a meeting of the Program Advisory Committee was convened from October 27–29, 2022. Serving as members of the committee were Margaret McCarthy, Stacey Ritz, Rebecca Shansky, Paula-Irene Villa, L. Zachary DuBois (Forum chair), Anelis Kaiser Trujillo (Forum chair), and Julia Lupp (Forum director). At this key meeting, members worked together to affix the scientific framework that would underpin the cutting-edge discussion and collaborative work at the Forum. In addition, they developed focal questions for each working group (detailed below), generated ideas for the background papers that were commissioned, selected the participants to be invited, and delineated the overarching goals for the Forum:

- to advance the conceptualization of sex, gender, and their entanglement,
- to align dialogue across disciplines, and
- to promote sound application in research, policy, medicine, and public health to engage in extended discussions and explore and apply their specific areas of expertise.

The ensuing discourse at the Forum was intense, focused, and invigorating. This volume—the final step in the Forum's process (see the Preface)—synthesizes the various strands of knowledge that emerged. We hope that it will inspire further interdisciplinary efforts, for much remains to be accomplished to advance and potentially transform how sex and gender are understood and used by science and society.

3 Points of Departure

3.1 Grappling with Binaries

Although sometimes they are treated as synonyms and used interchangeably in casual usage, the concepts of sex and gender are frequently distinguished from one another in ways that are especially meaningful in scientific research: sex is often used to refer to the binary categorization of individuals, bodies, and body parts into female and male based on a constellation of anatomical traits and/or physiological mechanisms linked to sexual reproduction, whereas gender is typically understood in reference to culturally embedded aspects of identity, experience, social interactions, norms, and power dynamics related to femininity and masculinity. There is, however, considerable diversity in the ways that sex and gender are defined, operationalized, and understood to relate to one another in different contexts, from scholarship to policy to public discourse. Although we emphasize their entanglement, at the same time we believe it remains important to articulate distinct conceptual definitions of sex and gender because of the problematic tendency to resort to biological essentialism and determinism in interpreting female–male difference (i.e., to invoke exclusively biological explanations without considering the ways that sociocultural conditions also produce and shape difference). This central tension motivated us to examine the issue of entanglement closely.

A key motivation behind this Forum was to contemplate how we might transform current scientific practices, which typically view sex and gender as conceptually separate, to a state where sex and gender are understood as being entangled and as co-constituting one another. The conceptual and terminological separation into sex and gender is often strongly connected to strictly binary notions of sex (and often gender as well), and this is often critiqued as failing to recognize complexity or for insufficient recognition that the categories themselves actually reflect many dynamic, nonbinary factors (as detailed further below). The phrase "sex is not binary" has thus become increasingly frequent in contemporary public discourses and can be contentious in some circles and disciplines. Many consider the binary sex categories of female and male to be self-evident common sense, reflecting a recognition of one component of reproduction: the requirement of a dyadic complementarity of gametes (i.e., sperm and ova). For many, this means that sex is binary, at least at this fundamental level. Discussions at the Forum made clear, however, that the claim "sex is not binary" does not necessarily require us to reject the categories of female and male altogether. Instead, it asks us to consider that other categorizations are possible, while drawing attention to the fact that many traits and factors associated with sex categories are not dimorphic. Such traits can have a high degree of intra- and interindividual variability across time and context, with overlapping distributions between the categories, and with an imperfect correlation of all sex-associated traits with the normative sex categories.

For many at the Forum, one important critique of categorizing sexually reproducing animals into female or male is that it is often unclear which criteria are being used to delineate the categories in any given case and that many unexamined assumptions are made about those categorizations. For example, it is common to reference chromosomal complement, gamete size, or the configuration of reproductive organs or genitalia as the traits that define an individual's membership in one sex category or the other. However, each of these has drawbacks, caveats, and limitations: there is no single trait that can be definitively used to determine *femaleness* or *maleness* that is universally appropriate. For any given sex-associated trait, there is some degree of heterogeneity within female and male categories as well as some degree of overlap between these categories. Application of these binary normative categories has implications for the pathologization of normal variation, including negative effects for individuals who may not conform with expected norms. This is a circular process, which some refer to as "biological normalcy" (DuBois and Shattuck-Heidorn 2021; Wiley 2023): scientifically generated, statistical norms reflect and reinforce sociocultural norms and, in turn, impact scientific practice, the development of policy, and the lives and health of individuals via embodied experiences of stigma and inequity (Brewis et al. 2011; DuBois et al. 2024).

A key topic for discussion at the Forum concerned the utility of binary sex/gender categories in scientific practice. Many argue that binary categorizations and concepts (sex vs. gender, female vs. male, woman vs. man, or femininity vs. masculinity) are insufficient to fully capture the breadth of human variation or understand embodied human experience. These arguments suggest that treating sex and gender as separate influences, and operationalized as a binary, obscures important insights about the complex and dynamic nature of the body's operations. For example, it is common in both scientific and lay discourses to talk about "sex hormones" in terms of "female hormones" (i.e., principally estrogens and progestogens) and "male hormones" (i.e., usually androgens). This is a flawed characterization in several ways. In fact, all hormones are present in all bodies, and there are dynamic patterns of natural hormonal variation across the lifespan, with considerable overlap in the expression patterns of most of these hormones between the female and male categories. In addition, hormone effects depend on patterns of hormone receptor expression and not just serum concentrations of those hormones. Moreover, there are marked effects of social experience and environmental exposure on the expression of hormone receptors and regulation of hormone synthesis (van Anders et al. 2011; van Anders and Watson 2006; Williams et al. 2023). The complexity of steroid hormones, their actions and interactions with physiology and behavior is vast and growing. Long-standing appreciation for the value of understanding hormones is evident in the founding of the journal *Hormones and Behavior* in 1969, the official journal of the Society for Behavioral Neuroendocrinology, as well as representative textbooks: *Introduction to Behavioral Endocrinology* (Nelson and Kriegsfeld 2022), now in its sixth edition, and *Behavioral Endocrinology*, in its second edition (Becker et al. 2002), among others. While it is not feasible for scientists from a range of disciplines studying sex or gender to be proficient in all the nuances regarding hormones, an awareness of the associated complexity will create guardrails and provide guidance in making conclusions around the causality of steroid action.

It is also noteworthy that the very use of the term gender is contentious within animal research. Here, the concept of gender identity is particularly challenging as it is not possible for an animal to report whether they feel that they are female, male, nonbinary, or other. Other aspects of gender may be reflected in animal societies that exhibit, for example, dominance hierarchies or matrilineal coalitions, but how much this mirrors the pervasive influence of gender in humans is highly debatable. Thus, when discussing results from animal researchers, it may be better to avoid using the term gender as this will help reduce the risk of anthropomorphizing findings (e.g., from rodents to women or men).

Given how deeply entrenched sex categories are, it is unlikely that the categories of female and male will disappear or cease to have utility in certain contexts. Importantly, our discussions at the Forum did not lead to a call for a proliferation of categories as a response to the challenge of seeing sex as something other than a binary. Although there are a range of views among Forum participants (and within our editorial team) about what this might mean or how it could be operationalized, there does seem to be general agreement that we must bring a more sophisticated conceptual understanding of those categories to our application, use, and interpretation of them, and to be rigorous and nuanced as conceptualizations of sex/gender continue to evolve. One possible alternative is presented in Chapter 10, where Yang et al. discuss a case study as part of their recommendation for sex to be considered a continuous variable in research. This approach could enable greater recognition of complexity, and has also been discussed by others (Joel 2020; Sanchis-Segura et al. 2022). In Chapter 5, Ritz et al. consider the possibility of taking a "hypothesis-free" approach to data analysis and to look "for clusters that emerge from the data rather than dividing the sample into sex categories a priori." (p. 96)

3.2 From Binaries to Sex/Gender Entanglement

The initial impetus for making a conceptual distinction between sex and gender was an explicitly feminist and historically important aim. Its deliberate intent was twofold: (a) to decouple the biological from the social so as to challenge and refute the biologically essentialist and determinist claims upon which discrimination against women is frequently based, and (b) to draw attention to the ways that social and cultural forces shape our lived realities and generate and sustain inequity. This distinction between sex and gender is valuable when trying to understand aspects of difference and inequity in humans. Yet over the past few decades, many scholars have drawn attention to the limitations of framing sex and gender as separate domains, with implications for what is possible to study in different disciplines. Distinguishing sex from gender in this way (i.e., to equate sex with the biological and gender with the social) can obscure their entanglement. For example, when gonadal hormones are understood to be a manifestation of "sex" because they are molecular substances, some scientists frame and interpret hormones as the *cause* of differences in physiology in health without considering the ways that social and environmental influences impact the expression of those hormones (Fausto-Sterling 2000; Oudshoorn 1994; van Anders et al. 2011; Williams et al. 2023). Similarly, it is problematic when interpretations of brain mapping data assume a biological basis for female–male differences without accounting for the ways that social experiences and behaviors alter our brains (Fine 2010; Rippon 2019; Zugman et al. 2023).

The complex, mutual influences of gender and sex have long been recognized to be centrally important (e.g., Bleier 1984; Fausto-Sterling 2000, 2005; Hrdy 1981/1999; Hubbard 1990; Krieger 2005; Schiebinger 2004; Schmitz 2010). Some now argue that distinguishing sex from gender (even when entangled) is futile or even potentially problematic (e.g., Ashley et al. 2024; Velocci 2024). Many authors in this volume grapple with these very issues: the challenge of acknowledging that the interactions between sex and gender are so intricate that operating as though they are two distinct entities may be conceptually flawed, while confronting the further challenge of operationalization of entanglement, particularly when studying nonhuman animals or isolated cells.

Another issue of contention concerns the terms gender and gender identity. Confusion can arise because contemporary public discourse about gender focuses disproportionate attention on gender identity; however, gender scholars typically understand gender to be comprised of multiple components (e.g., gender identity, gender expression, gendered behavior, gender norms, gender roles, and gender ideologies). Similarly, sex is comprised of multiple components (e.g., chromosomes, hormones, sexual behavior, and secondary sex characteristics). At the Forum, we confronted the problem of how entanglement views of sex and gender could be implemented in research contexts involving nonhuman animals or cells. In this context, it is unclear whether the culturally embedded human experiences of gender can be accounted for in research design and, if it can, how we would interpret this.

By referring to sex and gender entanglement, in an effort to avoid the primary division into sex and gender, we may overlook opportunities to highlight the multiplicity of components within each. There is a danger that people will nod to entanglement without engaging the complexity of elements that are part of the system, thereby failing to capture how the numerous subcomponents of sex and gender are intricately interconnected and entangled with one another (Duchesne and Kaiser Trujillo 2021). This is where new terminology, such as sex/gender entanglement or gender/sex entanglement, might facilitate reconceptualization in support of more nuanced and rigorous approaches to the study of sex and gender in a novel interdisciplinary way. This is the spirit in which these terms and their varied formulations (Table 1.1) are used throughout this volume.

The trajectory of sex/gender and gender/sex as hybrid terms has different genealogies at the intersection of biology and gender studies. Both address important aspects of this embeddedness, including inseparability and co-constitution, as well as the importance of word order when communicating these ideas. New terminology is an initial step toward a much-needed common interdisciplinary vocabulary and has been regarded by some as an indispensable and ethically crucial aspect in bridging disciplines (Barad 2007; Haraway 2003) and thus represents more than terminological innovation.

The need for new, interconnected terms related to sex and gender in gender-informed biomedicine, in particular, varies depending on the language in use. For example, in German, *Geschlecht* can mean both sex and gender, allowing for an entangled understanding of the two concepts due to its homonymous meaning. By contrast in Spanish, the term *sexo/género* has traditionally been used in an additive manner in feminist scholarship and scientific research, whereby distinct factors related to sex and gender would be added up as separate but interacting factors (e.g., Juárez-Herrera y Cairo et al. 2021). More recently, *sexo/género* has also been understood in a more entangled way (Ciccia 2023). Relatedly, when Canada established the Institute of Gender and Health at CIHR about two decades ago, the French translation was *Institut de la Santé des Femmes et des Hommes*: the French concept of *genre* is not used or understood in the way that gender is in English, thus leading to a translation into "women" (*femmes*) and "men" (*hommes*). Similar nuances, contextual interpretations, and meanings of sex, gender, and sex/ gender entanglement are likely applicable across many other languages. We have highlighted only a narrow subset of languages with which we are familiar. These examples make clear how crucial it is for the interdisciplinary and international research framework of sex/gender to be explicit about how terms are used in different specific contexts.

4 Areas of Discussion

To enable effective discussion, each working group faced the challenge of definitions and application of terms (sex, gender, sex/gender entanglement) for their topic area. Framing the discussions around the concept of entanglement provided a unifying focus for each group (Figure 1.2), yet it must be noted that emergent understandings were predominantly framed from a Euro-Anglocentric perspective. This prompts the question of how diverse our considerations can "really" be, even when we aim, for instance, to include cultural, geopolitical, and class-related norms. In addition, we note that our discussions did not *systematically* account for intersectional embeddedness, a point brought up by Bauer (Chapter 7) and Bowleg et al. (Chapter 9) in their constructive critiques. Both chapters stress the need for an intersectional approach, as disentangling sex, gender, or sex/gender from racialized and other minoritized intersections overlooks the role of the white racial frame in perpetuating this separation.

Below, we highlight some of the conceptual challenges that emerged within and between groups as well as how these can advance dialogue and thinking. Although each group reached an understanding concerning terminology, conceptual meanings,

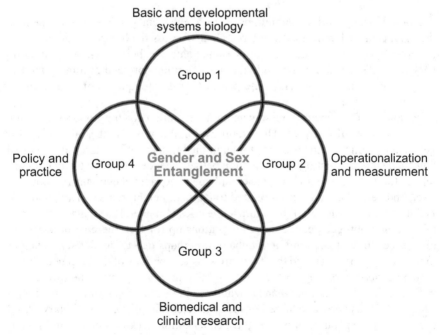

Figure 1.2   Research and policy fields in which sex/gender entanglement emerges, addressed by the four working groups at the 36th Ernst Strüngmann Forum.

and practice implications, the perspective of one group may not necessarily align with others. As editors, we purposely did not attempt to enforce consensus in terminology, as we believe that such inconsistencies mirror the complexity of the topic and our efforts. We view this as a strength of the volume, as it exposes issues that remain while highlighting a diversity of "solutions."

4.1 Entanglement of Gender/Sex Dynamics in Basic and Developmental Systems Biology

The first working group (Chapter 2) was comprised of a mix of bench scientists who study traditional animal models (e.g., rats and mice) and contributed in-depth expertise in neuroendocrinology as well as biological anthropologists who study evolutionary medicine in women's reproductive health, metabolism, and growth. Guiding questions set forth for their discussion included the following:

- How might the concept of "entanglement" of gender and sex transform research?
- What do we know about how gender/sex development varies across the life course?
- What do we know about gender/sex during critical periods of development (pre- and perinatal, juvenile/childhood, adolescence, adulthood, and senescence)?
- How can we better operationalize gender and sex entanglement across the life course?

After establishing working definitions of sex, gender, and gender/sex entanglement, they grappled with how gender and sex change across the lifespan. Saldanha et al. (Chapter 2) conceptualized gender/sex entanglement as being least relevant during very early life (i.e., from conception to sex determination) and becoming increasingly complex and influential as life proceeds as a consequence of gender and its accumulated impact.

Parallel to this dynamic view was an appreciation for the importance of the biological level of analyses. The group explored sex differences at the level of gene expression (i.e., the transcriptome, in the brain and elsewhere). It struggled, however, with articulating the distinction between gender/sex entanglement in sex-related traits (e.g., sex chromosome complement, gonadal phenotype, reproductive tract, and genitalia) and sex-correlated traits (e.g., height, bone density, body fat, facial hair, and breast size). In contrast, DeVries (Chapter 3) articulates the value of a whole-body perspective and the relationship (or lack thereof) between sex differences in neuroanatomy and behavior, perhaps due to gender/sex entanglement. The group recognized that there are sensitive periods of development during which gender/sex entanglement can be particularly emergent: some of these are unique to females (e.g., pregnancy) whereas others may be culturally specific. In Chapter 4, Dunsworth and Ware remind us to reject simplistic evolutionary-based, deterministic explanations of sex and gender differences in humans and nonhuman primates and cite multiple instances of alternative explanations for what has previously been viewed as reproductively adaptive. The group challenged itself to question how gender/sex entanglement might be incorporated into preclinical research using animal models, and how this would be operationalized and measured. Pros and cons of attempting to incorporate gender/sex entanglement into animal model research were considered. Finally, Saldanha et al. (Chapter 2) propose guidelines for operationalization and measurement, complementing suggestions made by Ritz et al. (Chapter 5).

4.2 Sex/Gender Entanglement and Issues of Operationalization and Measurement

The second working group (Chapter 5) was composed of a range of scholars from women, gender, and sexuality studies as well as psychology, biology, immunology, public health, epidemiology, and the history and philosophy of science. The following questions helped initiate their discussion:

- What does the use of categories of binary sex (female/male) and gender (man/woman) enable in research? What harms and biases may result? What is rendered invisible by using these categories and concepts?
- What does sex/gender entanglement mean for animal and *in vitro* studies?
- How can we get beyond biological essentialism (e.g., centering sex assigned at birth) and individualized gender fixation (e.g., centering self-reported gender identity only) in research?
- How can scientists account for the complex nature and social reality of gender, sex, and their entanglement?

In their discussions, Ritz et al. (Chapter 5) focused on biomedical, health, and neuroscience research. In spite of mandates to incorporate sex and gender considerations into research and policy, it is striking that there is considerable variation across disciplines and contexts in what is understood to constitute sex or gender. Regardless of the specific policy requirements, by far the most common way that health researchers take up the consideration of sex and gender is to include both female and male subjects in the research, to treat "sex category" as a simple binary variable, and then to compare them to one another. Although some argue that a female–male binary approach is better than nothing, or at least a "step in the right direction," others countered that an overreliance on the female–male binary may inadvertently *contribute* to data misinterpretation, reinforce inaccurate and stereotyped perceptions of human difference, and neglect the needs of individuals whose lived experiences do not align with that simple binary construct (DuBois and Shattuck-Heidorn 2021; Garcia-Sifuentes and Maney 2021; Haverfield and Tannenbaum 2021; Joel and Fausto-Sterling 2016; Maney 2016; Pape et al. 2024; Richardson et al. 2015; Ritz and Greaves 2022).

Ritz et al. grappled with how sex and gender categorization schemes function in scientific research: from *in vitro* cell culture work to animal experimentation, clinical and experimental research in humans, clinical trials, and epidemiologic research. In particular, they were concerned with how researchers might operationalize the entanglement of sex and gender, as the complexity inherent in sex, gender, and entanglement can be challenging to incorporate in forms of scientific practice that rely on the ability to control variables carefully to establish causal inferences. They also distinguished between *interaction* and *entanglement*, and discussed implications for scientific practice. Another significant dilemma confronted by Ritz et al. was the problem of how gender can be understood in the context of research in nonhuman animals and *in vitro* work on human cells and tissues, isolated from a social context.

Such challenges become even more complicated when viewed from an intersectional perspective, thus raising questions about whether we can legitimately prioritize sex and gender as a primary frame of analysis (see also Chapter 9). Bauer (Chapter 7) further emphasizes that sex and gender are not only entangled with one another but also simultaneously entangled with all other forms of intersectional considerations, and thus irretrievably bound up in relations of social power. Bauer highlights the limitations of categorical approaches and the value of intersectional perspectives for expanding conceptualizations of sex and gender in understanding health. In Chapter 6, Duchesne illustrates the transformative potential of neurofeminist and entanglement perspectives for scientific practice. Like Bauer, Duchesne considers the ways that sex and gender are closely connected to a broader social context. She contends that an overreliance on binary conceptualizations of sex and gender obscures the multiplicity and complexity of experiences and processes that shape human experience and health.

The core message of Ritz et al. (Chapter 5) is a call for scientists to engage the concepts of sex, gender, and sex/gender entanglement with a more critical lens, and to resist the tendency to slide into simplistic interpretations based on unexamined assumptions. They urge researchers to move to a more mechanism-focused approach rather than rely on categories. They also highlight the need to revisit and refine the conceptual and policy frameworks of sex and gender to better account for their inherent complexity.

4.3 Sex/Gender Entanglement Dynamics in Human Biomedical and Clinical Research

The third working group (Chapter 8) consisted of scientists and scholars with expertise in psychology, intersectional research, biomedical science, feminist studies, transgender health, HIV research, neuroscience, oncology, public health, and genetics. They focused on addressing the following set of questions:

- What is the meaning of sex, gender, and intersectionality in the context of clinical/biological sciences and what are our own personal biases in the context of research related to these topics?
- What do the categories of binary sex (male, female) and gender (man, woman, trans, nonbinary) enable us to achieve? What harms and biases can result?
- How can gender and sex be better integrated with biomedical research (e.g., neuropsychiatry, epigenetics, clinical research)?

- How can sex and gender be integrated into social sciences (e.g., public health, sociology, psychology)?
- How can gender experience and intersectionality, involving the dynamism within individuals and across historical and cultural contexts, be better integrated into health research?

From their discussions, Juster et al. (Chapter 8) highlight the necessity of including more social sciences-related research in biomedicine and the importance of recognizing gender/sex entanglement. Further, in addressing sex and gender, they stress that an inclusive, globally aware approach is necessary to ensure that biomedical research and clinical practices are sensitive to diverse cultural, social, and global experiences and perspectives. Juster et al. discussed further aspects of importance in the context of biomedical research and clinical application, such as including individual phenotyping, developing measurement tools for gender and sex that move beyond the binary in humans, statistical considerations such as continuum scales versus categorical measures, and the relevance of comparative-based versus mechanisms-based approaches.

One of the most significant insights and recommendations reached by this group reflects the position taken by Bowleg et al. (Chapter 9); namely, that it is futile to attempt to consider sex/gender entanglement separately from other intersectional positions, including racialized as well as sexual and gender minoritized status. Juster et al. (Chapter 8) highlight the importance of adopting an intersectional approach when conducting clinical research and providing medical care. They argue that this is crucial to understanding and addressing health inequities, particularly among marginalized groups. Further, the crosscutting perspective and its focus on interlocking systems of oppression of intersectional approaches underscores the importance of moving beyond generalizations to address the lived experiences of individuals in future biomedical research into gender/sex. Key examples are provided by Poteat and Ciccia in Chapter 11. They outline barriers and challenges to transgender health equity and highlight how reducing recognition of sex/gender complexity applies essentialist, cisheteronormative bias. This bias undermines important advancements needed in this area of biomedicine and clinical practice. In addition, they provide a critical overview of several research frameworks currently in use and suggest emerging approaches to advance transgender health equity research.

The case for how sex and gender can be better integrated within biomedical research was informed by Yang et al. (Chapter 10), who advocate for sex to be considered a continuous variable in the context of cancer. The group addressed individual phenotyping of both adult and pediatric cancers and stressed the need for further development of measurement tools that move beyond the binary in humans, statistical considerations (e.g., continuous scales vs. categorical measures), and the relevance of comparative- versus mechanisms-based approaches.

4.4 Entanglement of Sex/Gender Dynamics in Policy and Practice

Connecting scientific practice with policy is never easy, which is why a dedicated group at the Forum (Chapter 12) addressed the following issues:

- What are the consequences when biological definitions of sex that exclude sex/gender entanglement are used in social policies?
- Where is scientific clarification of sex-linked biologies, gendered behaviors, and health outcomes most needed?
- What alternative concepts/frameworks might more effectively account for human variation and diversity?
- How can the complexities of sex/gender entanglement be leveraged to foster positive change?

As in the other groups, this group's expertise was wide-ranging: from anthropology, gender studies and global health, to law and policy. Drawing on examples from sports, such as ACL injury, Brewis et al. (Chapter 12) highlight how gender and sex are often disentangled. Historically, cause of injury, for instance, has been assumed to be rooted in biology (i.e., sex-related), devoid of any social effects of gender. Brewis et al. stress that this must be corrected as it leads to detrimental policy decisions, such as gender-segregated practices and the underfunding of women's sports. They also explain how policies and practices impact differently gendered bodies in diverse ways and discuss how policies and practices produce individual bodies as gendered and at risk, at multiple analytic levels.

Varied definitions and usage were uncovered for the terms gender and sex as well as approaches to conceptualizing entanglement, thus further exposing the challenge of operationalizing the terms on their own or as entangled constructs. Through summarizing the key concepts relevant to each definition, Brewis et al. (Chapter 12 p. 247) stress the importance of sex/gender entanglement approaches for practice and policy and encourage the "investigation of potentially gendered factors and their constitutive interactions with the anatomy, physiology, biomechanics of bodies" and enable recognition of the dynamic interaction of biology and social and physical environments through which sex/gender interactions become embodied. They highlight the challenges involved in engaging sex, gender, and/or entanglement approaches, particularly concerning disciplinary differences and dichotomies of "social" versus lab-based or "hard" sciences, and support the emergence of scholarship across multiple disciplines. To this end, Brewis et al. suggest ways to address future policy and practice and advocate for greater precision and transparency in conceptualization and terminology when articulating what aspect(s) of sex/gender are under study, and why, particularly in research reports and publications.

Contributing to the discussion is Pape's review of the US National Institutes of Health policy on sex as a biological variable (NIH 2015). In Chapter 13, Pape reflects on its history and implementation, and considers how the idea of sex/gender entanglement could inform future policies and "open up the universe of sex and gender research" (p. 280) to bring about a better understanding of the complex ways that bodies and their sex-linked traits interact with their environment. Pape points to the possibility of bridging the positions of those arguing for the correction of male-only clinical trials with those who call for an augmentation of concepts and not just the inclusion of females. In addition, Greaves (Chapter 14) looks at ways to foster the integration of sex/gender into more equitable policy and practice. She argues that to redress gender inequities in health, we must address their root causes. Drawing on a feminist framework, Greaves describes how sex/gender entanglement can be an innovative add-on to the "gender transformative approach." She calls on all scholars and policy makers to engage in critically understanding the drivers of gender inequity from their own context and positionality. In Chapter 15, Currah considers further implications of the concept of sex and characterizes it in a way that is quite distinct. Building on arguments from his recent book *Sex Is as Sex Does: Governing Transgender Identity* (Currah 2022), he conceives of sex as an effect of a particular state action, a position that substantially shifts the focus of questions about sex/gender to the function and use of the categories themselves at the state level. Accordingly, Currah suggests that scientists should refrain from extending interpretations of biology to entangled and contested concepts and frameworks (e.g., sex, gender, male, and female) and to recognize the impact these have on policy making.

5 Next Steps

Our discussions at this Forum clearly revealed that there are no tidy answers or a one-size-fits-all template to conceptualize or operationalize sex/gender entanglement. This may, in part, reflect a history of scholarship and theorization rooted in diverse disciplines—areas of science that do not always share foundational assumptions or discourses—and the inherent challenges encountered when we try to bridge and translate across different fields. It may also reflect our reticence toward universalizing perspectives from the Global North, which do not necessarily include perspectives from the Global South or Indigenous knowledge and is not particularly cognizant of variation across cultures, geographies, and time. We would posit that the absence of clear resolution and the pressing areas of substantial divergence (and even some areas of dissensus) accurately reflect the messy, complex reality of sex/gender entanglement in human life. To paraphrase H. L. Mencken, anyone who claims to have a neat-and-tidy answer to a complex set of questions is probably wrong.

Nonetheless, looking back, it is truly remarkable how views converged throughout the entire process of this Forum. In our experience, this rarely occurs among such a diverse group of scholars. Such interdisciplinary dialogue is perhaps even more challenging when the focal point stems from such a highly politicized and contentious topic. Still, the trajectories taken since the Forum clearly demonstrate that this attempt at interdisciplinary dialogue was fruitful and must continue. Given the centrality of sex and gender as organizing social principles across so many cultures and contexts, approaches to conceptualization and operationalization of sex, gender, and sex/gender entanglement in research must continue to evolve, and for this, we need to pool our expertise and collaborate.

Acknowledgments This book and the Forum itself were several years in the making and would not have been possible without the intellectual contributions of many. We thank Amber Wutich, Heather Shattuck-Heidorn, and Alex Brewis for their formative contributions at early stages of the proposal development as well as the many scholars and scientists who participated in short interviews with Zachary DuBois, as ideas for the Forum were first being drafted. Those interviewed include Malin Ah-King, Sari van Anders, Niko Besnier, Lisa Bowleg, Paisley Currah, Lise Eliot, Anne Fausto-Sterling, Janet Hyde, Daphna Joel, Katrina Karkazis, Donna Maney, Sarah Richardson, Stacey Ritz, and Banu Subramaniam. We are grateful to Sarah Richardson for her contributions to the submitted proposal. We appreciate the dedication and hard work of the Scientific Advisory Board of the Ernst Strüngmann Forum (ESF); their feedback on several rounds of the proposal were crucial to the Forum's success as a whole. The program for the Forum, including the questions that guided the group discussions and the list of invitees, could not have been generated without the multiple days of in-person work by the Program Advisory Committee: Margaret McCarthy, Stacey Ritz, Paula-Irene Villa, Rebecca Shansky, Anelis Kaiser Trujillo (chair), L. Zachary DuBois (chair), and Julia Lupp (ESF director). Finally, our thanks to ESF's director, Julia Lupp. Without Julia's extraordinary efforts and expert guidance along with those of her staff, neither the Forum nor this book would have been possible. We are grateful to all of the participants who engaged in the week-long Forum, as those connections and interactions are the core of this book and each chapter within it.

References

Ashley F, Brightly-Brown S, and Rider GN (2024) Beyond the Trans/Cis Binary: Introducing New Terms Will Enrich Gender Research. Nature 630 (8016):293–295. https://doi.org/10.1038/d41586-024-01719-9

Barad K (2007) Meeting the Universe Halfway: Quantum Physics and the Entanglement of Matter and Meaning. Duke Univ Press, Durham

Bateup HS, Booth A, Shirtcliff EA, and Granger DA (2002) Testosterone, Cortisol, and Women's Competition. Evol Hum Behav 23 (3):181–192. https://doi.org/10.1016/S1090-5138(01)00100-3

Becker JB, Breedlove SM, and Crews D (eds) (2002) Behavioral Endocrinology. 2nd edn. MIT Press, Cambridge, MA

Becker JB, Prendergast BJ, and Liang JW (2016) Female Rats Are Not More Variable Than Male Rats: A Meta-Analysis of Neuroscience Studies. Biol Sex Differ 7:34. https://doi.org/10.1186/s13293-016-0087-5

Berg SJ, and Wynne-Edwards KE (2001) Changes in Testosterone, Cortisol, and Estradiol Levels in Men Becoming Fathers. Mayo Clin Proc 76 (6):582–592. https://doi.org/10.4065/76.6.582

Bernhardt PC, Dabbs Jr JM, Fielden JA, and Lutter CD (1998) Testosterone Changes during Vicarious Experiences of Winning and Losing among Fans at Sporting Events. Physiol Behav 65 (1):59–62. https://doi.org/10.1016/S0031-9384(98)00147-4

Bleier R (1984) Science and Gender: A Critique of Biology and Its Theories on Women. Pergamon Press, Oxford

Brewis AA, Wutich A, Falletta-Cowden A, and Rodriguez-Soto I (2011) Body Norms and Fat Stigma in Global Perspective. Curr Anthropol 52 (2):269–276. https://doi.org/10.1086/659309

Casto KV, and Edwards DA (2016) Testosterone and Reconciliation among Women: After-Competition Testosterone Predicts Prosocial Attitudes Towards Opponents. Adapt Hum Behav Physiol 2 (3):220–233. https://doi.org/10.1007/s40750-015-0037-1

Ciccia L (2023) Sexo/Género. In: Zahonero LA, Pérez Sedeño E, Sánchez N (eds) Enciclopedia Crítica del Género. Arpa, Barcelona, pp 281–290

Criado-Perez C (2019) Invisible Women: Data Bias in a World Designed for Men. Abrams, New York

Currah P (2022) Sex Is as Sex Does: Governing Transgender Identity. New York Univ Press, New York

Czarna AZ, Ziemiańska M, Pawlicki P, Carré JM, and Sedikides C (2022) Narcissism Moderates the Association between Basal Testosterone and Generosity in Men. Horm Behav 146:105265. https://doi.org/10.1016/j.yhbeh.2022.105265

DuBois LZ, Puckett JA, Jolly D, et al. (2024) Gender Minority Stress and Diurnal Cortisol Profiles among Transgender and Gender Diverse People in the United States. Horm Behav 159:105473. https://doi.org/10.1016/j.yhbeh.2023.105473

DuBois LZ, and Shattuck-Heidorn H (2021) Challenging the Binary: Gender/Sex and the Bio-Logics of Normalcy. Am J Hum Biol 33 (5):e23623. https://doi.org/10.1002/ajhb.23623

Duchesne A, and Kaiser Trujillo A (2021) Reflections on Neurofeminism and Intersectionality Using Insights from Psychology. Front Hum Neurosci 15:684412. https://doi.org/10.3389/fnhum.2021.684412

European Commission (2021) Directorate General for Research and Innovation. Horizon Europe Guidance on Gender Equality Plans. https://data.europa.eu/doi/10.2777/876509

European Research Executive Agency (n.d.) Gender in EU Research and Innovation— European Commission. https://rea.ec.europa.eu/gender-eu-research-and-innovation_en

Fausto-Sterling A (2000) The Five Sexes, Revisited. Sciences (New York) 40 (4):18–25. https://doi.org/10.1002/j.2326-1951.2000.tb03504.x

Fausto-Sterling A (2005) The Bare Bones of Sex: Part 1—Sex and Gender. Signs 302 (2):1491–1527. https://doi.org/10.1086/424932

Fine C (2010) Delusions of Gender: How Our Minds, Society, and Neurosexism Create Difference. W. W. Norton, New York

Garcia-Sifuentes Y, and Maney DL (2021) Reporting and Misreporting of Sex Differences in the Biological Sciences. eLife 10:e70817. https://doi.org/10.7554/eLife.70817

Gettler LT, McDade TW, Feranil AB, and Kuzawa CW (2011) Longitudinal Evidence That Fatherhood Decreases Testosterone in Human Males. PNAS 108 (39):16194–16199. https://doi.org/10.1073/pnas.1105403108

Government of Canada (2018) Sex and Gender in Health Research – CIHR. https://cihr-irsc.gc.ca/e/50833.html

Haraway DJ (2003) The Companion Species Manifesto: Dogs, People, and Significant Otherness. Prickly Paradigm Press, Chicago

Haverfield J, and Tannenbaum Cc (2021) A 10-Year Longitudinal Evaluation of Science Policy Interventions to Promote Sex and Gender in Health Research. Health Res Policy Syst 19 (1):94. https://doi.org/10.1186/s12961-021-00741-x

Hrdy SB (1981/1999) The Woman That Never Evolved. Harvard Univ Press, Cambridge, MA

Hubbard R (1990) The Politics of Women's Biology. Rutgers Univ Press, New Brunswick

Joel D (2020) Beyond Sex Differences and a Male–Female Continuum: Mosaic Brains in a Multidimensional Space. Handb Clin Neurol 175:13–24. https://doi.org/10.1016/b978-0-444-64123-6.00002-3

Joel D, and Fausto-Sterling A (2016) Beyond Sex Differences: New Approaches for Thinking about Variation in Brain Structure and Function. Philos Trans R Soc Lond B Biol Sci 371 (1688):20150451. https://doi.org/10.1098/rstb.2015.0451

Juárez-Herrera y Cairo LA, Muñóz IEJ, and Chew AGM (2021) El Análisis Sexo/Género en la Enseñanza/Aprendizaje de la Medicina. El Caso de Las Enfermedades Cardiovasculares. Investigación educ méd 10 (37):78–87. https://doi.org/10.22201/fm.20075057e.2021.37.20262

Kaiser A (2012) Re-Conceptualizing "Sex" and "Gender" in the Human Brain. Z Psychol 220 (2):130–136. https://doi.org/10.1027/2151-2604/a000104

Kaiser A, Haller S, Schmitz S, and Nitsch C (2009) On Sex/Gender Related Similarities and Differences in fMRI Language Research. Brain Res Rev 61 (2):49–59. https://doi.org/10.1016/j.brainresrev.2009.03.005

Kaiser A, Kuenzli E, Zappatore D, and Nitsch C (2007) On Females' Lateral and Males' Bilateral Activation during Language Production: A fMRI Study. Int J Psychophysiol 63 (2):192–198. https://doi.org/10.1016/j.ijpsycho.2006.03.008

Krieger N (2005) Embodiment: A Conceptual Glossary for Epidemiology. J Epidemiol Community Health 59 (5):350–355. https://doi.org/10.1136/jech.2004.024562

Levy DR, Hunter N, Lin S, et al. (2023) Mouse Spontaneous Behavior Reflects Individual Variation Rather Than Estrous State. Curr Biol 33 (7):1358–1364. https://doi.org/10.1016/j.cub.2023.02.035

Maney DL (2016) Perils and Pitfalls of Reporting Sex Differences. Philos Trans R Soc Lond B Biol Sci 371 (1688):20150119. https://doi.org/10.1098/rstb.2015.0119

Nelson RJ, and Kriegsfeld LJ (2022) An Introduction to Behavioral Endocrinology (Sixth edition). Oxford Univ Press/Sinauer Assoc., Oxford

NIH (2001) NIH Policy and Guidelines on the Inclusion of Women and Minorities as Subjects in Clinical Research. https://grants.nih.gov/policy-and-compliance/policy-topics/inclusion/women-and-minorities/guideline

—— (2015) NOT-OD-15-102: Consideration of Sex as a Biological Variable in NIH-Funded Research. https://grants.nih.gov/grants/guide/notice-files/not-od-15-102.html

Oliveira T, Gouveia MJ, and Oliveira RF (2009) Testosterone Responsiveness to Winning and Losing Experiences in Female Soccer Players. Psychoneuroendocrinology 34 (7):1056–1064. https://doi.org/10.1016/j.psyneuen.2009.02.006

ORWH (2015) Consideration of Sex as a Biological Variable in NIH-Funded Research. https://orwh.od.nih.gov/sites/orwh/files/docs/NOT-OD-15-102_Guidance_508.pdf

Oudshoorn N (1994) Beyond the Natural Body: An Archaeology of Sex Hormones. Routledge, London/New York

Pape M, Miyagi M, Ritz SA, et al. (2024) Sex Contextualism in Laboratory Research: Enhancing Rigor and Precision in the Study of Sex-Related Variables. Cell 187 (6):1316–1326. https://doi.org/10.1016/j.cell.2024.02.008

Prendergast BJ, Onishi KG, and Zucker I (2014) Female Mice Liberated for Inclusion in Neuroscience and Biomedical Research. Neurosci Biobehav Rev 40:1–5. https://doi.org/10.1016/j.neubiorev.2014.01.001

Richardson SS, Reiches M, Shattuck-Heidorn H, LaBonte ML, and Consoli T (2015) Opinion: Focus on Preclinical Sex Differences Will Not Address Women's and Men's Health Disparities. PNAS 112 (44):13419–13420. https://doi.org/10.1073/pnas.1516958112

Rippon G (2019) The Gendered Brain: The New Neuroscience That Shatters the Myth of the Female Brain. Random, New York

Ritz SA, and Greaves L (2022) Transcending the Male–Female Binary in Biomedical Research: Constellations, Heterogeneity, and Mechanism When Considering Sex and Gender. Int J Environ Res Public Health 19 (7):4083. https://doi.org/10.3390/ijerph19074083

Sanchis-Segura C, Aguirre N, Cruz-Gómez ÁJ, Félix S, and Forn C (2022) Beyond "Sex Prediction": Estimating and Interpreting Multivariate Sex Differences and Similarities in the Brain. Neuroimage 257:119343. https://doi.org/10.1016/j.neuroimage.2022.119343

Schiebinger LL (2004) Nature's Body: Gender in the Making of Modern Science. Rutgers Univ Press, New Brunswick

Schmitz S (2010) Sex, Gender, and the Brain: Biological Determinism versus Socio-Cultural Constructivism. In: Klinge I, Wiesemann C (eds) Sex and Gender in Biomedicine: Theories, Methodologies, Results. Universitätsverlag, Göttingen

Scott JW (2010) Gender: Still a Useful Category of Analysis? Diogenes 57 (1):7–14. https://doi.org/10.1177/0392192110369316

Shansky RM (2019) Are Hormones a "Female Problem" for Animal Research? Science 364 (6443):825–826. https://doi.org/10.1126/science.aaw7570

van Anders SM (2009) Androgens and Diversity in Adult Human Partnering. In: Endocrinology of Social Relationships. Harvard Univ Press, Boston, pp 340–363

—— (2015) Beyond Sexual Orientation: Integrating Gender/Sex and Diverse Sexualities via Sexual Configurations Theory. Arch Sex Behav 44 (5):1177–1213. https://doi.org/10.1007/s10508-015-0490-8

—— (2024) Gender/sex/ual Diversity and Biobehavioral Research. Psychol Sex Orientat Gend Divers 11 (3):471–487. https://doi.org/10.1037/sgd0000609

van Anders SM, and Dunn EJ (2009) Are Gonadal Steroids Linked with Orgasm Perceptions and Sexual Assertiveness in Women and Men? Horm Behav 56 (2):206–213. https://doi.org/10.1016/j.yhbeh.2009.04.007

van Anders SM, Goldey KL, and Kuo PX (2011) The Steroid/Peptide Theory of Social Bonds: Integrating Testosterone and Peptide Responses for Classifying Social Behavioral Contexts. Psychoneuroendocrinology 36 (9):1265–1275. https://doi.org/10.1016/j.psyneuen.2011.06.001

van Anders SM, Tolman RM, and Volling BL (2012) Baby Cries and Nurturance Affect Testosterone in Men. Horm Behav 61 (1):31–36. https://doi.org/10.1016/j.yhbeh.2011.09.012

van Anders SM, and Watson NV (2006) Social Neuroendocrinology: Effects of Social Contexts and Behaviors on Sex Steroids in Humans. Hum Nat 17 (2):212–237. https://doi.org/10.1007/s12110-006-1018-7

——— (2007) Effects of Ability- and Chance-Determined Competition Outcome on Testosterone. Physiol Behav 90 (4):634–642. https://doi.org/10.1016/j.physbeh.2006.11.017

Velocci B (2024) The History of Sex Research: Is "Sex" a Useful Category? Cell 187 (6):1343–1346. https://doi.org/10.1016/j.cell.2024.02.001

Wiley AS (2023) Biological Normalcy. Annu Rev Anthropol 52 (1):223–238. https://doi.org/10.1146/annurev-anthro-052721-090632

Williams JS, Fattori MR, Honeyborne IR, and Ritz SA (2023) Considering Hormones as Sex- and Gender-Related Factors in Biomedical Research: Challenging False Dichotomies and Embracing Complexity. Horm Behav 156:105442. https://doi.org/10.1016/j.yhbeh.2023.105442

Wingfield JC, Hegner RE, Dufty AM, and Ball GF (1990) The "Challenge Hypothesis": Theoretical Implications for Patterns of Testosterone Secretion, Mating Systems, and Breeding Strategies. Am Nat 136 (6):829–846. https://doi.org/10.1086/285134

Zugman A, Alliende LM, Medel V, et al. (2023) Country-Level Gender Inequality Is Associated with Structural Differences in the Brains of Women and Men. PNAS 120 (20):e2218782120. https://doi.org/10.1073/pnas.2218782120

2

Entanglement of Gender/Sex Dynamics in Basic and Developmental Systems Biology

Colin J. Saldanha, Gillian R. Bentley, Charlotte A. Cornil, Geert J. de Vries, Holly Dunsworth, Margaret M. McCarthy, Rebecca M. Shansky, Lynnette Leidy Sievert, and Catherine S. Woolley

Abstract Establishing a common language in research is essential but challenging in the context of gender/sex entanglement. Without agreement and understanding of specific terms, study design and analysis of findings risk bias, misinterpretation, or misrepresentation. This chapter scrutinizes working definitions and argues that animal research can incorporate the concept of gender by modeling components of human gendered expectations or interactions in ethological contexts relevant to these expectations and interactions for animals. It looks at the considerable variation in physiological and social factors that contribute to sex and gender, and analyzes gender/sex entanglement across multiple levels of biological organization and different life stages. Particular attention is given to critical periods of brain development. It advocates for incorporating gender/sex entanglement into the conception, conduct, and communication of science, and discusses new initiatives and resources that can support studies of gender/sex entanglement. Illuminating how deeply sex and gender are entangled—and all the attendant complexity—holds promise for expanding the acceptance of that complexity in all corners of society.

Keywords Gender/sex entanglement, animal research, critical periods of brain development, life stages, neuroendocrinology

Colin J. Saldanha (✉)
Dept. of Neuroscience, American University, Washington, DC 20016, USA
Email: saldanha@american.edu

Affiliations for the coauthors are available in the List of Contributors

© Frankfurt Institute for Advanced Studies (FIAS) 2025
L. Z. DuBois et al. (eds.), *Sex and Gender*, Strüngmann Forum Reports,
https://doi.org/10.1007/978-3-031-91371-6_2

Group photos (top left to bottom right)   Colin Saldanha, Catherine Woolley, Gillian Bentley, Rebecca Shansky, Charlotte Cornil, Geert de Vries, Gillian Bentley, Margaret (Peg) McCarthy, Lynnette Sievert, Holly Dunsworth, Colin Saldanha, Charlotte Cornil, Rebecca Shansky, Geert de Vries, Holly Dunsworth, Lynnette Sievert, Margaret (Peg) McCarthy, Catherine Woolley. Photos by Norbert Miguletz.

1 Introduction

This Ernst Strüngmann Forum was convened "to advance conceptualizations of *gender* and *sex*, to align dialogue across disciplines, and to promote sound application in research, policy, medicine, and public health." The focus of our working group was on basic and developmental systems biology, which accordingly restricted our focus to research and medicine as opposed to policy or public health writ large. We embraced the central goals of the Forum itself and further asked how the concept of sex and gender entanglement could transform research. More specifically, we asked: What might be gained (or lost) if sex and gender are viewed as entangled? How might the entanglement vary across the lifespan or during critical periods of development? How can we better operationalize gender and sex entanglement? Our working group consisted of scientists who work in areas ranging from neurogenomics through behavioral neuroendocrinology and endocrinology to biological anthropology. This set of expertise conferred a range and depth of knowledge about sex, gender, and their entanglement. It also created challenges as variation in discipline-specific principles and vocabularies require mutual understanding to allow fruitful discussion and debate. Establishing a common language in research is essential but uniquely challenging in the context of gender/sex entanglement. Without mutual agreement and understanding of specific terms, there is risk of confounds to study design, biases in analyses, and misinterpretation or misrepresentation of experimental findings.

The complexity of gender/sex entanglement makes it difficult to generate clean and unassailable definitions of sex and gender due to the nature of life itself, in which there are always exceptions to any "rule" that further erode absolutism. Moreover, critical differences among observations for the "majority" should not be labeled as the "the norm" or "veridical." Exclusive reliance on mean differences between groups can also obscure the actual difference in terms of effect size. Inclusion of information about variance goes a long way toward tempering the potentially misleading nature of data presentation and consequent interpretations. As researchers, we must carefully assign language to a messy distributional space keeping in mind that "*it is not that words have meanings, but it is meanings to which we assign words. That is why we have so many languages*" (adapted from Raman 2018). Thus, we recognize that terminology can and should evolve as our understanding of gender/sex entanglement matures.

With this context in mind, we grappled with the definition of "sex" and collectively identified differences in gamete size as the most consistent sex difference across eukaryotic species. However, gamete size is a function of gonadal phenotype, which is in turn a function of chromosomal complement, hormone synthesis and secretion, and so on, resulting in the emergence of a constellation of features that contribute to a sexually differentiated phenotype. Nonetheless, gamete size by itself remains an imperfect correlate of sex because (a) there is no internal consistency between the various components of the genotype and phenotype of sex within some individuals and (b) during some phases of the lifespan there are few or no gametes (i.e., pregonadally differentiated embryos or greatly reduced and nonviable gametes in postmenopausal women). Indeed, the elusive nature of sex has been pointed out previously, with Lillie (1932) famously declaring: "There is no such biological entity as sex. What exists in nature is a dimorphism within species into male and female individuals…" Thus, we invite further discussion as to the meaning of the term "sex," and fully acknowledge that the "proper" definition may vary by context and purpose.

To date, the majority of scientific work using animal models, including humans, has failed to recognize the diversity associated with sex. More specifically, until relatively recently, only a small minority of preclinical and clinical studies in the life sciences collected and analyzed data on female subjects, while some disciplines purposefully excluded males (Beery and Zucker 2011; Becker et al. 2016; Shansky and Woolley 2016). This inherent biological bias in the generation of knowledge is being corrected due to new mandates, guidelines, and encouragements from biomedical funding agencies across the world (Miller et al. 2017; Clayton and Collins 2014; McCullough et al. 2014). However, a new challenge has emerged in that explanations for variations in phenotype across the lifespan are constrained by the need to operationalize the biosocial development of an individual within its environment. This includes the development of individual gender identity within humans as well as the aspects of human gender in the larger social and cultural environment, which can influence well-being across the lifespan. Moreover, to our knowledge, there is no evidence to support the concept that nonhuman animals experience gender identity nor that they have similar structures of gender as they are commonly understood in human societies (see Chapter 4). That said, some components of sex do not exist independently of an animal's social experience (more on this later). Thus we argue there is an opportunity to incorporate the concept of gender into animal research by modeling components of human gendered expectations or gendered interactions in ethological contexts relevant to these expectations and interactions for animals. Examples include manipulating resource allocation, social context (e.g., parental care, dominance hierarchies, social defeat), or social stress. Modeling components of gendered experiences and behaviors in nonhuman animals that resemble those in humans might allow for an assessment of the biological impact of those experiences and thereby improve the quality of research and the translation of science related to sex and gender to health and well-being.

2 Working Definitions of Sex, Gender, and Gender/Sex Entanglement

Consider the following broad topics: (a) the variation of sex, gender, and their entanglement over time (the lifespan) and within biological levels of analysis (molecules to behavior) and (b) sex, gender, and their entanglement in biomedical research and their applications in the clinic and policy. Engaging with these topics requires working definitions of sex and gender as well as an understanding of how they do and do not generalize across different species and research questions. Yet, as noted above, there is no single definition for sex or gender that universally fits across all contexts and intended usages. For example, the optimal definition of sex is likely to differ based on the application (e.g., between a project that seeks to simulate sexual selection on genetic variation as compared to one that seeks to understand health outcomes among cis- and transgender individuals and populations). Similarly, the optimal approach to operationalizing gender is likely to differ between biomedical versus anthropological or sociological research contexts. Notwithstanding this complexity, effective communication requires clarity regarding how the terms sex and gender are being used in the context at hand. With this principle in mind, this chapter builds on the definitions of sex and gender that have been proposed by the Canadian Institutes of Health Research Institute (CIHR) of Gender and Health, outlined below.

- *Sex* refers to a set of biological attributes in humans and animals. It is primarily associated with physical and physiological features including chromosomes, gene expression, hormone levels and function, and reproductive/sexual anatomy. Sex is usually categorized as female or male, but there is variation in the biological attributes that comprise sex and how those attributes are expressed.
- *Gender* refers to the socially constructed roles, behaviors, expressions, and identities of people. It influences how people perceive themselves and each other, how they act and interact, and the distribution of power and resources in society. Gender is not confined to a binary (girl/woman, boy/man) nor is it static; it exists along a continuum and can change over time. There is considerable diversity in how individuals and groups understand, experience, and express gender through the roles they take on, the expectations placed on them, relations with others, and the complex ways that gender is institutionalized in society.
- *Gender identity*, while not explicitly defined by CIHR, is a personal and internal sense of oneself as male, female, or other.

While the definitions above apply well to humans, the nature of our group and its charge necessitated consideration of sex and gender in both human and nonhuman animals. Thus, we suggest the following:

- *Sex* is a multidimensional construct that for most mammals is primarily based on differences in sex chromosome complement and anatomy of the genitourinary tract. Natural variation encompasses diverse combinations of these features, but the majority of mammals have either XY chromosomes and testes (males) or XX chromosomes and ovaries (females), accompanied by a corresponding visible dimorphism of the external genitalia into male or female phenotype, with considerable variation therein. We will use the terms male and female in this way while acknowledging that a substantial number of human individuals (estimated at ~1% of the population or about 80 million people globally) have noncanonical combinations of sex chromosomes and genitourinary anatomy.
- *Gender and gender identity* are operationalized in this chapter as experiences that differ in magnitude or form across variations in groups based on the perception of the self or others. We will use gender in this way while acknowledging that there is considerable diversity in what gender means across different professional and societal communities. This includes the term "gender fluidity," which can be used to describe individuals switching between different gender modes (e.g., trans and cis conditions). Here, it may prove useful to make a clear distinction between attributes of gender in

humans compared to nonhuman animals. Querying the gender of a nonhuman animal is, of course, difficult at this time since, put plainly, we do not speak the same language. However, we can certainly talk about "gendered experiences," in animals. Since we refer to and study "anxiety-like" and "depressive-like" behaviors, there may be a space for discussing "gendered experiences" when referring to nonhuman animals.

• The theory of *gender/sex entanglement* is a topic of scholarly debate and a central topic for this Forum (see Chapter 4; DuBois and Shattuck-Heidorn 2021; Viuff et al. 2023; Fausto-Sterling 2019; Greaves and Ritz 2022; Kaiser 2012). For our purposes here, we acknowledge that although sex and gender (or gendered experience) are defined as different terms, they are deeply intertwined across the life course, as humans and other animals are shaped by and shape their environment. As such, individual attributes that differ as a function of gender or sex will variably reflect accumulated effects of a reiterative interaction (or entanglement) between sex and gender over development. Disentangling the contributions of sex and gender to an outcome of interest is both conceptually challenging and logistically difficult, especially in humans, where the space for gender/sex entanglement is large and often beyond experimental analysis. Indeed, in humans, gender cannot exist without sex and vice versa. However, as detailed throughout this chapter, it is nevertheless possible to propose and study different modes of gender/sex entanglement.

3 Gender/Sex Entanglement across the Lifespan and Biological Levels

There is considerable variation in the nature of the physiological and social factors that contribute to sex and gender. In gender/sex entanglement, an individual develops through interaction with its environment such that at any given moment, the traits of the organism represent the cumulative history of these interactions. Importantly, these interactions occur across multiple levels of biological organization and different stages of the lifespan. While some factors are essentially fixed and immutable (e.g., sex chromosomal complement), others are considered plastic or mutable (e.g., morphology, physiology, and behavior). The degree to which gender/sex entanglement is entrenched within the fixed and plastic factors varies across life. This variation may be envisioned along the two axes of space and time. By "space," we mean levels of biological analysis: from molecules, genes, cells, and organs onward to behavior and systems.

Phenotypic variation within an individual occurs in molecules, cells, tissues, and organs, as well as in the orchestration of behavior by complex ensembles of communication across organs. At the molecular level, phenotypic variation begins with genes; the X and Y chromosomes are immediate sources of sex differences in genes and are essentially immutable. As one moves away from genes, there is an increase in plasticity and influence of the environment. The greatest plasticity may be reflected in the redundant communication systems that orchestrate complex physiology, including behavior, but the degree of gender/sex entanglement also varies along this axis, as aspects of gender can become encoded in or entangled with components of sex. Put more simply, the nature of gender/sex entanglement increases in complexity with greater biological organization.

3.1 Sex, Gender, and Their Entanglement across the Lifespan

In eukaryotes, after fertilization, and with a few notable exceptions, the resulting zygote has a sex chromosome complement, which in mammals is XX for female and XY for male. Individuals with a multitude of sex chromosomes are also categorized into a single sex such that XO is female and XXY is male and so on. Within weeks following fertilization in humans and nonhuman primates, and within days in rodents, the SRY (sex determining region of the Y chromosome) gene will initiate a gene expression cascade that will direct the formation of a testis from the primordial gonadal anlage. Conversely, the action of COUP-TFII and other transcription factors directs the embryo toward developing female gonads and genitalia. Maintenance of the sex-typical phenotypes requires functional signaling pathways throughout the life course (e.g., SOX9 in the male and WTNT4 in the female) (Lin and Capel 2015; Chassot et al. 2014). In mammals, the germ cells within a testis versus an ovary will subsequently be directed toward the generation of sperm, which can be expelled from the body, versus oocytes, which are expelled from the ovary into the fallopian tubes and uterus. Hormonal secretions specific to gonadal phenotype will direct formation of the reproductive tract to match the demands of internal versus external gamete release. These events occur prenatally and are therefore largely, but not entirely, independent of gender. The developing embryos of various vertebrate species are influenced by changes in the maternal or developmental environment in a sex-biased manner. For example, when a pregnant rat or mouse experiences an inflammatory condition, such as the mimicking of a viral or bacterial infection, the developing fetus often develops brain anomalies, and these are significantly more likely and more severe in male fetuses than female (Arambula and McCarthy 2020).

As gestation proceeds, the potential for gender/sex entanglement in the fetus increases due to the gendered cultural world in which the biological mother lives. This could include restriction of nutritional and other resources or heightened psychological stress if a woman gives birth to a child of a culturally less desired sex. There is a step-function increase in gender/sex entanglement at birth when a child is born into the gendered world. Our group agreed that the closer we are to conception, the larger the contribution of sex compared to gender is for any phenotype. In other words, the degree of entanglement between the impact of sex versus gender is minimal early in life but entanglement increases as an individual progresses through the life course, ultimately blurring the contributions of sex and gender to almost any measurable endpoint. One example of the impact of such interaction is how being a female (descriptor intentionally chosen to refer to genital sex) in a given environmental context that is not favorable to women (descriptor intentionally chosen to refer to the female gender) may generate psychosocial stress, which may in turn alter gonadal, adrenal, and/or neural steroidogenesis (Dedovic et al. 2009). The multiplication of such interactions (or the cumulative experience of gendered situations) as one progresses through life can result in further divergence from the initial, lower degree of sex-typical variation.

3.2 Gender, Sex, and Their Entanglement: Critical Periods

Although gender/sex entanglement increases with age, this should not be taken to imply a linear accumulation of the impact of gender. Brain development is characterized by epochs during which specific stimuli or entrained genetic programming dictate the formation of neural circuits or enduring neuronal activity patterns relevant to function. These functions vary from detection and encoding of sensory stimuli such as light and touch, to adult reproductive physiology and behavior. If the appropriate stimuli are not experienced during that critical period, the adaptive developmental process will not occur. Some critical periods are sex-specific whereas others are influenced by sex, gender, and gender/sex entanglement (McCarthy et al. 2018; Steensma et al. 2013). As a result, age-dependent amplification of the entanglement between sex and gender occurs in a stepwise fashion, with some steps bigger than others.

3.2.1 Hormonally Mediated Sexual Differentiation of the Mammalian Brain

Early in gestation, the process of sex determination proceeds with the differentiation of the testes and ovaries, as we have discussed. By mid to late gestation, the fetal testis begins producing androgens, which induces a process generally referred to as "sexual differentiation of the brain" but is more appropriately thought of as early life programming by androgens to masculinize the brain in males. One reason for this is that in the absence of androgen synthesis, the brain develops as female (McCarthy and Arnold 2011). The onset of androgen production in the testis in males defines the beginning of the critical period. The closing of the critical period is defined as the time at which the fetus is no longer sensitive to the developmental programming effects of androgens. The timing of the loss of sensitivity is not the same for the two sexes because once males are exposed to androgens the window closes; for females, however, the window of opportunity for effects of androgens (and in some cases estrogens) endures for several more days (in rodents) and potentially weeks in humans, although we do not really know. This is nonetheless important in that if developing females are exposed to androgens, they are as responsive as males to the programming effects. During the time that the fetal testis is producing androgens, the ovary is not steroidogenic, but in rodents, at least, there is a later developmental period when the ovary produces estrogen and this contributes to early life programming of the female brain (Bakker and Brock 2010). It is important to acknowledge that while there may be other female-specific signals that could program the female brain, they remain largely undescribed.

Puberty constitutes an additional sensitive period during which androgens in males act on the brain to further modify and complete the developmental process started perinatally in a manner that will align adult physiology and behavior with the presence of testes while, similarly, in females estrogens act on the brain to align behavior with the presence of an ovary (Schulz et al. 2009). More specifically, in the realm of physiology, this means in females that the pulsatile pattern and release of the gonadotropins—luteinizing hormone (LH) and follicle-stimulating hormone—from the anterior pituitary differs in frequency and amplitude across the menstrual cycle and induces ovulation at mid-cycle. In contrast, in males the daily pulsatile pattern of the gonadotropins contributes to testosterone production and spermatogenesis. The pattern of LH release by the pituitary is controlled by the brain and programmed by early life hormone exposure.

Behaviorally, the motor patterns of both courting and copulatory behaviors are different between the sexes in most mammals. Experiments in rodents have greatly expanded our understanding of the process of hormonally mediated sexual differentiation of the brain and, importantly, determined that both males and females are responsive to the impact of androgens and bioactive metabolites such as estrogens. In humans, various conditions can result when females are exposed to higher than typical levels of androgens, either from their own adrenals, as in the case of congenital adrenal hyperplasia, or from their mothers due to polycystic ovarian syndrome, or even from the *in utero* presence of a male twin (Hines 2011; Cohen-Bendahan et al. 2005). This can impact the phenotype of girls in subtle or dramatic ways, each of which will have subsequent entanglements with gender (reviewed in McCarthy et al. 2017).

3.2.2 Brain Development in Adolescence and Puberty

Adolescence is a particularly critical period for the impact of peers on an individual's psychosocial development and that impact is entangled by sex and gender, including the complexity of puberty. Relatively little preclinical work focuses on the adolescent period, in large part because of the reliance on rodents, which have a short span between independence from the dam and reproductive capacity. Humans are notable for an extended childhood compared to other long-lived species (Bogin 2020), which greatly increases the potential for gender/sex entanglement prior to full adulthood.

Puberty is marked by dynamic changes in both sex- and gender-related factors, as well as sex-biased risks for multiple health-related outcomes including mental health (Alderman et al. 2003). It is also a critical period during which males and females are exposed to the same gonadal hormones but at different levels and patterns. The developmental timing of this critical period also differs between the sexes: the onset of puberty is defined differently for boys and girls and occurs on average at a younger age in females than males across a wide range of mammalian species, including humans (Cheng et al. 2021). This important difference during a major life transition is undoubtedly entangled with gender, especially how the world genders children, but differently for girls and boys who are becoming young women and men. Such gendering mechanisms include, but are not limited to, cultural, social, and familial pressures that may impact gender differentially across both age and sex. Multiple sex-related factors first emerge or undergo dynamic changes during puberty, including sex differences in height, emergence of secondary sexual characteristics (Cheng et al. 2021), and motivation toward sexual behavior (Feldmann and Middleman 2002). The stereotypic, consistent emergence of these physical and behavioral changes across diverse sociocultural and historical settings strongly suggests multiple sex-biased developmental programs that "switch on" during puberty (Bordini and Rosenfield 2011). It is difficult, however, to recognize a strict pubertal influence on gender/sex entanglement, as there is substantial interindividual variability in the timing of puberty (Marceau et al. 2011) and abundant evidence that this timing can be influenced by environmental factors (Fisher and Eugster 2014).

Sex-biased behavior and mental health risk in adolescence are well documented. Specifically, with the onset and progression of puberty in humans, there is a rapid and disproportionately male increase in accidental deaths, suicide, substance abuse, and violent offenses (Federal Interagency Forum on Child and Family Statistics 2009) alongside a disproportionately female increase in mood, anxiety, and eating disorders (Green et al. 2005). Teenage girls are also five times more likely to experience sexual assault than their male peers (Bentivegna and Patalay 2022). The expression of these conditions and their prognosis often differs between males and females. However, very little is understood regarding the potential for gender, such as the aforementioned examples, to interact with brain-based pubertal changes. Puberty takes place over multiple years and is associated with cognitive, emotional, and social development, gender differences in stress, as well as sexual development (Copeland et al. 2019).

3.2.3 Critical Periods Specific to Females

Pregnancy and menopause are developmental transitions involving intense and documented neuroplasticity (Lonstein 2003). Potential changes in the brain related to menopause, as opposed to aging per se, remain relatively poorly documented and hotly debated (Morrison et al. 2006; Sherwin 2007; Maki and Sundermann 2009). There are clear increased health risks in both pregnant and postmenopausal women, which entangle with gender in terms of perception of need and delivery of appropriate health care. Neither of these is well addressed in preclinical research. Taken together, the accumulation of gendered experiences superimposed on the effects of biological sex at critical/sensitive/transition periods results in an increasing gender/sex entanglement throughout life (Figure 2.1).

3.2.4 Sex, Gender, and Their Entanglement across Generations

Humans differ from mice, songbirds, and many other animals (but not all) in having two pools of inherited conditions: a gene pool and a cultural pool. The interaction between these two pools contributes perhaps most strongly to the entanglement of sex and gender. Culture moves across time from generation to generation, slowly or quickly. It can move from parents to children, children to parents, or horizontally within generations, one to many or many to one (Boyd and Richerson 1988). The gene pool also moves across time from generation to generation. While changes in the gene pool come about slowly relative to cultural changes, and depend on mutations, natural selection, genetic drift, and gene flow, epigenetic changes in gene expression are becoming increasingly recognized as having the potential to accelerate processes of change (Muyle et al. 2021). Furthermore, epigenetic changes in gene expression in the reproductive system can occur within the order of one generation

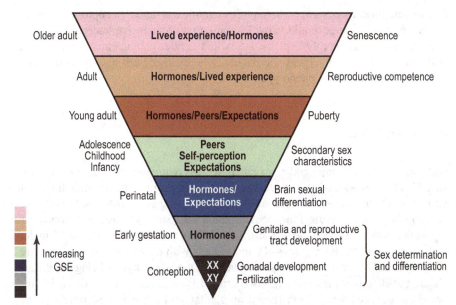

Figure 2.1 A conceptualization of increased gender/sex entanglement (GSE) across the lifespan that occurs in epochs as opposed to linear progression, with each building upon the past. The degree of GSE varies with the biological level: entanglement increases as processes move beyond the molecular and cellular level, as occurs early in life. Expectations based on gender begin as early as birth and accumulate into lived experience gradually over the life course.

(Bar-Sadeh et al. 2020). The interaction between genes and culture continues to be transformed across thousands, even millions of years (Durham 1991) and may influence each other. For example, domestication of dairy animals led to a rapid spread of a mutation in the lactase gene that allowed humans to digest milk beyond infancy (Gerbault et al. 2013). Other examples of coevolution of culture and gene pools in humans could include increased brain size (genetic change) and tool complexity (Heldstab et al. 2016) or cooking (cultural change) (Carmody et al. 2016). Since tool use and cooking are engendered behaviors, this coevolution may provide examples of gender/sex entanglement.

4 Gender, Sex, and Their Entanglement in Biomedical Research

To advance conceptualizations of gender/sex entanglement needed to promote sound application in research, policy, medicine, and public health, we sought to define gender/sex entanglement so that testable hypotheses could be generated to guide biomedical research. Challenges arise from the sheer complexity of the variables associated with sex, gender, and their entanglement, as well as the fact that the very species in which gender is most developed is one on which we cannot experiment, at least in the sense to which preclinical researchers are accustomed. There is additional tension between the efficiency of gathering data and an accurate modeling of the complexity of the world from which the data are collected. Hypothesis-driven research attempts to eliminate variables down to known causal entities and thus, by design, oversimplifies complex systems. In contrast, other scientific approaches seek to add variables until a comprehensive understanding of a phenomenon is acceptably explained. Sex and gender entanglement arguably sits at the nexus of these divergent approaches, which either limit or expand the set of potential explanations of phenotype.

4.1 Measuring Sex

Quantitatively analyzing gender/sex entanglement requires defined and measurable variables for sex and gender as well as a statistical framework for modeling these variables. There are several approaches for measuring sex. Historically and especially in large-scale studies, the most common approach is to ask an individual to report their own sex, code the responses as a binary variable of "male" or "female," and assume that "male" captures XY individuals with testes and "female" captures XX individuals with ovaries. This approach has the advantage of being simple and cheap, but it fails to acknowledge two critical issues. First, reported sex is often based on assigned sex at birth (based on genital sex) but may also reflect gender identity. Second, this approach ignores the complexity of sex-related variables and can introduce measurement errors, the size and import of which will vary depending on the research setting. Definitive prevention of this kind of error requires measuring (a) the sex chromosome complement by genetic analysis and (b) the gonad type by physical exam and/or imaging (e.g., ultrasound). Further, untangling the confounds of genital sex and gender identity cannot be done without explicitly asking about one or the other (Bauer et al. 2017; Beischel et al. 2022).

Sex chromosome complement and gonadal type at birth are core sex-related traits. Both traits share two key properties: (a) typically they are developmentally static and fixed in the absence of medical intervention or disease; (b) for the vast majority of individuals, the features can be measured as binary variables (XX or XY, testes or ovaries) that are highly correlated with each other across individuals (Sanchez et al. 2023; Garcia-Acero et al. 2020). It is the combination of these two phenomena that enables males to be defined as individuals with XY chromosomes and testes, and females as individuals with XX chromosomes and ovaries. However, these binary variables fail to distinguish individuals with noncanonical sex chromosome dosages (e.g., sex chromosome aneuploidies), gonadal types (e.g., gonadal dysgenesis), or uncommon endocrine conditions (e.g., androgen insensitivity, congenital adrenal hyperplasia) (Garcia-Acero et al. 2020). Therefore, it is crucial that researchers describe precisely how they measured sex and (if relevant) core sex-related variables.

These concerns with measurement error are particularly relevant in the study of populations that are enriched for noncanonical sex chromosome dosages and/ or gonadal types, or in cases where the research question is focused on such variations. In practice, the strong autocorrelations among sex chromosome complement, gonadal type, and self-reported sex combined with the logistic and resource considerations that apply in most research settings mean that risk of measurement error is brought within a tolerable range for most applications. However, the broad statistical applicability of a binary scale for measuring the phenotypes that comprise sex in most research settings coexists with there being (a) diverse naturally occurring sex chromosome and gonadal complements represented in the population and (b) an ethical and scientifically important reason to engage with this diversity (Goetz et al. 2023).

Unlike core sex-related traits (i.e., chromosomal complement and gonadal type), other sex-associated phenotypic features exhibit continuous variation across individuals of a given species. Thus, while the vast majority of individuals for any given mammalian species falls into one of two distinct groups in an unsupervised clustering based on core sex-related variables (males and females), such binarizations would not emerge from clustering based on most other organismal features related to sex, such as height or weight. However, organismal features do vary in their statistical correlation and causal relationship with core sex-related traits such that some sex-associated phenotypic features are highly colinear with the binary variable of sex. Examples of such features include expression levels of X- or Y-linked genes, circulating levels of gonadal steroids, and secondary sexual characteristics.

Sex-associated phenotypes (e.g., stature, muscle mass, body fat, facial and body hair, pelvic dimensions) have a range within each sex group and can overlap between sex groups, but will covary strongly with the binary variable of sex at the population level (Figure 2.2). There are also sex-associated phenotypes that only vary within each sex group, such as age at menarche or oigarche (menarche is only possible given ovaries and oigarche given testes). Importantly, almost all organismal features, sex-associated or otherwise, are potentially sensitive to common environmental exposures—including gendered ones—in ways that the core sex-related traits of sex chromosomal dosage and gonadal type are not.

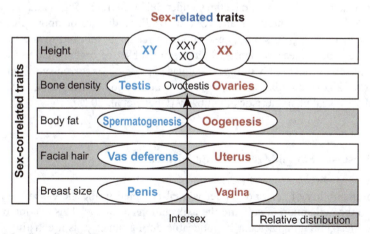

Figure 2.2 Sex-related traits versus sex-correlated traits. Sex-related traits are unique to males and females and form the definition therein. Generally, but not exclusively, they are immutable and covary within one individual; this means that gonadal type, reproductive tract, and genitalia are consistent with sex chromosome complement and gamete type. Sex-correlated traits are common to males and females but influenced by sex-related traits. They vary along a continuum and overlap in range between males and females. Thus, they have little predictive capacity for sex. Another way to frame this concept would be to refer to the axes as "sexually dimorphic phenotypes" and "sex differences in phenotypic traits."

4.2 Measuring Gender

The definition of gender varies widely between and within different academic disciplines (for further discussion, see Chapters 3 and 4). For the specific purpose of quantitatively analyzing gender/sex entanglement, we propose that gender variables be defined as those that index *environmentally mediated influences which correlate with sex-related traits through the perception of sex-correlated phenotypes by others or oneself*. These correlations can arise in an active manner (e.g., a person actively seeks out female/women-associated environments because they perceive themselves to be a female/woman) or in a passive, evoked manner (e.g., a person is treated differently at work because they are perceived to be a female/woman). Gender variables could also refer to an individual's internal mental environment, such as mental experiences of one's gender identity, sexual orientation, the state of pregnancy, having menstrual periods, or experiencing nocturnal emissions. The potential for these experiences is strongly correlated with core sex-related traits and can vary across individuals due to biological, psychological, or social factors. Just as sex-correlated phenotypes can vary in their causal proximity and statistical correlation with core sex-related traits, so too can gender variables. For example, for cis individuals, the gendered pronouns by which a person is addressed will be very highly correlated (albeit imperfectly) with that person's core sex-related traits, whereas a person's experience of gendered social media content is likely to be more weakly correlated with their core sex-related traits.

The measurement of gender is a rapidly evolving field, which presents substantial complexity given the profound heterogeneity in how different academic disciplines and practitioners operationalize gender. From the developmental biology perspective represented by our group, some examples of tools for measurement of gender-related variables would be the Gender Self-Report (Strang et al. 2023) or the UN Women's Model Questionnaire (UN 2016). For a discussion about phenotypes other than sex and core sex-related traits, the reader is referred to the following articles regarding measurement of that special subset of organismal features that represent sex-correlated phenotypes, such as expression of sex chromosome genes (Liu et al. 2023), circulating levels of gonadal steroids (Casals et al. 2023), and secondary sexual characteristics in puberty (Cheng et al. 2021).

4.3 Measuring Sex and Gender Entanglement

The inseparability of sex and gender is well-reflected in the increased interaction between the perception of self and the expectations of others. This occurs throughout the lifespan but is particularly noticeable during puberty, as it constitutes a period of tremendous change particularly in the infiltration of external cues and social evaluation. For a description of interactions between sex, gender, and their entanglement with attention to the interaction of self and others, consider the following anecdote from Gillian Bentley, concerning cultural differences in the entanglement of sex and gender.

Each of us have our own assumptions about how particular items of clothing are either "male" or "female" in their associations. Of course, none of these cultural associations have any meaning in the context of the Ituri Forest. They can, however, reveal how specific assumptions about appropriate dress can be foisted onto young children, who may not have assimilated these specific gendered associations.

In the Ituri Forest in what was Zaire, now Democratic Republic of Congo, lives a horticulturalist group of Sudanic origin called the Lese. They live in symbiosis with a Pygmy group, the Efe, with whom they exchange food items and labor. This area is isolated but there is a market about one hour's walk from the village where I was living, where second-hand clothing from Europe and the United States could be purchased, presumably originating from global charity networks (e.g., Oxfam) in the wider African continent. One day, while walking along the only road (i.e., dirt track) to find women participating in our project, I saw one male Lese whom I knew walking toward me, wearing what was obviously a newly acquired item of clothing: a white frilly nylon petticoat. Leon (pseudonym) was clearly delighted with his new purchase, telling me how happy he was while I was desperately trying to keep a straight face. After he went on his way, I dashed back to our own village to tell my fellow field companions about Leon's "new look" and to keep an eye open for him.

Another item of Western clothing that was highly prized among the Efe male hunters were nylon undies usually purchased to cover female genitalia in Western countries (called "slippis" in Kingwana, a local dialect of Swahili familiar to the local people). Acting rather like jock straps, I imagine, and worn underneath loin cloths made of bark cloth, a Pygmy man who was able to obtain a slippi of this kind (and red ones appeared particularly desired) was very happy indeed, although I wouldn't describe nylon as an optimal fabric in the hot and humid tropical rain forest.

The blurring of sex and gender makes it difficult to speak of pure sex or gender differences in human research. One can, however, define parameters important in disentangling factors that contribute to sex/gender differences in human traits. For example, discomfort experienced by some people during menses may illustrate sex/gender entanglement. A combination of physical (e.g., cramps) and environmental conditions (e.g., climate, social conditions) contributes to discomfort. At first glance, physical conditions appear to be directly related to sex: having ovaries and the accompanying cyclical hormonal release cause women to have menses. However, social situations play a role as well. For instance, the presence of others may cause embarrassment and lead a woman to avoid specific activities (e.g., swimming, gym, other athletic events); in certain religious traditions, a woman may be deemed unclean whereas in specific societies, women may have to use a menstrual hut. These situations may cause emotional hardship. It should be remembered that the experience and frequency of menses have differed substantially across time and in different settings, depending on how humans have been able, or have chosen, to regulate their fertility (Riddle 1994; Strassmann 1997).

Although the experience of menses is based on a biological process atypical in males, these are examples where sex and gender interact, making the dysphoria associated with menses altogether an engendered condition. In addition to the obvious end result, through the subjective level of unpleasantness and ensuing changes in behavior (e.g., withdrawal from social events), one can identify and measure parameters linked to the physical conditions, and therefore more or less directly to sex, such as menstrual cycle profile, timing and size of hormonal changes. Underlying variables such as levels of inflammation or psychological stress can further influence mood states. One can also identify environmental parameters that are physical (e.g., temperature, humidity) or social (e.g., presence of peers, interruption of social interactions, religious mandates) in nature and clearly linked to gender.

4.4 Statistical Modeling of Sex and Gender Variables

Standard statistical methods, including correlation, regression, principal component, and cluster analyses, can be used to interrogate the contribution of sex and gender variables and how they relate to a given endpoint. These efforts can help to identify correlated aspects of sex and gender to inform theory, improve measurement tools, prioritize variables for analysis with complementary (e.g., qualitative) methodologies, and optimize the predictive power of machine learning algorithms. Statistical frameworks for interrelating sex and gender variables can also help to articulate, formalize, and test hypothesized phenomena across biological and social sciences alike. For example:

1. The potential for effects that can be reliably separated as reflecting sex or gender can be tested by including all as predictors of target outcomes. Colinearity is likely to be a challenge when including both sex and gender as independent variables in regression. This challenge can be mitigated by increasing sample sizes, running stepwise regressions, and (if compatible with the question at hand) intentionally selecting decorrelated subcomponents of gender.
2. The potential for varying effects of gendered variables as a function of demographic variables (e.g., "ethnicity") can be addressed by modeling outcomes of interest through using gender and other variables in interaction with each other. The estimated interactions are important quantitative references for the rich social science field of intersectionality research.
3. The potential for developmental dynamism in sex and gender effects—or for effects that are contingent on the relative timing and duration of differed gendered exposures—can be measured by time series and longitudinal analyses allowing nonlinear modeling of age effects.
4. The relative role of different components of gender for a variable of interest can be measured through multivariate models.

When conducting statistical analyses of this sort, investigators should consider whether sex and gender (as defined in this chapter) are separable or inextricably entwined. If the latter, consider whether they blur into each other or maintain a distinction. The use of longitudinal data, whenever possible, will increase relevance to long-lived humans that accumulate the impacts of gender/sex entanglement over decades. For some, explanations based on sex are commonly accepted as true, whereas those based on gender are viewed as suspect or even untrue, perhaps because of the inherent uncertainty involved. By improving the operation, measurement, and experimental design of studies involving sex and gender, such implicit and explicit biases can be mitigated.

5 Use of Nonhuman Animals in Gender/Sex Entanglement Research

As noted earlier, nonhuman animals are unlikely to experience gender or gender identity as it is understood in human societies. That said, whether they may possess something that can be called "gender" or "gender-like" remains an open question and depends heavily on how gender is defined; for instance, how one behaves or is treated by others in a way that is correlated with sex may be a "gender-like" experience (Cortes et al. 2019). Regardless, researchers interested in understanding the influence of sex have an obligation to control environmental and social factors to avoid confounds that are typically considered to be outside of what is defined as "sex." One common issue illustrates this point. In many veterinary resource facilities at universities, research institutes, or industry, there is often a sex difference in animal housing density. Male mice are often housed alone to prevent fighting while females are housed in groups, which saves on animal care costs. When males of any species are housed together, robust dominance hierarchies are established which influence a wide range of physiological and behavioral parameters. Thus sex is confounded by the parameter of housing. Researchers using animals should be alert to other parameters that are secondary to sex and yet impact its effects on a given biological parameter.

For most researchers who work with animals, the need to consider gender/sex entanglement comes at the stage of interpretation and generalization to humans. The problem with generalization is not that nonhuman animals necessarily lack gender, but that humans can hardly escape it. All biomedical research results must be interpreted with care when translating to humans, but in the case of sex differences, the additional role of gender should be incorporated. Most if not all sex differences in humans are confounded or entangled in some way by gender.

6 Incorporating Gender/Sex Entanglement in Nonhuman Animal Research

As discussed above, there is a tension between the reductionist approach of most biomedical scientific investigation and the complexity of the lived experience of individual humans. Attempting to model gender-like experiences in nonhuman animals is challenged by the inescapable fact that animal models lack the characteristic cumulative experience of gender and that the goal of most researchers is to limit variables, not add them. Whether naturally occurring behavioral traits that vary on average in animals have validity to the human experience is a further challenge. For example, it is clinically well established that women are more likely to be treated for anxiety, whereas female rodents generally display lower rates of "anxiety-like" behaviors than their male counterparts (Rodgers and Cole 1993; Scholl et al. 2019). There are multiple plausible reasons for this incongruity: (a) anxiety-like behaviors in rodents are sexually monomorphic as opposed to those in humans, (b) the tests used to model anxiety-like behavior in rodents are not relevant to human anxiety, and (c) animal models do not recapitulate the cumulative exposure to the sex-biased/gendered experiences that occur in humans.

Indeed, one major difference between animal and human studies is the degree of variability that exists, both in the sex component (e.g., variability in exposure to hormones or in genetic diversity) and in the gender component (e.g., variability in the extent of exposure or sensitivity to gendered situations) as well as in their accumulation throughout life. Based on this difference in variability, two alternative approaches to testing gender-like situations are proposed:

1. With machine learning and artificial intelligence, simultaneously measure and quantify large numbers of variables and test for covariance with the category of sex.
2. Focus only on sex differences for which a function has been identified in controlled experiments and introduce contextual variability (e.g., by complexifying the social environment, applying stressful stimuli) to determine the influence on physiology or behavior and whether the outcomes model what is observed in humans.

Another possibility is to complexify the environment and determine whether sex differences arise under more variable conditions.

7 Incorporating Gender/Sex Entanglement into the Conception, Conduct, and Communication of Science

The importance of understanding the pervasive effects of sex, gender, and gender/sex entanglement on scientific study and on the dissemination of scientific information cannot be overstated. To develop inclusive and equitable policies, the incorporation of gender/sex entanglement into the conduct and communication of science must be enhanced. While the benefits seem obvious, there are also risks (see Table 2.1). Not all scientific enterprises are of unassailable quality nor are they all well-intentioned. Preordained ideologies can lead to misinterpretation or intentional designation of findings. Such distortions of evidence may lead, for example, to assertions that there are "only two sexes" when the data support a continuous distributional space, or a claim that there are no sex differences when data distributions are, in fact, different between groups. The framing of findings as evidence for what is natural, normal, or divinely intended is just one of the many ways in which research on animal models that consider aspects of gender can be misappropriated. However, the notion of not studying something because it might do harm is not new and should not be used as an excuse to avoid the topic of gender/sex entanglement.

New initiatives have recently been established: Sex as a Biological Variable by the US National Institutes of Health; sex reporting requirements in Canada; the Sex and Gender Equity in Research Guidelines for publishing statements on definitions of sex and gender by the World Health Organization, the US Institute of Medicine, and the CIHR. They have generated a firestorm of opinion pieces, a variety of "how to" guidelines, and more recently, thoughtful analytical approaches to determine whether these new mandates are working. A similar effort around the entanglement of gender with sex would provide much needed light in a space that has had little

Table 2.1 Potential benefits and concerns of studying gender in nonhumans and incorporating gender/sex entanglement.

Benefits to Research on:	Concerns:
Mother/infant interaction	Limitations of lab practices and feasibility
Dominance hierarchies and play behavior	Only one component is studied at a time
Anxiety- and depression-like behaviors	Limited abstract reasoning in animals
Habit formation/addiction/telescoping	Inability to measure cumulative effects
Hormone therapy in both sexes	Difficult to assess transgenerational effects
Sex-related hormone therapy	Increased reinforcement of stereotypes
Sex-based precision medicine	Poorly conducted studies in gender/sex entanglement
Improvements in science itself	Lack of diversity in subjects and researchers
Increased societal understanding	Lack of diversity in research topics
Understanding gender biases in neuroscience and neuropsychiatry	

illumination, and it is hoped that the results of this Ernst Strüngmann Forum will be a first step in that direction. Already there are some potential resources available to support studies of gender/sex entanglement. We include this information below to demonstrate how researchers can collect and analyze relevant data on both sex and gender from humans and also have the potential to influence what kinds of data (retrospective and prospective) should be collected from participants:

1. In 1976, the Nurses' Health Study was conducted in the United States. It recruited married registered nurses, aged 30 to 55, from 11 states: baseline n = 121,700. The first phase studied contraceptive methods, smoking, cancer, and heart disease and later expanded to include additional diseases, lifestyle factors, and behaviors. A follow-up questionnaire was issued every two years, including food-frequency questionnaires. Subsamples of participants have provided toenail, blood, and urine samples for hormone levels and genetic markers as well as DNA from cheek cells. The second phase, Nurses' Health Study II, began in 1989 with women aged 25 to 42; baseline n = 116,430. The research began as a study of oral contraceptives, diet, and lifestyle risk factors (physical activity and diet). Every two years, there are questionnaires about diseases and health-related topics, and every four years there are food-frequency questionnaires. Subsamples of participants have provided blood and urine samples, as well as DNA from cheek cells. The third phase, Nurses' Health Study III, began in 2010 with female and male nurses in the United States and Canada, aged 19 to 46. This study includes nurses with more diverse ethnic backgrounds. Enrollment is still open. Questionnaires include dietary patterns, lifestyle, environment, and nursing occupational exposures. Investigators interested in collaborations are invited to fill out a simple form that asks about the details of the collaboration.

2. The National Health and Nutrition Examination Survey began in the early 1960s and examines a nationally representative sample of ~5,000 people

each year, selected to represent the US population of all ages. Interviews include demographic, socioeconomic, dietary, and health-related questions. In addition, there are medical visits, dental screenings, and physiological measurements. For data users and researchers throughout the world, survey data are available online or via easy-to-use CD-ROMs.

3. The UK BioBank began in 2006 as a longitudinal study of half a million participants aged 40 to 69 at the time of recruitment. Baseline data were collected and there are periodic data sweeps. Data include fMRI, nutritional surveys, blood work, etc.

4. The Avon Longitudinal Study of Parents and Children (ALSPAC) was established to understand how genetic and environmental characteristics influence health and development in parents and children (Fraser et al. 2013). From April 1991 to December 1993, ALSPAC recruited over 14,000 women from a specific area in Southwest England and has followed up with them intensely over the next two decades.

5. Born in Bradford is a more fluid and much less well-funded study than ALSPAC, but it contains a higher diversity of ethnic groups since Bradford has a high proportion of people from South Asia and especially Pakistan. It has a birth cohort component (about 30,000 participants).

6. The data collected by Understanding Society (2024) offer a cross-sectional study, conducted annually, that collects data from entire households with longitudinal elements and biometric data. The United Kingdom excels at longitudinal cohort studies and provides strong financial support. Recently, a new longitudinal study, Our Future Health (2024), was initiated; similar to the UK BioBank, it aims to recruit five million people to assess how life course events and experiences influence health in later life.

8 Future Directions

Gender is an inextricable factor in human development yet does not clearly translate easily to any other species. Nonetheless, given the biosocial similarities that we share with nonhuman animals, there is an opportunity to learn more about animal development, including our own, if we incorporate and account for gender/sex entanglement. Not only will an understanding of developmental and evolutionary biology improve, so will the translation of research in animal models to the improvement of human health and well-being. Recognizing that the culture of science, scientists, and researchers across disciplines has biosocially developed within the context of gender/sex entanglement improves science and its applications.

In nonhuman animals, given the lack of evidence supporting the existence of gender (and therefore gender/sex entanglement) in the way we understand it in humans, it may be tempting to overemphasize the challenges of studying gender and gender/sex entanglement in animal models for research. However, the scientific community is well-versed at operationalizing and approximating human constructs to further research on a number of topics. For example, animal model researchers refer to "depressive-like," and "anxiety-like" behaviors when trying to untangle the origins of human depressive and anxiety disorders. Toward that end, they have constructed assays and measurements that have strong face validity with humans and reliability in cross-species effects. Indeed, pharmaceuticals that reduce anxiety in animals invariably reduce anxiety in a subset of humans as well.

Similarly, researchers who explore the origins of neurodevelopmental disorders (e.g., autism spectrum, attention and hyperactivity, and early onset schizophrenia) have identified multiple factors in animal models, from gene transcription to neuroanatomical changes, that are also evident in these disorders in humans. All of these exhibit strong gender biases in rates of diagnosis, with males being overwhelmingly represented. Emerging evidence reveals a contribution of gender/sex entanglement to this bias, with clinicians quicker to diagnose boys than girls, with girls displaying behaviors that hide core symptoms, and the original diagnostic criteria being based on symptoms frequently observed in boys. This has led to inherent biases in what are considered core symptoms. For instance "systemizing," where boys may obsessively collect model trains or other objects, is considered a core symptom, but girls who obsessively collect dolls or other "feminine" objects are viewed as unremarkable. Despite these inherent biases, there remains strong evidence of a biological contribution to the higher prevalence in boys, with girls not reaching the criterion for diagnoses until older and with a heavier mutational burden than boys.

As with humans, animals experience social stimuli that are explicitly or implicitly gendered. This includes parental care by just one parent, learning to produce courtship songs in species where one sex sings more than the other, or aggressive sex-specific behavior. These experiences necessarily involve an interaction between two individuals, one of whom is the recipient of a gender-like social cue. Similar to gender in humans, the reciprocal nature of this interaction involves the assessment of social cues and their accuracy in terms of predicting future behaviors. Ideally, studies of gender-like conditions in nonhuman animals would be longitudinal to mimic the cumulative effect of gender/sex entanglement. Thus, it may be possible to study the influence of these gender-like experiences on behavioral phenotypes. Importantly, conducting such research and extrapolating experimental findings to human populations comes with caveats, limitations, and the warnings of mistakes which we, as a scientific community, have made in the past. Yet, we also stand at a point in history where there is an escalation of inclusion and thoughtfulness, with increasing awareness of variable ways in which individuals express themselves and how they experience the world around them. Illuminating how deeply sex and gender are entangled and all the attendant complexity holds promise for expanding the acceptance of that complexity in all corners of society. Now is the time to capitalize on the opportunity to effect change.

References

Alderman EM, Rieder J, and Cohen MI (2003) The History of Adolescent Medicine. Pediatr Res 54 (1):137–147. https://doi.org/10.1203/01.PDR.0000069697.17980.7C

Arambula CE, and McCarthy MM (2020) Neuroendocrine-Immune Crosstalk Shapes Sex-Specific Brain Development. Endocrinology 161 (6):bqaa055. https://doi.org/10.1210/endocr/bqaa055

Bakker J, and Brock O (2010) Early Oestrogens in Shaping Reproductive Networks: Evidence for a Potential Organisational Role of Oestradiol in Female Brain Development. J Neuroendocrinol 22 (7):728–735. https://doi.org/10.1111/j.1365-2826.2010.02016.x

Bar-Sadeh B, Rudnizky S, Pnueli L, et al. (2020) Unravelling the Role of Epigenetics in Reproductive Adaptations to Early-Life Environment. Nat Rev Endocrinol 16 (9):519–533. https://doi.org/10.1038/s41574-020-0370-8

Bauer GR, Braimoh J, Scheim AI, and Dharma C (2017) Transgender-Inclusive Measures of Sex/Gender for Population Surveys: Mixed-Methods Evaluation and Recommendations. PLoS One 12 (5):e0178043. https://doi.org/10.1371/journal.pone.0178043

Becker JB, Prendergast BJ, and Liang JW (2016) Female Rats Are Not More Variable Than Male Rats: A Meta-Analysis of Neuroscience Studies. Biol Sex Differ 7:34. https://doi.org/10.1186/s13293-016-0087-5

Beery AK, and Zucker I (2011) Sex Bias in Neuroscience and Biomedical Research. Neurosci Biobehav Rev 35 (3):565–572. https://doi.org/10.1016/j.neubiorev.2010.07.002

Beischel WJ, Gauvin SEM, and van Anders SM (2022) "A Little Shiny Gender Breakthrough": Community Understandings of Gender Euphoria. Int J Transgend Health 23 (3):274–294. https://doi.org/10.1080/26895269.2021.1915223

Bentivegna F, and Patalay P (2022) The Impact of Sexual Violence in Mid-Adolescence on Mental Health: A UK Population-Based Longitudinal Study. Lancet Psychiatry 9 (11):874–883. https://doi.org/10.1016/S2215-0366(22)00271-1

Bogin B (2020) Patterns of Human Growth, vol 88. Cambridge Univ Press, Cambridge

Bordini B, and Rosenfield RL (2011) Normal Pubertal Development: Part I: The Endocrine Basis of Puberty. Pediatr Rev 32 (6):223–229. https://doi.org/10.1542/pir.32-6-223

Boyd R, and Richerson PJ (1988) Culture and the Evolutionary Process. Univ Chicago Press, Chicago

Carmody RN, Dannemann M, Briggs AW, et al. (2016) Genetic Evidence of Human Adaptation to a Cooked Diet. Genome Biol Evol 8 (4):1091–1103. https://doi.org/10.1093/gbe/evw059

Casals G, Costa RF, Rull EU, et al. (2023) Recommendations for the Measurement of Sexual Steroids in Clinical Practice. A Position Statement of SEQC(ML)/SEEN/SEEP. Adv Lab Med 4 (1):52–69. https://doi.org/10.1515/almed-2023-0020

Chassot AA, Gillot I, and Chaboissier MC (2014) R-Spondin1, WNT4, and the CTNNB1 Signaling Pathway: Strict Control over Ovarian Differentiation. Reproduction 148 (6):R97–110. https://doi.org/10.1530/REP-14-0177

Cheng TW, Magis-Weinberg L, Guazzelli Williamson V, et al. (2021) A Researcher's Guide to the Measurement and Modeling of Puberty in the ABCD Study® at Baseline. Front Endocrinol 12:608575. https://doi.org/10.3389/fendo.2021.608575

Clayton JA, and Collins FS (2014) Policy: NIH to Balance Sex in Cell and Animal Studies. Nature 509 (7500):282–283. https://doi.org/10.1038/509282a

Cohen-Bendahan CC, van de Beek C, and Berenbaum SA (2005) Prenatal Sex Hormone Effects on Child and Adult Sex-Typed Behavior: Methods and Findings. Neurosci Biobehav Rev 29 (2):353–384. https://doi.org/10.1016/j.neubiorev.2004.11.004

Copeland WE, C. W, L. S, Costello EJ, and Angold A (2019) Early Pubertal Timing and Testosterone Associated with Higher Levels of Adolescent Depression in Girls. J Am Acad Child Adolesc Psychiatry 58 (12):1197–1206. https://doi.org/10.1016/j.jaac.2019.02.007

Cortes LR, Cisternas CD, and Forger NG (2019) Does Gender Leave an Epigenetic Imprint on the Brain? Front Neurosci 13:173. https://doi.org/10.3389/fnins.2019.00173

Dedovic K, Wadiwalla M, Engert V, and Pruessner JC (2009) The Role of Sex and Gender Socialization in Stress Reactivity. Dev Psychol 45 (1):45–55. https://doi.org/10.1037/a0014433

DuBois LZ, and Shattuck-Heidorn H (2021) Challenging the Binary: Gender/Sex and the Bio-Logics of Normalcy. Am J Hum Biol 33 (5):e23623. https://doi.org/10.1002/ajhb.23623

Durham WH (1991) Coevolution: Genes, Culture, and Human Diversity. Stanford Univ Press, Stanford

Fausto-Sterling A (2019) Gender/Sex, Sexual Orientation, and Identity Are in the Body: How Did They Get There? J Sex Res 56 (4-5):529–555. https://doi.org/10.1080/00224499.2019.1581883

Federal Interagency Forum on Child and Family Statistics (2009) America's Children: Key National Indicators of Well-Being. https://www.childstats.gov/pubs/pubs.asp?PlacementID=2&SlpgID=20

Feldmann J, and Middleman AB (2002) Adolescent Sexuality and Sexual Behavior. Curr Opin Obstet Gynecol 14 (5):489–493. https://doi.org/10.1097/00001703-200210000-00008

Fisher MM, and Eugster EA (2014) What Is in Our Environment That Effects Puberty? Reprod Toxicol 44:7–14. https://doi.org/10.1016/j.reprotox.2013.03.012

Fraser A, Macdonald-Wallis C, Tilling K, et al. (2013) Cohort Profile: The Avon Longitudinal Study of Parents and Children: ALSPAC Mothers Cohort. Int J Epidemiol 42 (1):97–110. https://doi.org/10.1093/ije/dys066

Garcia-Acero M, Moreno O, Suarez F, and Rojas A (2020) Disorders of Sexual Development: Current Status and Progress in the Diagnostic Approach. Curr Urol 13 (4):169–178. https://doi.org/10.1159/000499274

Gerbault P, Roffet-Salque M, Evershed RP, and Thomas MG (2013) How Long Have Adult Humans Been Consuming Milk? IUBMB Life 65 (12):983–990. https://doi.org/10.1002/iub.1227

Goetz TG, Aghi K, Anacker C, et al. (2023) Perspective on Equitable Translational Studies and Clinical Support for an Unbiased Inclusion of the LGBTQIA2S+community. Neuropsychopharmacology 48 (6):852–856. https://doi.org/10.1038/s41386-023-01558-8

Greaves L, and Ritz SA (2022) Sex, Gender and Health: Mapping the Landscape of Research and Policy. Int J Environ Res Public Health 19 (5):2563. https://doi.org/10.3390/ijerph19052563

Green H, McGinnity Á, Meltzer H, Ford T, and Goodman R (2005) Mental Health of Children and Young People in Great Britain, 2004, vol 175. Palgrave Macmillan, Basingstoke

Heldstab SA, Kosonen ZK, Koski SE, et al. (2016) Manipulation Complexity in Primates Coevolved with Brain Size and Terrestriality. Sci Rep 6:24528. https://doi.org/10.1038/srep24528

Hines M (2011) Gender Development and the Human Brain. Annu Rev Neurosci 34:69–88. https://doi.org/10.1146/annurev-neuro-061010-113654

Kaiser A (2012) Re-Conceptualizing "Sex" and "Gender" in the Human Brain. Z Psychol 220 (2):130–136. https://doi.org/10.1027/2151-2604/a000104

Lillie FR (1932) General Biological Introduction. In: Allen E (ed) Sex and Internal Secretions: A Survey of Recent Research. Williams & Wilkins, Baltimore, pp 1–11

Lin YT, and Capel B (2015) Cell Fate Commitment during Mammalian Sex Determination. Curr Opin Genet Dev 32:144–152. https://doi.org/10.1016/j.gde.2015.03.003

Liu S, Akula N, Reardon PK, et al. (2023) Aneuploidy Effects on Human Gene Expression across Three Cell Types. PNAS 120 (21):e2218478120. https://doi.org/10.1073/pnas.2218478120

Lonstein J (2003) Individual Difference in Maternal Care Reveal the Neural Mechanisms of Nurturance. Endocrinology 144 (11):4718–4719. https://doi.org/10.1210/en.2003-0984

Maki PM, and Sundermann E (2009) Hormone Therapy and Cognitive Function. Hum Reprod Update 15 (6):667–681. https://doi.org/10.1093/humupd/dmp022

Marceau K, Ram N, Houts RM, Grimm KJ, and Susman EJ (2011) Individual Differences in Boys' and Girls' Timing and Tempo of Puberty: Modeling Development with Nonlinear Growth Models. Dev Psychol 47 (5):1389–1409. https://doi.org/10.1037/a0023838

McCarthy MM, and Arnold AP (2011) Reframing Sexual Differentiation of the Brain. Nat Neurosci 14 (6):677–683. https://doi.org/10.1038/nn.2834

McCarthy MM, De Vries GJ, and Forger NG (2017) Sexual Differentiation of the Brain: A Fresh Look at Mode, Mechanisms and Meaning. In: Pfaff DW, Joels M (eds) Hormones, Brain and Behavior, vol 3. Elsevier, San Diego, pp 3–32

McCarthy MM, Herold K, and Stockman SL (2018) Fast, Furious and Enduring: Sensitive versus Critical Periods in Sexual Differentiation of the Brain. Physiol Behav 187:13–19. https://doi.org/10.1016/j.physbeh.2017.10.030

McCullough LD, de Vries GJ, Miller VM, et al. (2014) NIH Initiative to Balance Sex of Animals in Preclinical Studies: Generative Questions to Guide Policy, Implementation, and Metrics. Biol Sex Differ 5:15. https://doi.org/10.1186/s13293-014-0015-5

Miller LR, Marks C, Becker JB, et al. (2017) Considering Sex as a Biological Variable in Preclinical Research. FASEB J 31 (1):29–34. https://doi.org/10.1096/fj.201600781R

Morrison JH, Brinton RD, Schmidt PJ, and Gore AC (2006) Estrogen, Menopause, and the Aging Brain: How Basic Neuroscience Can Inform Hormone Therapy in Women. J Neurosci 26 (41):10332–10348. https://doi.org/10.1523/jneurosci.3369-06.2006

Muyle A, Bachtrog D, Marais GAB, and Turner JMA (2021) Epigenetics Drive the Evolution of Sex Chromosomes in Animals and Plants. Philos Trans R Soc Lond B Biol Sci 376:20200124. https://doi.org/10.1098/rstb.2020.0124

Our Future Health (2024). https://ourfuturehealth.org.uk/

Raman I (2018) Workshop on Strategies to Promote Scientific Rigor, October 22, 2018. NINDS Office of Research Quality.

Riddle JM (1994) Contraception and Abortion from the Ancient World to the Renaissance (Illustrated and Revised ed.). Harvard Univ Press, Cambridge, MA

Rodgers RJ, and Cole JC (1993) Anxiety Enhancement in the Murine Elevated Plus Maze by Immediate Prior Exposure to Social Stressors. Physiol Behav 53 (2):383–388. https://doi.org/10.1016/0031-9384(93)90222-2

Sanchez XC, Montalbano S, Vaez M, et al. (2023) Associations of Psychiatric Disorders with Sex Chromosome Aneuploidies in the Danish iPSYCH2015 Dataset: A Case-Cohort Study. Lancet Psychiatry 10 (2):129–138. https://doi.org/10.1016/S2215-0366(23)00004-4

Scholl JL, Afzal A, Fox LC, Watt MJ, and Forster GL (2019) Sex Differences in Anxiety-Like Behavior in Rats. Physiol Behav 211:112670. https://doi.org/10.1016/j.physbeh.2019.112670

Schulz KM, Molenda-Figueira HA, and Sisk CL (2009) Back to the Future: The Organizational-Activational Hypothesis Adapted to Puberty and Adolescence. Horm Behav 55 (5):597–604. https://doi.org/10.1016/j.yhbeh.2009.03.010

Shansky RM, and Woolley CS (2016) Considering Sex as a Biological Variable Will Be Valuable for Neuroscience Research. J Neurosci 36 (47):11817–11822. https://doi.org/10.1523/JNEUROSCI.1390-16.2016

Sherwin BB (2007) The Critical Period Hypothesis: Can It Explain Discrepancies in the Oestrogen-Cognition Literature? J Neuroendocrinol 19 (2):77–81. https://doi.org/10.1111/j.1365-2826.2006.01508.x

Steensma TD, Kreukels BP, de Vries AL, and Cohen-Kettenis PT (2013) Gender Identity Development in Adolescence. Horm Behav 64 (2):288–297. https://doi.org/10.1016/j.yhbeh.2013.02.020

Strang JF, Wallace GL, Michaelson JJ, et al. (2023) The Gender Self-Report: A Multidimensional Gender Characterization Tool for Gender-Diverse and Cisgender Youth and Adults. Am Psychol 78 (7):886–900. https://doi.org/10.1037/amp0001117

Strassmann BI (1997) The Biology of Menstruation in Homo sapiens: Total Lifetime Menses, Fecundity, and Nonsynchrony in a Natural Fertility Population. Current Anthropology. Curr Anthropol 38 (1):123–129. https://doi.org/10.1086/204592

UN (2016) UN Women Model Questionnaire: Measuring the Nexus between Gender and Environment. https://data.unwomen.org/sites/default/files/documents/Publications/Model-Questionnaire-Gender-Environment.pdf

Understanding Society (2024) Data & Documentation: User Guides, Fieldwork Documents, Questionnaires, Technical Reports. https://www.understandingsociety.ac.uk/documentation/

Viuff M, Skakkebaek A, Johannsen EB, et al. (2023) X Chromosome Dosage and the Genetic Impact across Human Tissues. Genome Med 15 (1):21. https://doi.org/10.1186/s13073-023-01169-4

3

How Do Sex Differences in the Brain Help Our Understanding of Sex and Gender in Humans?

Geert J. de Vries

Abstract Although myriad sex differences have been found in the brain, the functional consequences of these differences are understood in only a few cases. This chapter presents new insights derived from broad-scale genomic approaches to studies that focus on sex differences in the expression of single genes. It illustrates the pervasiveness and interconnectedness of sex differences in the brain. Consideration is also given to the development and function of sex differences in brain function. A case will be made that sex differences in the brain can only be fully understood against a whole-body perspective. Although it is too early for a full-scale translation of the observed sex differences in structure into sex differences in overt functions, current knowledge underscores the need to consider sex as a biological variable. Doing this will improve scientific discourse, and considering sex differences will refine medical practice.

Keywords Behavior, vasopressin, sex chromosomes, gender-affirming treatment

1 Introduction

Over the past sixty years, thousands of papers have reported on the sex differences in the brains of animals as well as humans. The sheer number of papers published annually has increased logarithmically, at least through 2023 (Figure 3.1). The reported sex differences involve many aspects of the brain, such as differences in the expression of specific genes, the size and number of specific neurons, density of synapses and projections, the size of brain regions, neuronal activity, and activation of brain regions as revealed by fMRI (McCarthy et al. 2017). Remarkably, however,

Geert J. de Vries (✉)
Dept. of Biology, Georgia State University, Atlanta, GA 30303, USA
Email: devries@gsu.edu

© Frankfurt Institute for Advanced Studies (FIAS) 2025
L. Z. DuBois et al. (eds.), *Sex and Gender*, Strüngmann Forum Reports,
https://doi.org/10.1007/978-3-031-91371-6_3

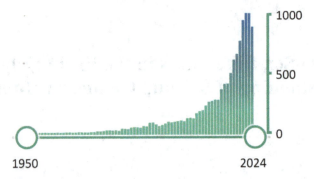

Figure 3.1 A PubMed search by year, using "sex differences" and "brain" as search terms, yielded over 8,000 hits for all years combined. As these terms are quite limited, the number of published articles on sex differences in the brain, shown above, is undoubtedly grossly underestimated.

the functional significance of most of these differences has remained obscure.

While still a graduate student in the late 1970s, I stumbled upon one such difference more or less by chance. Studying the development of vasopressin innervation in the brains of rats, one of the earliest detected neuropeptide systems in the brain, I noticed that some rats developed robust networks of vasopressin axons much earlier than others. My observation occurred only a couple of years after a sex difference had been observed that was so large, it could be seen with the naked eye in stained sections of the brain (Gorski et al. 1978). Unaware of such differences at the time, I followed a path that is still taken all too often; namely, using experimental subjects without keeping track of their sex (Beery and Zucker 2011). A follow-up study showed that males develop vasopressin innervation much earlier, and to a far greater extent, than females (de Vries et al. 1981). This was the first report of what has become one of most consistently found sex differences in the brain, present in all vertebrate classes except for fish (de Vries and Panzica 2006). Apparently, this difference was important enough to be preserved through evolution.

Hunting down the *function* of this sex difference, however, turned out to be surprisingly difficult. One problem involved overinterpreting the functional significance of neural sex differences that were known at the time. Invariably, structural brain differences were linked to sex differences in behavior or other centrally regulated functions. There was also a palpable unease with studies on sex differences in the brain. In 1983, during one of the first international meetings on sex differences in the brain in Amsterdam (de Vries 1984), the headlines of a local newspaper read: "Brain researchers go on a sex tour. Is one [sex] dumber than the other after all?" Concern over public perception coupled by the tendency to overinterpret the functional significance of sex differences in brain structure may have blinded researchers to what is arguably the most prominent reason for variability in structure: it is likely an adaptation to variability in conditions (Dobzhansky 1964). Put simply, sex differences were viewed within a far too limited scope. Brains do not operate in a vacuum. Just as their actions affect every major organ system, so too must it be assumed that sex differences (in form and function of any organ system) impact directly or indirectly the brain. Therefore, a sex difference in the brain may be an *adaptation* to differences elsewhere in the body and not necessarily the *cause* of physiological or behaviorial differences. Figure 3.2 illustrates various interactions

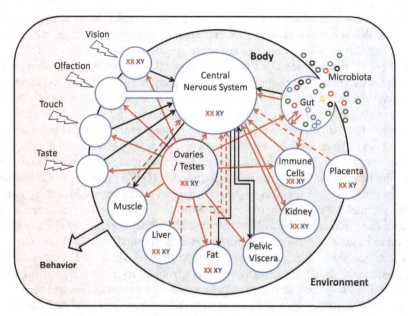

Figure 3.2 Sex differences in peripheral influences on the central nervous system. Solid arrows indicate a sex influence from one organ to another. Dashed arrows show an influence inferred from circumstantial evidence. Black and red arrows indicate neural and humoral communication, respectively. "XX XY" indicates organs in which the sex chromosome complement has a direct effect; in most cases, it is not known whether this effect is mediated within that organ or indirectly via effects on other organs. The small colored circles (upper right) represent the many species of microorganisms that live commensally in our gut or on our skin. Adapted from de Vries and Forger (2015) (http://creativecommons.org/licenses/by/4.0/).

between the central nervous system and sexually differentiated systems elsewhere in the body, as reported in the literature (reviewed in de Vries and Forger 2015). It presents the central nervous system as an organ embedded in a sexually differentiated body, which itself is embedded in an environment that may interact with the body in a manner that varies by sex and gender (Cortes et al. 2019).

In this chapter, I review evidence that sex differences in the body are pervasive and can be found in any tissue studied, including the brain. I discuss strategies for uncovering the functional significance of such differences, contrasting insights derived from studies that reveal sex differences in cohorts of genes with studies that focus on differences in the expression of single genes. I conclude with a brief discussion of the medical implications when considering sex and gender differences in the brain.

2 Disease Patterns and Genome-Wide Studies Suggest Pervasive and Significant Sex Differences Throughout the Body

Compelling arguments suggest that the presence of sex differences in human behavior and cognitive abilities are often exaggerated or even nonexistent (Eliot et al. 2021; Fine 2012). Making the same argument for the presence of sex differences in disordered brain function, however, is much harder to make. There is a growing realization that sex differences play an important role in the prevalence, progression, and treatability of disease across the disease spectrum (Mauvais-Jarvis et al. 2020) (Figure 3.3). Environmental as well as biological factors are likely to contribute to these differences. For example, gendered experiences of patients and caretakers alike can interact with biological factors to influence disease outcomes (Mauvais-Jarvis et al. 2020; for further discussion, see Chapter 2). In addition, broad genomic analyses of sex differences in gene expression support the idea that there are significant sex differences in homeostatic processes that keep brain and bodily systems functioning within physiological range. Of particular interest are studies associated with the large-scale Genotype-Tissue Expression (GTEx) project. In GTEx, multiple institutions work together to generate data on gene expression and regulation; tissue samples are taken from multiple sites in the body, gathered at different stages of life, from males as well as females. One of the major goals in GTEx is to link variation in gene expression with variation in health and disease (Lonsdale et al. 2013). Using these data, several studies report pervasive sex differences in gene expression and regulatory network in tissues from all over the body. In one study, Oliva et al. (2020) compared gene expression in 44 different tissues and found over 13,000 differentially expressed genes (DEGs). In subcutaneous adipose tissue, for example, they found 2,954 DEGs and in the skin, an astonishing 4,558 DEGs. In addition, Oliva et al. (2020) sampled many areas of the brain and found, in each, thousands of DEGs: 2,416 are differentially expressed in only one single tissue and 1,628 are expressed in only two tissues, with the number steadily decreasing until one reaches 30 DEGs in every type of tissue sampled. Importantly, while the numbers decrease, the proportion of X-linked DEGs increases (Figures 3.4 and 3.5).

The expression of 30 DEGs, found in every tissue sampled, is overwhelmingly female biased (Figure 3.5) and is probably related to sex differences in the expression of genes on the X and Y chromosomes. X and Y chromosomes carry mostly different genes, except for a small portion of genes found at the poles of these chromosomes, the so-called "pseudoautosomal region" (PAR), named as such because these are the only regions of the sex chromosomes that exchange, in males, genetic material during meiosis. Arguably the best-known gene in the non-PAR on the Y chromosome is SRY—the gene that instructs the developing gonad in males to become a testis—which starts producing testosterone, a major differentiating factor of brain and body (Arnold 2012).

An equally notable gene in the non-PAR of the X chromosome is XIST, which is transcribed from only one of the two X chromosomes. Instead of being translated into protein, XIST mRNA causes inactivation of the other X chromosome by epigenetic mechanisms, presumably to prevent dosage differences in X-chromosomal genes, which may be deleterious (Lyon 1999). This process is not without its flaws, as illustrated elegantly in another study from the GTEx project (Tukiainen et al. 2017). Here, Tukiainen et al. compared the expression of X-chromosomal genes in 29 tissues, including brain regions, and found that many genes on the PAR showed a male bias. Presumably, inactivation of one of the two X chromosomes in females inhibits expression of genes on the PAR of the X chromosome to some extent, something that does not happen to the PAR of the X chromosome in males. Many genes on the non-PAR of the X chromosome, however, showed a female bias.

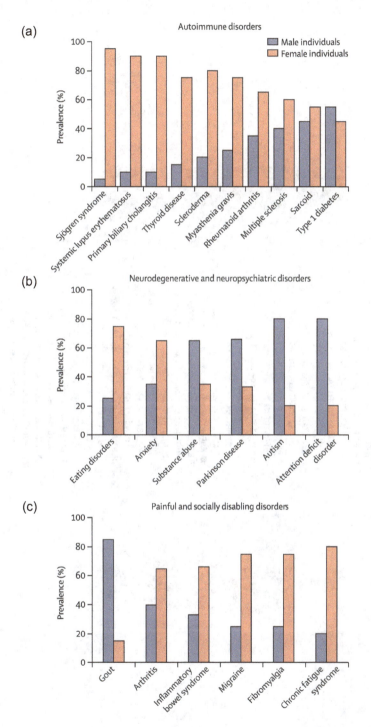

Figure 3.3 Prevalence of disorders in male and female individuals: (a) autoimmune disorders, (b) neurodegenerative and neuropsychiatric disorders, and (c) painful and socially disabling disorders. Adapted from Mauvais-Jarvis et al. (2020).

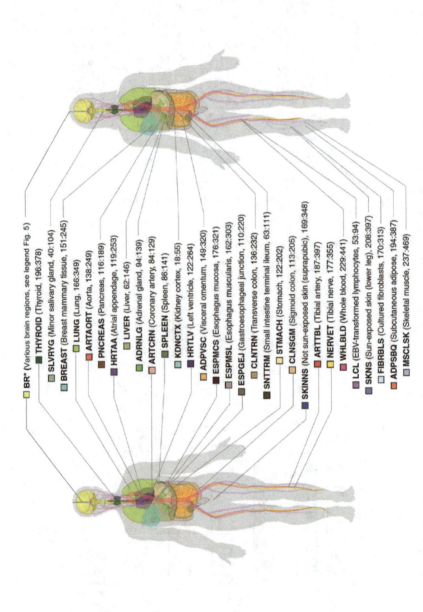

Fig. 3.4 Tissue types analyzed by Oliva et al. (2020): sample numbers from genotyped donors are given in parentheses (females:males) and color coded according to type of tissue. Tissue name abbreviations are shown in boldface. Specific labels and numbers for brain tissues (BR*) are listed in Figure 3.5. Adapted from Oliva et al. (2020).

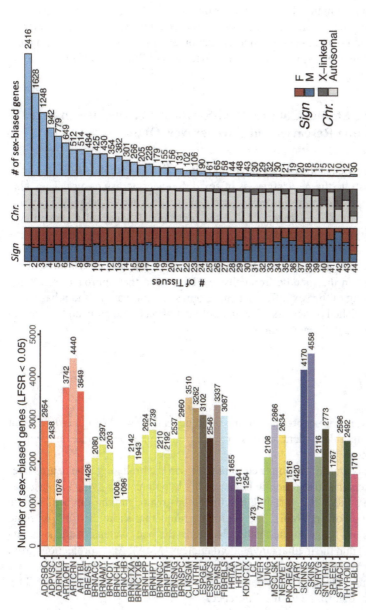

Figure 3.5 (a) Number of differentially expressed genes (sex-biased genes) per tissue; tissue colors correspond with those used in Figure 3.4. (b) Number of sex-biased genes as a function of the number of tissues that express these genes (histogram) and characteristics of these sex-biased genes (stacked bars in the center). Sign: proportions of female (F)- and male (M)-biased genes; Chr.: proportion of X-linked and autosomal sex-biased genes. Labels for brain tissue samples (BR* in Figure 3.4): BRNACC, anterior cingulate cortex; BRNAMY, amygdala; BRNCDT, caudate; BRNCHA/BRNCHB, cerebellum/cerebellar hemisphere; BRNCTXA/BRNCTXB, cortex/frontal cortex; BRNHPP, hippocampus; BRNHPT, hypothalamus; BRNCC, nucleus accumbens; BRNPTM, putamen; BRNSNG, substantia nigra; BRNSPC, spinal cord (cervical c-1). Adapted from Oliva et al. (2020).

One explanation is that this region contains several genes that have a homologous gene on the non-PAR of the Y chromosome. These homologous genes on X and Y, sometimes called XY genes, are not identical but often have similar functions. Inactivation of one of those genes on X would automatically lead to an overexpression of XY genes in males. Presumably for that reason, XY genes escape inactivation. Once again, however, this process is imprecise, and genes in the vicinity of XY genes tend to escape inactivation to greater or lesser degree as well, leading to a female bias in the expression of those genes (Carrel and Willard 2005).

3 Single Gene Studies Confirm Well-Established Principles in Sexual Differentiation Research and Uncover New Ones

Although differences in the expression of individual DEGs were mostly quite small, GTEx data illustrate the pervasiveness of sex differences across tissues. They also support the idea that X-chromosomal genes that escape inactivation may be an important factor in sexual differentiation of the molecular makeup of all organ systems, including the brain (Arnold 2022). It is more difficult to link differential expression of large sets of genes to specific functional outcomes. Studies that employ broad genomic analyses typically follow up by looking at the effects of specific DEGs identified in their study. Ironically, we have used that approach ever since the chance finding of the sex difference in vasopressin innervation, and it has led to some well-established principles of sexual differentiation of the brain and suggested some new ones, as delineated below.

3.1 Principle 1

Sex differences in gene expression in the brain depend on sex hormone-dependent and sex hormone-independent actions of sex chromosomes as well as on environmental influences.

Vasopressin is made in various neuronal groups in the brain (e.g., the neurosecretory paraventricular and supraoptic nuclei of the hypothalamus, which release vasopressin as an antidiuretic hormone into the bloodstream, and the suprachiasmatic nucleus, the clock of the brain). The neuronal groups that show the most extreme sex differences are found in the telencephalon, in the bed nucleus of the stria terminalis (BNST) and the medial amygdaloid nucleus (MeA). Axon projections from these nuclei provide the majority of vasopressin innervation of the brain and are much denser in males than in females (de Vries and Panzica 2006) (Figure 3.6). Multiple factors contribute to this difference.

Many sex differences in brain and behavior have been shown to depend on early organizing effects of differences in gonadal hormone levels (most notably the higher levels of testosterone in males) that permanently set brain development on a male or female track, as well as on acute effects of circulating gonadal hormones in adulthood (McCarthy et al. 2017). This is true for the sex difference in

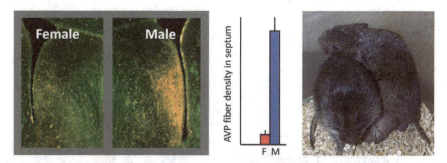

Figure 3.6 (a) A sex difference in the density of vasopressin-immunoreactive projections from BNST and MeA to the lateral septum of the rat, which is much higher in males than in females. (b) The bar graph shows the extreme difference in vasopressin (AVP) innervation between male and female prairie voles, even though they provide similar levels of parental care to a pup. Modified from de Vries (2004).

vasopressin expression as well. For example, male rats castrated on the day of birth develop a fiber density similar to what is found in intact females. Males castrated at three weeks of age develop a fiber density similar to that found in intact males. Males castrated at one week of age develop an intermediate density; this suggests that around this time, testosterone naturally programs the system to develop male characteristics. To see these differences, however, males had to be treated acutely with testosterone in adulthood; without testosterone in circulation, the system does not produce vasopressin. In this respect it resembles male sexual behavior in rats, the circuitry of which is programmed around birth by higher testosterone levels in males. For the sex difference in the propensity to show male sexual behavior in the presence of a receptive female, testosterone must be present in adulthood as well (McCarthy et al. 2017).

Differences in sex-chromosomal constitution determine whether an animal develops a testis or an ovary. In case of a testis, the animal gets exposed to the masculinizing effects of early and later circulating levels of testosterone. For many decades, sex differences in gonadal hormone levels were seen as the primary drivers of sexual differentiation in the brain. After noticing that these factors cannot explain some instances of sexual differentiation (e.g., sex differences found in the brains of songbirds), Art Arnold and colleagues developed a novel mouse model, the Four Core Genotype model, to test whether sex differences were caused primarily by sex hormones or directly by sex chromosomes (de Vries et al. 2002). In these mice, the Sry gene has been lost from the Y chromosome and now resides as a transgene in an autosomal location. XX and XY mice with this transgene develop testes and an ensuing male phenotype. XX and XY mice without this transgene develop a female phenotype. This allowed us to compare the effects of XX- and XY-chromosomal complements within phenotypic males or females. In the initial cohort we studied the effects on a number of well-known sex differences in the brain. Of these, only the sex difference in vasopressin was affected directly by sex-chromosomal conditions: XY males and females had overall denser vasopressin projections than XX males and females. The Four Core Genotype model is now widely used to study sex-chromosomal effects throughout the body and has uncovered many traits that are directly affected by sex chromosomes (Arnold et al. 2023). For example, XX

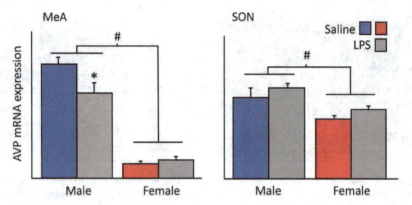

Figure 3.7   Levels of vasopressin (AVP) mRNA expression in the medial amygdala (MeA) and supraoptic nucleus of the hypothalamus (SON) in mothers treated with lipopolysaccharides (LPS) or saline during pregnancy. Notice that the size of the sex difference in the MeA is reduced in the offspring of LPS-treated rats. The smaller sex difference in the SON, however, is unaffected. Modified from Taylor et al. (2012). (creativecommons.org/licenses/by/2.0)

mice become heavier and develop more adipose tissue than XY mice (Arnold 2020).

The size and sometimes even the direction of some sex differences in the brain and behavior depend on the circumstances under which individuals are raised. We found this to be true for the sex difference in vasopressin as well. For example, rats born to mothers that had been given an immune challenge during pregnancy (i.e., exposure to lipopolysaccharides, a proxy for a bacterial infection) developed a much smaller sex difference in the number of vasopressin cells in the telencephalon than offspring from unexposed dams. This was mainly caused by a reduction in the number of vasopressin cells in the male offspring of the exposed dams (Figure 3.7). Other areas that produce vasopressin were not affected, even though the supraoptic nucleus showed a similar, albeit much more modest, sex difference in vasopressin expression, which I will discuss below.

3.2 Principle 2

Sex differences in gene expression in the brain may cause as well as prevent sex differences in overt functions and behaviors.

The hormone-dependent and hormone-independent effects of sex-chromosomal composition as well as environmental effects makes the vasopressin system an exemplary model for understanding the biological basis of sexual differentiation of the brain. Interestingly, however, chasing down the function of sex differences in this system has not been easy. A breakthrough came from studying the vasopressin innervation in prairie voles. Prairie vole males and females show remarkably similar behavior. For example, apart from nursing, males and females spend equal amounts of time caring for their young, yet the sex difference in vasopressin expression is much higher in voles (Figure 3.6) than has been reported for any other mammal. After finding that becoming parents activated the vasopressin system, but only in males, we tested the effects of vasopressin on parental behavior and confirmed that vasopressin drives parental behavior in male voles (Wang et al. 1994). The irony was that we had just linked one of the most dimorphic neuropeptide systems (see Figure 3.6) to a behavior that does not differ that much in males and females (i.e., parental care of pups). This actually made sense. Like other rodents, female voles must undergo pregnancy and give birth before they exhibit parental behavior. Males, of course, do not get pregnant, let alone give birth. Therefore, to become equally parental, they must follow a different strategy, part of which may involve recruitment of the vasopressin system.

This realization led us to propose that sex differences in the brain may induce sex differences in behaviors and other centrally regulated functions, but it may also compensate for sex differences in physiology and hormonal state to reach similar endpoints in behaviors and other centrally regulated functions (de Vries 2004). This hypothesis is perfectly testable. For example, in the first case, one would predict that reducing or removing a specific sex difference in a transmitter system would eliminate or reduce a sex difference in functions modulated by that system. In the second case, one would predict that the same manipulation would create novel sex differences in function. The literature suggested that this was true for vasopressin. For example, the denser vasopressin projections to telencephalic areas may drive higher levels of aggressive behavior in males while preventing sex differences in social recognition memory (the ability to distinguish familiar from novel conspecifics). Indeed, blocking vasopressin transmission reduced the sex difference in aggressive behavior, while creating a sex difference in social recognition memory, in both cases by reducing these modalities in males and leaving them unchanged in females (de Vries 2004).

One way to explain the hypothesis that sex differences cause and prevent sex differences in function (the Dual Function Hypothesis) is by pointing out that there are no circuits in the brain that are exclusively dedicated to one specific behavior. For example, neural circuitry needed for male sexual behavior invariably shares nodes with circuitry for functions that are not conspicuously dimorphic. If sex differences in such a circuit are needed to induce sex differences in male sexual behavior, for instance, other sex differences may be required to prevent unnecessary sex differences from occurring in other functions served by that circuitry. This is probably true for every level of organization (from molecules to organs) and is likely to occur each time a sex difference is needed for a specific function but may cause maladaptive sex differences in another. Some of the most clear-cut examples are found at the molecular level. Consider, for example, the largest sex difference found in the brain, as well as in every cell of the body with the exclusion of gametes: the expression of the Xist gene, which takes place exclusively in female cells. The primary reason for this sex difference is that it prevents sex differences in function that may result from dosage differences in the expression of X-chromosomal genes, many of which serve functions that do not obviously differ between the sexes. This very conspicuous sex difference in gene expression is actually there to prevent sex differences in cellular function.

Testing the Dual Function Hypothesis directly for the sex difference in vasopressin innervation was made possible by the development of powerful genetic tools, which allowed us to specifically manipulate sexually dimorphic vasopressin cells. This approach showed that these cells cause as well as prevent sex differences in function, most notably in social behavior. For example, we injected Cre-dependent viral vectors into the brain of vasopressin-iCre mice—mice that express Cre exclusively in vasopressin cells. Using Cre-dependent caspase viral vectors, which cause expression of the cell death-signaling protein caspase in cells where Cre is present, we were able to delete specifically the sexually differentiated vasopressin cells. Under normal conditions, both males and females spend more time investigating novel mice than familiar mice. Removing the vasopressin cells eliminated this bias in males but not in females, thereby creating a sex difference that did not previously exist (Rigney et al. 2019). Similar experiments indicate that these cells have a stronger impact on certain social behaviors in males than in females (Rigney et al. 2023).

3.3 Principle 3

Sex differences in gene expression in the brain may explain sex differences in the vulnerability for behavioral and other disorders.

If a specific neurotransmitter system has a more prominent role in controlling a physiological process or behavior in one sex over the other, it is not difficult to imagine that dysfunction in that system will affect one sex more than the other. For vasopressin transmission, this has been done artificially. Larry Young and colleagues, for example, found that deletion of the vasopressin 1 receptor gene specifically affected anxiety-like behaviors in male but not female mice (Bielsky et al. 2005). Similar scenarios may come into play each time a function is driven by mechanisms that differ between the sexes.

At the molecular level, this may be very common in mice and is likely to apply to humans as well, as demonstrated in a study that used material generated by the GTEx project to compare gene regulatory networks across 29 different tissues in humans (Lopes-Ramos et al. 2020). Interestingly, although the study did not find many differences in the expression of transcription factors (the molecular signals that regulate expression levels of target genes), it discovered that these transcription factors often targeted different sets of genes and, correspondingly, that target genes were controlled by different sets of transcription factors in males and females (Figure 3.8). Such an arrangement suggests that dysfunction in any of these differently connected genes will result in different health consequences in males and females, a factor that likely contributes to the impressive sex differences in the prevalence of specific disorders mentioned earlier. This undoubtedly applies to brain disorders as well. For example, genes linked to Parkinson disease, which has a greater prevalence and develops at an earlier age in males (Gillies et al. 2014), were found to be targeted by different sets of transcription factors in males and females (Lopes-Ramos et al. 2020).

3.4 Principle 4

Sex differences in gene expression in the brain can only be understood from a whole-body perspective.

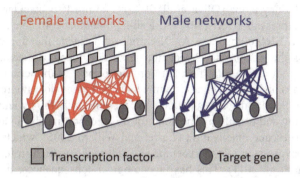

Figure 3.8 Part of the graphical abstract used in the Lopes-Ramos et al. (2020) paper showing sets of transcription factors influencing expression of target genes differently in males and females. Not shown here is that the data also indicated that each of the transcription factors were influenced by a different network of genes in males and females.

The study by Lopes-Ramos et al. (2020) maps quite well to the idea of "the sexome," proposed by Arnold and Lusis (2012), who pointed out that the function of every cell in the body is defined by a network of interactions among all molecules that make up a cell. They defined the sexome as "the sum of all sex-specific and sex-biased modulatory interactions that operate within [these] networks" (Arnold and Lusis 2012:2552). Because cells, including cells in brains, do not live in a vacuum, it must be assumed that intracellular molecular networks are influenced by gene products from elsewhere in the body. Given the pervasiveness of sex differences in gene expression throughout the body (Lopes-Ramos et al. 2020; Oliva et al. 2020), many of these influences will have a sex bias. The sexome, therefore, covers networks that reach every cell in the body. The drivers of the sex bias in the interactions are the same as mentioned above for the brain: the direct and indirect effects of sex-chromosomal complement, programming and acute effects of gonadal hormones, and environmental effects, which may include gender-based treatment of self and others (see below). These drivers will affect some of the nodes in this sexome more than others. For example, genes with a so-called estrogen receptor response element in a cell that expresses estrogen receptors are more likely to be first responders to changes in estrogen levels than genes without such response elements. Interconnectedness, however, ensures that all genes in the network are affected to a greater or lesser degree by the sex bias. Although the nature of the connections between the nodes in the sexome proposed by Arnold and Lusis, or in the networks discussed in the GTEx studies, is often unclear, one could envision, at a higher level of organization, a network of functional interactions between the different organs that make up a body. As the aforementioned sex factors have wide-ranging effects that affect many of these organs directly, we referred to this network of organs depicted in Figure 3.2, somewhat tongue-in-cheek, as the "sexorganome" (de Vries and Forger 2015).

To illustrate how sex effects on one organ may affect another, let us look at the functional connections between the kidney and the brain. Upon superficial inspection, it is difficult to tell male and female kidneys apart, yet there are remarkable sex differences in gene expression. For example, the GTEx study on sex differences in gene expression across the human body found over 1,200 differentially expressed genes in the kidney alone (Oliva et al. 2020). The most spectacular difference is found in the expression of Xist mRNA, leading to almost complete inactivation of one of the two X chromosomes. An X-chromosomal gene that appears to escape inactivation is the vasopressin 2 receptor gene. This receptor mediates vasopressin's action as an antidiuretic hormone on the kidney. Malfunction of that gene leads to diabetes insipidus, a condition in which the kidney does not reabsorb water filtered out of blood; this causes patients to feel constantly thirsty and leads to excessive drinking and urination. The gene for this receptor resides in an area that is more likely to escape inactivation (Carrel and Willard 2005), which may explain why, in rats, expression of this receptor is about twice as high in females as it is in males (Liu et al. 2011). This has physiological consequences. In females, the vasopressin analog desmopressin has a stronger antidiuretic effect when given in the same dosage than in males (Liu et al. 2011). Clinical research suggests similar sex differences in effectiveness in adults who are prescribed desmopressin to treat an abnormally high need to urinate at night (nocturia), and in children who are prescribed desmopressin to treat bedwetting (nocturnal enuresis) (Schroeder et al. 2018). Nature seems to compensate for the higher expression of the vasopressin 2 receptor in females, as males have higher plasma levels of vasopressin (antidiuretic hormone) than females (Asplund and Aberg 1991), perhaps to keep osmoregulation and control of water balance similar between the sexes. This difference may directly explain a sex difference in vasopressin expression that we and others have found in the supraoptic nucleus of the hypothalamus, which expresses more vasopressin in males than in females (Figure 3.7; Taylor et al. 2012). Without considering the possibility that sex differences in the brain may compensate for sex differences in the body, one might be tempted to link this difference to sex differences in overt functions or behaviors. Data on vasopressin 2 receptor expression, however, suggest that this neural sex difference may primarily function to avoid sex differences in water balance. Interestingly, the sex difference in sensitivity to desmopressin did not go unnoticed by Ferring Pharmaceuticals, which produces a drug to treat nocturia in older individuals. Since women suffer more serious side effects (Juul et al. 2011), Ferring Pharmaceuticals proposed and got permission to market the drug in two different dosages: a higher dose geared toward men and a lower dose for women, under the trade name NOCDURNA®. Currently, this is one of the very few examples of sex-specific dosing in medicine.

The most notorious medicine with sex-based dosing is zolpidem (AMBIEN®), a drug that targets the brain to induce sleep. Based on reports that women were more likely to suffer serious side effects (e.g., impairment in driving a vehicle the next day), the FDA approved, in 2013, zolpidem to be prescribed to women at half the dose as that for men. Unlike desmopressin, however, there was no compelling biological explanation to warrant this change, and many argued that the evidence was thin or downright misguided to justify prescribing different dosages in males and females (Greenblatt et al. 2019). Interestingly, public perception and media coverage may have played a role in blinding researchers and regulatory agencies to the negative consequences of a binary treatment of sex in developing medications, leading to the "concretization of a sex difference fact" (Zhao et al. 2023). In practice, the change in dosing may have subjected women to ineffective treatment (Greenblatt et al. 2019). One of the challenges in determining an appropriate course of action for certain conditions—in this case, sleep—is that the physiological processes that control sleep are far less well understood than, for example, urine production. This makes it harder to make a compelling case for sex-specific dosages.

4 Consequences of Sex Differences in Brain Organization for Understanding Issues Related to Gender Differences in Brain Function

The studies discussed above address primarily "sex," used in this chapter to refer to immediate biological factors: the composition of sex chromosomes (XX vs. XY), nature of the gonads and the hormones they secrete (ovaries vs. testes), and other body features intimately related to chromosomal sex and gonadal constitution (secondary sex characteristics). Although "gender" is ultimately influenced by these factors, gender involves an important social aspect and is related to the perception of, and treatment by, oneself and others. As a result, physical characteristics that are typically used to define sex are more bimodal whereas gender appears more fluid. Currently, we are far from a comprehensive understanding of how sex differences in the brain contribute to sex differences in brain function. A valid criticism of this field is that many reports of sex differences in human brain structure or physiology link these differences to purported differences in cognitive function and behavior, often without providing strong scientific arguments to back up these links (Eliot et al. 2021; Fine 2012; Fine et al. 2013). Given the fluidity of gender, we are even further away from explaining how sex differences in the brain contribute to gender differences in brain function.

We have, however, enough data to develop strategies for incorporating "sex" and "gender" in the fight against disease. For example, genome-wide studies, such as the one discussed above (Lopes-Ramos et al. 2020), indicate widespread (but small) differences in the molecular regulation of cellular processes across systems, including the brain. They also indicate that genes correlated with specific diseases are targeted by different networks of regulatory transcription factors in males and females. Any medical intervention based on targeting such genes should therefore take sex into account and, if warranted, should include treatment strategies that vary by sex, such as prescribing different dosages of desmopressin.

It is far more challenging to address "gender," as we lack comprehensive genome-wide studies, like the GTEx data, that are stratified according to the various forms of gender. Nevertheless, one can point to issues that must be considered in developing safe medical interventions to manage, for example, gender dysphoria in cases of gender-affirming care, which may include (a) hormonal treatment to slow down or block pubertal development and thereby the development of secondary sexual characteristics, (b)hormonal treatment to stimulate the development of secondary sexual characteristics that align with an individual's gender identity, and (c) gender-affirming surgical interventions. The Netherlands, one of the first countries to adopt a standard practice of gender affirmation, has accumulated several decades of experience and thus provides a rich source for outcome research. As reported by van der Loos et al. (2023), detransition is uncommon and in most cases, the effects on mental health have been positive. In a review of 23 studies that addressed the effects of gender-affirming treatment on suicide-related outcomes, Jackson (2023) suggests that gender-affirming treatment reduces the risk of suicide and suicide ideation in the majority of cases. In addition, a recent study involving a large cohort of responders suggests that gender-affirming surgery was associated with lower psychological distress, reduced smoking, and less suicidal ideation (Almazan and Keuroghlian 2021).

Given the obvious benefits of gender-affirming treatment for mental health, priority should be given to the development of treatments that are effective and safe. This may involve studying whether sex differences in gene regulatory networks and in the linkage of such networks to diseases, found by Lopes-Ramos et al. (2020), are affected by gender-affirming care, and if so, what the consequences are of such changes for optimizing treatment.

References

Almazan AN, and Keuroghlian AS (2021) Association between Gender-Affirming Surgeries and Mental Health Outcomes. JAMA Surg 156 (7):611–618. https://doi.org/10.1001/jamasurg.2021.0952

Arnold AP (2012) The End of Gonad-Centric Sex Determination in Mammals. Trends Genet 28 (2):55–61. https://doi.org/10.1016/j.tig.2011.10.004

———— (2020) Four Core Genotypes and XY* Mouse Models: Update on Impact on SABV Research. Neurosci Biobehav Rev 119:1–8. https://doi.org/10.1016/j.neubiorev.2020.09.021

———— (2022) X Chromosome Agents of Sexual Differentiation. Nat Rev Endocrinol 18 (9):574–583. https://doi.org/10.1038/s41574-022-00697-0

Arnold AP, Abdulai-Saiku S, M.-F. C, et al. (2023) Sex Differences in Neurological and Psychiatric Diseases. In: Zigmond MJ, Wiley CA, M.-F. C (eds) Neurobiology of Brain Disorders. Academic Press, New York, pp 933–952

Arnold AP, and Lusis AJ (2012) Understanding the Sexome: Measuring and Reporting Sex Differences in Gene Systems. Endocrinology 153 (6):2551–2555. https://doi.org/10.1210/en.2011-2134

Asplund R, and Aberg H (1991) Diurnal Variation in the Levels of Antidiuretic Hormone in the Elderly. J Intern Med 229 (2):131–134. https://doi.org/10.1111/j.1365-2796.1991.tb00320.x

Beery AK, and Zucker I (2011) Sex Bias in Neuroscience and Biomedical Research. Neurosci Biobehav Rev 35 (3):565–572. https://doi.org/10.1016/j.neubiorev.2010.07.002

Bielsky IF, Hu SB, and Young LJ (2005) Sexual Dimorphism in the Vasopressin System: Lack of an Altered Behavioral Phenotype in Female V1a Receptor Knockout Mice. Behav Brain Res 164 (1):132–136. https://doi.org/10.1016/j.bbr.2005.06.005

Carrel L, and Willard HF (2005) X-Inactivation Profile Reveals Extensive Variability in X-Linked Gene Expression in Females. Nature 434 (7031):400–404. https://doi.org/10.1038/nature03479

Cortes LR, Cisternas CD, and Forger NG (2019) Does Gender Leave an Epigenetic Imprint on the Brain? Front Neurosci 13:173. https://doi.org/10.3389/fnins.2019.00173

de Vries GJ (1984) Sex Differences in the Brain: The Relation between Structure and Function. In: Proceedings of the 13th International Summer School of Brain Research, Amsterdam, the Netherlands, 1983. Elsevier, Amsterdam, pp 589–596

———— (2004) Minireview: Sex Differences in Adult and Developing Brains: Compensation, Compensation, Compensation. Endocrinology 145 (3):1063–1068. https://doi.org/10.1210/en.2003-1504

de Vries GJ, Buds RM, and Swaab DF (1981) Ontogeny of the Vasopressinergic Neurons of the Suprachiasmatic Nucleus and Their Extrahypothalamic Projections in the Rat Brain—Presence of a Sex Difference in the Lateral Septum. Brain Res 218 (1–2):67–78. https://doi.org/10.1016/0006-8993(81)90989-6

de Vries GJ, and Forger NG (2015) Sex Differences in the Brain: A Whole Body Perspective. Biol Sex Differ 6:15. https://doi.org/10.1186/s13293-015-0032-z

de Vries GJ, and Panzica GC (2006) Sexual Differentiation of Central Vasopressin and Vasotocin Systems in Vertebrates: Different Mechanisms, Similar Endpoints. Neuroscience 138 (3):947–955. https://doi.org/10.1016/j.neuroscience.2005.07.050

de Vries GJ, Rissman EF, Simerly RB, et al. (2002) A Model System for Study of Sex Chromosome Effects on Sexually Dimorphic Neural and Behavioral Traits. J Neurosci 22 (20):9005–9014. https://doi.org/10.1523/JNEUROSCI.22-20-09005.2002

Dobzhansky T (1964) Biology, Molecular and Organismic. Am Zool 4 (4):443–452. https://doi.org/10.1093/icb/4.4.443

Eliot L, Ahmed A, Khan H, and Patel J (2021) Dump the "Dimorphism": Comprehensive Synthesis of Human Brain Studies Reveals Few Male-Female Differences Beyond Size. Neurosci Biobehav Rev 125:667–697. https://doi.org/10.1016/j.neubiorev.2021.02.026

Fine C (2012) Explaining, or Sustaining, the Status Quo? The Potentially Self-Fulfilling Effects of "Hardwired" Accounts of Sex Differences. Neuroethics 5 (3):285–294. https://doi.org/10.1007/s12152-011-9118-4

Fine C, Jordan-Young R, Kaiser A, and Rippon G (2013) Plasticity, Plasticity, Plasticity...And the Rigid Problem of Sex. Trends Cogn Sci 17 (11):550–551. https://doi.org/10.1016/j.tics.2013.08.010

Gillies GE, Pienaar IS, Vohra S, and Qamhawi Z (2014) Sex Differences in Parkinson's Disease. Front Endocrinol 35 (3):370–384. https://doi.org/10.1016/j.yfrne.2014.02.002

Gorski RA, Gordon JH, Shryne JE, and Southam AM (1978) Evidence for a Morphological Sex Difference within the Medial Preoptic Area of the Rat Brain. Brain Res 148 (2):333–346. https://doi.org/10.1016/0006-8993(78)90723-0

Greenblatt DJ, Harmatz JS, and Roth T (2019) Zolpidem and Gender: Are Women Really at Risk? J Clin Psychopharmacol 39 (3):189. https://doi.org/10.1097/JCP.0000000000001026

Jackson D (2023) Suicide-Related Outcomes Following Gender-Affirming Treatment: A Review. Cureus 15 (3):e36425. https://doi.org/10.7759/cureus.36425

Juul KV, Klein BM, Sandstrom R, Erichsen L, and Norgaard JP (2011) Gender Difference in Antidiuretic Response to Desmopressin. Am J Physiol Renal Physiol 300 (5):F1116–F1122. https://doi.org/10.1152/ajprenal.00741.2010

Liu J, Sharma N, Zheng W, et al. (2011) Sex Differences in Vasopressin V(2) Receptor Expression and Vasopressin-Induced Antidiuresis. Am J Physiol Renal Physiol 300 (2):F433–440. https://doi.org/10.1152/ajprenal.00199.2010

Lonsdale J, Thomas J, Salvatore M, et al. (2013) The Genotype-Tissue Expression (Gtex) Project. Nat Genet 45 (6):580–585. https://doi.org/10.1038/ng.2653

Lopes-Ramos CM, Chen C-Y, Kuijjer ML, et al. (2020) Sex Differences in Gene Expression and Regulatory Networks across 29 Human Tissues. Cell Rep 31 (12):107795. https://doi.org/10.1016/j.celrep.2020.107795

Lyon MF (1999) X-Chromosome Inactivation. Curr Biol 9 (7):R235–237. https://doi.org/10.1016/s0960-9822(99)80151-1

Mauvais-Jarvis F, Bairey Merz N, Barnes PJ, et al. (2020) Sex and Gender: Modifiers of Health, Disease, and Medicine. Lancet 396 (10250):565–582. https://doi.org/10.1016/S0140-6736(20)31561-0

McCarthy MM, De Vries GJ, and Forger NG (2017) Sexual Differentiation of the Brain: A Fresh Look at Mode, Mechanisms and Meaning. In: Pfaff DW, Joels M (eds) Hormones, Brain and Behavior, vol 3. Elsevier, San Diego, pp 3–32

Oliva M, Muñoz-Aguirre M, Kim-Hellmuth S, et al. (2020) The Impact of Sex on Gene Expression across Human Tissues. Science 369 (6509):aba3066. https://doi.org/10.1126/science.aba3066

Rigney N, de Vries GJ, and Petrulis A (2023) Modulation of Social Behavior by Distinct Vasopressin Sources. Front Endocrinol 14:1127792. https://doi.org/10.3389/fendo.2023.1127792

Rigney N, Whylings J, Mieda M, de Vries G, and Petrulis A (2019) Sexually Dimorphic Vasopressin Cells Modulate Social Investigation and Communication in Sex-Specific Ways. eNeuro 6 (1):ENEURO.0415–0418.2019. https://doi.org/10.1523/eneuro.0415-18.2019

Schroeder MK, Juul KV, Mahler B, Nørgaard JP, and Rittig S (2018) Desmopressin Use in Pediatric Nocturnal Enuresis Patients: Is There a Sex Difference in Prescription Patterns? Eur J Pediatr 177 (3):389–394. https://doi.org/10.1007/s00431-017-3074-x

Taylor PV, Veenema AH, Paul MJ, et al. (2012) Sexually Dimorphic Effects of a Prenatal Immune Challenge on Social Play and Vasopressin Expression in Juvenile Rats. Biol Sex Differ 3 (1):15. https://doi.org/10.1186/2042-6410-3-15

Tukiainen T, Villani A-C, Yen A, et al. (2017) Landscape of X Chromosome Inactivation across Human Tissues. Nature 550 (7675):244–248. https://doi.org/10.1038/nature24265

van der Loos MATC, Klink DT, Hannema SE, et al. (2023) Children and Adolescents in the Amsterdam Cohort of Gender Dysphoria: Trends in Diagnostic- and Treatment Trajectories during the First 20 Years of the Dutch Protocol. J Sex Med 20 (3):398–409. https://doi.org/10.1093/jsxmed/qdac029

Wang Z, Ferris CF, and De Vries GJ (1994) Role of Septal Vasopressin Innervation in Paternal Behavior in Prairie Voles (*Microtus ochrogaster*). PNAS 91 (1):400–404. https://doi.org/10.1073/pnas.91.1.400

Zhao H, DiMarco M, Ichikawa K, et al. (2023) Making a 'Sex-Difference Fact': Ambien Dosing at the Interface of Policy, Regulation, Women's Health, and Biology. Soc Stud Sci 53 (4):475–494. https://doi.org/10.1177/03063127231168371

4

How Can Gender/Sex Entanglement Inform Our Understanding of Human Evolutionary Biology?

Holly Dunsworth and Libby Ware

Abstract Few who study human evolutionary biology would defend a view of life that pits nature versus nurture. However, moving on without continuing to seek the primary driver of the evolution of a trait, or without aiming to disentangle the relative importance of each factor in a biocultural phenomenon, has been a challenge for researchers, writers, and educators because scientists are interested in measurable causes. When evolving traits of scientific interest are related to sex and/or gender, then gender/sex entanglement theory will be imperative for escaping the legacy of nature versus nurture. But that progress means rethinking research questions; creating clever, careful, and ever-complex study designs; employing new data collection and analytic methods; and incorporating diverse qualitative, theoretical, and philosophical perspectives. Humbly, and in lieu of providing instructions for such a path forward, this chapter argues that practitioners, educators, and communicators of human evolutionary biology should (continue to) endeavor to carve such a path. To contribute, this chapter applies the gender/sex entanglement lens to sex differences in primate behavior and bones that relate to body size and male dominance.

Keywords Human evolutionary biology, apes, body size, embodiment, growth and development, male dominance, sexual selection

Holly Dunsworth (✉)
Dept. of Sociology and Anthropology, University of Rhode Island, Kingston, RI 02881, USA
Email: holly_dunsworth@uri.edu

Libby Ware
Dept. of Anthropology, The George Washington University,
Washington, DC 20052, USA

L. Z. DuBois et al. (eds.), *Sex and Gender*, Strüngmann Forum Reports,
https://doi.org/10.1007/978-3-031-91371-6_4

1 Introduction

As a rigorous approach to human evolutionary biology, gender/sex entanglement theory will require changes to the research process and to how researchers, communicators, and educators comprehend biology and conceptualize evolution. Here we attempt to energize a paradigm shift by holding up the gender/sex lens to key overlapping, entangled areas in human evolutionary biology concerning sex differences in behavior and bone development that are related to body size and male dominance.

2 Background

There are presently two main approaches to sex and gender in human evolutionary biology. The adoption of either approach depends on the researcher, educator, or communicator and the trait or behavior in question. It also depends on conceptions of sex and gender.

DuBois and Shattuck-Heidorn (2021) describe sex and gender as it is used in human evolutionary biology: sex "broadly refers to biological characteristics generally related to reproductive anatomy or physiology" whereas gender is "culturally contextualized social and structural experience as well as expressions of identity" (DuBois and Shattuck-Heidorn 2021:3; see also Sobo 2020). Though DuBois and Shattuck-Heidorn do not describe it as such, we take their definition of "sex" to represent the field's working definition. Note that an explicit mention of behavior is absent from this definition of "sex," which in human evolutionary biology reflects a tradition of equating patterned sex differences in reproductive anatomy and physiology (e.g., the hormones involved) with a biologically determined concept of gender to explain behavior. In other words, the behavioral differences between the sexes are assumed to be caused by the same factors, in the same ways, that cause sex differences in anatomy and physiology. In addition, because it is culturally contextualized and includes expressions of identity, "gender" only applies to humans. In human evolutionary biology, this perspective has created room to assume that sex differences in behavior in nonhuman primates are biological characteristics, or just sex, and has led many to interpret similar or analogous sex differences in human behavior as also being biologically based.

These conceptions of sex and gender are reflected in two different approaches to sex and gender in human evolutionary biology: one equates sex differences in behavior to gender whereas the other distinguishes sex and gender, layering gender on top of biology. Yet, when nonhuman primates are the focus, both approaches encourage the same outcome: scientists and communicators essentialize sex differences in behavior, rendering them sex not gender. *Within the enduring, traditional evolutionary perspective* (i.e., biologically deterministic and not particularly modern or feminist), the result has been that gender describes a smaller, more circumscribed, more superficial realm of human existence than its broader incarnation, with greater, variable, and fluid possibilities, as gender functions among an increasing proportion of society. This, in turn, has caused tension and conflict between scientific and progressive sociopolitical views of gender, despite many progressive views espoused by human evolutionary biology's scientists, educators, and communicators.

The crux of the problem is that when it comes to sex and gender, and even when multiple, complex variables are considered together, human evolutionary biology is languishing in the outdated nature (gender = sex) versus nurture (gender is separate from sex) framework. Meredith (2015:72) spotlighted the living legacy of nature versus nurture in human evolutionary biology and called for "a dynamic systems approach that focuses on understanding how the interactions of social, environmental, somatic, and historical factors work to produce sex-typed [i.e., sex-related] behaviors."

The integrated approach of gender/sex entanglement theory (DuBois and Shattuck-Heidorn 2021; Fausto-Sterling 2019) offers a more realistic (albeit more complicated) view from which to ask and answer evolutionary questions for humans (and potentially other species). Since few aspects of adult behavior, emotions, sexual orientation, or identity can be "sourced purely to sex or purely to gender," and because none of those qualities are fixed over a lifetime, and "gendered structures" change biological function and structure, gender/sex entanglement neither synonymizes nor separates sex and gender (Fausto-Sterling 2019:4). It should be the paradigm for humans as well as for nonhuman primates, even though their cultural dynamics are qualitatively different from ours.

Fausto-Sterling describes gender/sex entanglement as "a softly assembled dynamic system that comes into being starting in infancy and is maintained through one-on-one interactions with other individuals and via cultural enforcement of gender/sex" (Fausto-Sterling 2019:4). One key aspect entails embodiment, "a neuromuscular habit, a nonconscious phenomenon that may entail both the central and autonomic nervous systems" (Fausto-Sterling 2020:280). Children and adults "choose consciously from among the many cultural features of gender to embed new bodily habits into [their] sensorimotor (neuromuscular) system" (Fausto-Sterling 2019:5). Cultural features of gender shape how our bodies function even without conscious choice.

Development is key as well. Gender/sex development is a "continuously evolving (both intra- and intergenerationally) set of habits resulting from ongoing interactions between the child and other humans and objects in their world" (Fausto-Sterling 2019:6). To use the gender/sex approach is to think about *becoming* a gender/sex, so gender/sex (from infancy to adulthood) would be understood to "sediment gradually in the body, seeming to arise 'naturally,' but in fact being a biosocial sediment built up over a lifetime" (Fausto-Sterling 2019:6). Sex and gender are two layers that "can only be understood in relation to each other" (Fausto-Sterling 2019:6). Because they are co-constructed, intermingled, and interwoven, the gender layer cannot be peeled off to reveal what lies underneath. Despite the challenges, gender/sex entanglement theory will enable human evolutionary biology to enact science that is no longer constrained by nature versus nurture perspectives.

Let us consider the following examples put forth by Anne Fausto-Sterling (pers. comm):

1. Weightlifting, a gendered behavior, affects anatomical and physiological traits that already differ by sex; their modification affects how a person is

gendered and how they behave.

2. Wearing a hijab, a gendered behavior, limits exposure to sunlight and affects vitamin D and bone density (Lips 2007); this trait differs by sex throughout life and any ensuing frailty can shape the gendering of a person.

3. Living in an abusive relationship is experienced in gendered ways and impacts stress physiology, steroid, and hormone systems (which already differ by sex); behaviors and biological effects that emerge can compound the gendering of the person.

Gender/sex entanglement is the inseparable development of anatomy, physiology, hormones, and genetics within a fluid sociocultural context, including identity, roles, norms, relations, and power. Gender/sex entanglement acknowledges that culture seizes on the baseline biological variation and that there is the potential to increase as well as decrease it. For instance, there are sex differences in nonhuman animals that are known to contribute to the same physiological response, or phenotype, in both sexes (see Chapter 3).

Gender/sex is entangled with race and other intersecting factors (Collins 1990; Crenshaw 1989; see Chapter 9); the very study of sex and gender in human evolutionary biology has a history of racism (e.g., Markowitz 2001). The legacy of the WEIRD (Western, educated, industrial, rich, democratic), monotheistic colonial cultural context (Henrich et al. 2010; see also Clancy and Davis 2019) has influenced how people understand human biological variation over time and space and enabled those ideas to spread globally.

Gender/sex entanglement is a process more than an attribute; thus, it poses a challenge to human evolutionary biology. Traditionally, the field carved up a spectrum of variation into mutually exclusive units (e.g., species, sexes, genders) and treated qualitative and quantitative variables as discrete traits under direct selection (e.g., height, strength). Often the goal was to disentangle such complex, interconnected traits to uncover the main driver of a phenomenon or to disentangle biological from social causes in the evolution of a trait. In addition, the field has focused primarily on adult members of a species. This suggests that selection on mature individuals drives human evolution more than at earlier stages of development, which implies that reproductive-aged adults are more valuable sources of evolutionary insight than any other age group. Although only a small portion of studies in the field actually measure reproductive success or fitness, all are intimately tied to reproduction. Therefore, the paradigm shifts required by gender/sex entanglement theory will impact the entire scope of human evolutionary biology, even when studies are not directly tied to gender/sex.

To our knowledge, gender/sex entanglement theory has not been incorporated into sex- and gender-attuned critiques of human evolutionary biology (including primatology) or sexual selection, but it has benefited from and built upon those contributions (Ah-King 2007, 2022; Cooke 2022; Fedigan 1986; Fisher et al. 2021; Fromonteil et al. 2023; Gowaty et al. 2012; Hoquet 2020; Hrdy 1981/1999; Khorasani and Lee 2019; Ocobock and Lacy 2023; Roughgarden 2004; Tang-Martinez 2016; Vandermassen 2004; Zihlman 1985). In psychology, Wood and Eagly's (2012) comprehensive discussion of the "biosocial construction of sex differences and similarities in behavior" may come closest to how gender/sex entanglement theory could be applied to the evolution and development of sex and gender differences in human behavior. Let us, therefore, apply the gender/sex approach to long-standing questions related to body size and male dominance: In primates, how do we make evolutionary sense of sex differences in long bone growth and behavioral development?

3 Evolution of Sex Differences in Human Height

Phenomena as basic as sex differences in anatomy and body size—and the fact that in primates females (not always) have sex with males and males (not always) have sex with females—have been at the root of profound claims made about innate, evolved (i.e., adaptive) sex differences in psychology. Sex-typed psychologies are hypothesized to explain, for instance, sex differences in chimpanzee infant behaviors (discussed below) and in long bone growth. Here, we use the lens of gender/sex entanglement to explore alternative evolutionary explanations for sex differences compared to the canon that relies on theorized sex differences in psychology.

We begin by considering evolutionary explanations for one such sex difference in anatomy: human height. Viewed through a gender/sex lens, traditional assumptions about height's evolutionary relationship to sex differences in behavior/psychology—namely, sexual selection (Darwin 1871)—become difficult to accept (Dunsworth 2020). Although scholars and scientists disagree on the exact definition of sexual selection and its applications, the assumption that it occurs within each sex (i.e., it is intrasexual), not between them, is a standout but contested concept. For recent reviews and applications of sexual selection in humans, see Lassek and Gaulin (2022) and Wilson et al. (2017).

If we simply consider what we know about long bone growth, a different answer emerges. Until puberty, all human children grow at about the same rate. With the onset of puberty, females seem to experience a slight bump in growth velocity for a short phase. After menarche, long bones in typical female bodies stop lengthening and the growth plates fuse. In typical male bodies, long bones continue to lengthen at the same rate for a few more years until their growth plates eventually fuse (Bogin et al. 2018). Estrogen is the main cause of this complex phenomenon (for further discussion, see Dunsworth 2020).

As a primary driver of long bone growth in all humans, estrogen is biphasic. At high levels, estrogen ends long bone growth with the fusion of the growth plates. As estrogen soars during routine female development, teenage girls stop gaining stature. Estrogen increase at puberty is fundamental to ovarian development and crucial to the initiation of regular ovulatory/menstrual cycling. Because typical teenage boys do not have ovaries, estrogen does not typically reach high enough levels to cause bones to stop growing at the same age as they do in girls. Boys stop gaining stature a few years after girls, because there is nothing causing them to stop sooner. Both male and female bodies depend on a delicate balance of estrogen (not too much, not too little) for gonad, genital, and gamete function. Bone growth is also affected by numerous other factors, hormonal and otherwise, that are involved in multiple functions and traits beyond body size.

Given the nutritional, energetic, metabolic, and locomotor costs of pregnancy, lactation, and mothering, there are context-specific limits to female body size, perhaps leading to the idea that biology "prioritizes" reproduction over growth. This is an evolutionary framework that refers to primate females being the "ecological sex." That framework has endured separately alongside sexual selection to explain height in males. Yet even within that narrative, with selection distinctively optimizing skeletal growth in both sexes, the sexual selection perspective on male height seems unnecessary.

Underneath all the factors that explain human height variation, there are hundreds (and maybe thousands) of genomic connections. To date, however, no male-specific genes have been identified that can account for the male-specific biology of height. There is only a common biology of skeletal growth shared with females, in which similar processes significantly controlled by estrogen play out differently in different bodies during development. These issues, and the remaining gaps in our knowledge, apply not only to human skeletal growth but to the great apes as well, as they share this growth pattern with us.

Still, according to conventions accepted by human evolutionary biologists, estrogen and the biology of long bone development do not provide a reasonable evolutionary answer to the question of sex differences in height. Since 1871, before anyone understood about hormones let alone their role in bone growth, Charles Darwin's explanation dominated the field: Males are taller than females because males compete for sex, and the taller males have been the winners.

As sexual selection theory goes, male height has been caused by what it is *for*, which is winning the competition for mates. Consider a textbook example of how this is described: "Sex differences in pubertal development are closely tied to sex differences in intrasexual competition and the corresponding sex differences in physical size" (Geary 2021:300). In this scenario, greater male height is assumed to be conspicuous evidence of inconspicuous, biological underpinnings of male (i.e., not female) behavior. Here is how one researcher explained the phenomenon: "The mere existence of the physical differences tells us that human males have been subject to stronger selection for aggression and violence than females" (Stewart-Williams 2019).

Darwin's sexual selection explanation for sex difference in human height has endured through the powerful influence of Ernst Mayr (1961). Mayr's framework distinguishes proximate and ultimate biological causation, and has led human evolutionary biologists to separate hormones and development (i.e., proximate) from evolutionary (i.e., ultimate) causes, such as sexual selection, with mate competition and mate choice. Interpreting and applying this framework in human evolutionary biology, when a biological explanation falls under the "proximate" category, it is considered a mere "mechanism" (too often a simplistic, black-boxed, and unidirectional cause) and is not deemed, rather surrealistically, to be "evolutionary" despite having to be adaptive if it is to typify a lineage's biology. In addition, the proximate category is typically where there is evidence to parse, whereas the ultimate explanations typically rest (at least when it comes to human anatomy, physiology, and *especially* behavior) on theory alone. That is how the proximate versus ultimate convention has contributed to the persistence of spurious correlations and untestable evolutionary "hypotheses," thereby baking assumptions about gender into human evolutionary biology by empowering a bioessentialized gender.

Sex differences in height are typically explained by asking "what for?" From this (usually implicit) perspective, the answer cannot simply be about the crucial function of estrogen in the human reproductive system because the sex difference is rendered purposeless. Indeed, the question implies that there is an end goal; for any given trait—even a patterned difference between sexes, which is not a trait—the default assumption is purpose. Sex differences in height are for winning male-male contests. The thought process which follows is that while the "proximate" biological mechanisms that underpin sex differences in height may be hormonal, the ultimate "reason" they exist is masculinity or, broadly, gender. That is, male behavior brought about differences in height as taller males were competitively more successful than shorter ones. With increased understanding of reproductive biology, including hormone variation, the behavioral (i.e., sexual selection) explanation for sex differences in bone growth still endures because it is "ultimate": it is *the* evolutionary one. Greater male height continues to serve not merely as evidence for sexual selection being its cause, but also as evidence that men and women are fundamentally distinct, with men being fundamentally more competitive and dominant in their evolved (i.e., adaptive) biology.

At present, human evolutionary biologists treat Mayr's theory like Darwin's, but it is only a convention. In 2011, Laland et al. wrote that "Ernst Mayr's formulation has acted to stabilize the dominant evolutionary paradigm against change but may now hamper progress in the biological sciences." Combined with other obstacles to progress, like adaptationism, teleology, and a reluctance to reckon with knowns and unknowns in developmental biology (e.g., what counts as evidence or defines causation), Laland et al.'s critique applies to human evolutionary biology, including applications of sexual selection theory (e.g., male competition, female deception).

Without the traditional story and with a gender/sex entanglement prerogative, it becomes easier to see how sex differences in body size (i.e., the by-product of egg- and sperm-related reproductive physiology) could contribute to how the sexes develop, learn, and enact behavior—how being bigger bodied than half the species could contribute to "masculine" male behavior. The sexual selection explanation tempts us to think unidirectionally, with behavior always being the cause of the anatomical evolution. Yet over time, as anatomical evolution occurs and the context for the behavior changes, behavior may change and the changes in anatomy may also follow. As lineages experience more or less sex difference in body size, for whatever evolutionary reason (e.g., the strict sexual selection route or the by-product route), it is not just behavior that is the context for the physical evolution, it is always also vice versa. The gender/sex framework exposes the shortcomings of sexual selection and helps us to question its relevancy. Natural selection, constraints, and by-products could be producing sex differences which help to create the context for sex differences in behaviors that may have incidental, weak, or strong biological connections to those physical adaptations.

Given that the evolution of human sex differences is understood by the field in comparison to other primate and mammalian species (see Geary 2021; reviewed in Lassek and Gaulin 2022), the gender/sex framework opens up new ways to view the evolution and development of sex differences in nonhuman primates and their relevance in explaining ours.

Though gender/sex is not the explicit lens, Cassini (2020) offers an alternative model to the classic Darwinian one for greater male body size and related dominance behavior, which echoes the embodied framework of gender/sex and rejects the unidirectional causation that is impossible in gender/sex entanglement. In the classic model, greater average height is caused by precopulatory competition between males. In Cassini's model, greater male body size exists, perhaps as a result of sex differences in hormone levels involved directly in gamete production (as described above). This initial condition enables larger-bodied males to sexually harass and coerce smaller-bodied females, which spurs females to aggregate and, in turn, enables males to monopolize "harems," thus explaining the well-known, widespread polygynous social condition across mammalian species where males are larger. According to Cassini (2020:115): "The fact that males fight for mates, and even that the largest males win combats, does not mean that sexual selection, understood as direct competition between males, plays a predominant role in the evolution of sexual [body size] dimorphism or in male reproductive behavior." The emphasis on intrasexual selection in males as the main or only process in sexual selection has acted to obscure important, similar processes within females and evolutionarily salient intersexual dynamics, not just in pinnipeds and nonhuman primates but also when narrating human evolution. It is crucial to acknowledge the role of body size in the development of behavior, rather than to continue to focus narrowly on the reverse, as this elevates behavior to the evolutionary cause.

Given what is currently known about skeletal biology, it is difficult to imagine how selection for tall males, but not also females, could work. If sexual selection could reduce estrogen in males in favor of a longer growth period and a taller outcome, then the estrogen required to pass on those height-enhancing genes to their offspring would be reduced. This poses an interesting problem that cannot easily be explained by sexual selection. Thus, it appears that sex differences in height evolved as a by-product of the evolution of genital and gonad function. Women stop growing earlier than men because their bodies have more estrogen; this causes growth plates to fuse and bones to stop getting longer. Men grow taller simply because nothing stops their growth early in life, as it does in women. There is no "male" or "female" skeleton; there is a human skeleton that develops in patterned ways. For a similar view of brains, see Joel (2021) and Eliot et al. (2021). Viewing sex differences in skeletal growth through a gender/sex lens opens up the possibility that what Savell et al. (2016) described across human populations may also apply between sexes, which is that trait differences may not directly reflect the forces of selection that shaped them (see also Auerbach et al. 2023).

Binary thinking—which views males and females as mutually exclusive or (in the extreme) as being separate species from conception—is hampering our scientific imagination. Gender/sex entanglement theory, however, will help propel us forward: We start with a human, who is nearly identical to every other human, and that human develops over time, in context. Conceptually, it is that simple, yet it is far more complicated to implement scientifically than male versus female. But if we are to break free of the delusions of nature versus nurture and proximate versus ultimate, then we must find a way.

Many evolutionary biologists have adopted "sex differences" instead of "sexual dimorphism" when describing traits that do not differ in quality, and whose quantitative range overlaps between the sexes (Astorino 2019). It seems absurd to continue to conceptualize height and other quantitative traits that are present in all humans as being "sexually dimorphic." Furthermore, the label "sexual dimorphism" earns an automatic sexual selection explanation; the example of height may serve as a warning against that practice. Genitals and gonads are exempt from sexual selection, but other "sexually dimorphic" traits, including height, should be as well until evidence demonstrates otherwise.

Perpetuating the proximate-ultimate convention, and enforcing it on others, is holding up another sort of binary or dimorphic construction that may be preventing scientists from describing evolutionary reality. Not only is it difficult to imagine how to falsify the enduring "ultimate" sexual selection explanation, it remains an obstacle to improving the scientific quality of the narratives of human evolution, and, thus, liberating all of us from gendered oppression in the name of "human nature." If science tells us that male bodies are "for" competition and dominance (or if that is what we hear as we make sense of the science), then that is the story we will enact, and in doing so, we will embody what has counted as evidence for the sexual selection explanation, and then we are back to the self-fulfilling prophecy of human evolutionary biology.

4 Evolution of Sex Differences in Nonhuman Primate Behavior

Studies of nonhuman primates are often associated with questions about human sex and gender, yet often these questions seek to reduce human behavior to bioessentialist categories of sex-based behavior. One example is "Sex Differences in Wild Chimpanzee Behavior Emerge during Infancy" by Lonsdorf et al. (2014) in which the authors analyze data compiled on 40 infants from long-term studies of chimpanzees at Gombe National Park, Tanzania. Here, we begin with a summary of this study and then discuss how the data could be reinterpreted using a gender/sex lens. As with the previous section, we interrogate the heavy reliance on theorized, evolved sex differences in psychology for explaining variation within a species.

Not all variables included in the study differed by sex. No sex differences were apparent in suckling or grooming behaviors or in the integration of solid foods into the diet. Some key variables across categories of motor development, spatial independence, and social behavior did differ by sex. Males switch from riding ventrally (considered to be the more immature form of travel) to riding dorsally at a younger age than females. Male infants begin to travel and spend more time traveling independently at a younger age than females. By three years of age, male infants maintain a farther distance from their mother and remain at longer distances than female chimpanzees do, up to the age of five. Sex differences were also apparent in social but not solitary play. Males dedicated more time to social play at an earlier age, whereas females peaked in the percent of time spent on social play later in development. One might wonder if this pattern is similar to what seems to be happening in humans: newborn and infant boys are slightly larger than girls and caregivers may accordingly treat them differently, and girls learn to speak and might mature gonadally at an earlier age than boys (Fausto-Sterling 2015).

Because play is related to locomotor independence, the analysis by Lonsdorf et al. (2014), which encompassed whole-body physical activity, is not surprising nor are their interpretations: "Sex differences were found for indicators of social behavior, motor development, and spatial independence with males being more physically precocious and peaking in play earlier than females. These results demonstrate early sex differentiation that may reflect adult reproductive strategies" (Lonsdorf et al. 2014:1). Yet what "adult reproductive strategies" exist to explain infant sex differences? For social play, combined with references to other studies, the authors write that sex differences "may reflect the relative importance of socialization for young males given the importance of social dominance in adulthood" (Lonsdorf et al. 2014:4). The logic here is that selection for dominant adult males includes selection for their relatively precocial independence and social play behavior in infancy and youth.

The authors write that their findings are "consistent with adult sex-specific social roles in chimpanzees and parallel similar patterns found in humans" and suggest "that some biologically based sex differences in behavior may have been present in the common ancestor of chimpanzees and humans, and operated independently from influences of modern sex-biased parental behavior and gender socialization" (Lonsdorf et al. 2014:7–8). This approach assumes that chimpanzees are useful models for our hominin ancestors prior to the emergence of complex language-based culture. Accordingly, when compared to us, chimpanzees may hold the keys to understanding what about our behavior is innate, what is socially learned, and what is uniquely, culturally constructed. This approach, however, de-emphasizes equifinality, or variable paths to the same end point. Are male infants doing similar behaviors as adult males? If so, do the behaviors actually have the same causes? Would having the same causes be necessary to link them causally in a developmental way? Does the timing difference between the sexes in these variables lead to a different endpoint between the sexes in other variables? The thread from development to endpoint would need to be studied (e.g., Karasik et al. 2023; Schneider and Iverson 2023).

If a reader is not careful, the analyses and interpretations by Lonsdorf et al. could be taken as a validation of existing, biased assumptions about humans. Interestingly, these very assumptions contributed to their interpretation and, if care is not taken, could become further embedded in science. Whether by emphasis or omission, their article implies that the best explanation for sex differences in infant chimpanzee behavior is that there is something adaptive and biologically essential in male chimps that is different from what is adaptive and biologically essential in female chimps, and that this difference accounts for the variation in spatial independence and gregariousness observed in young chimps. Accordingly, this raises the possibility (already widely assumed in science) of inherent sex differences in psychology, regarding behaviors such as dominance, adventurousness, confidence, independence, risk-taking, and leadership, and from there it is easy to redirect such assumptions onto humans. These assumptions exemplify the challenges and pitfalls that have impacted human evolutionary biology, knowingly, for decades.

Yet when we view Lonsdorf et al. (2014) through the lens of gender/sex entanglement—within a relentlessly developmental, embodied, and relational context—we do not arrive at their conclusion. Humans who care for infants know that as children grow, they become increasingly heavy and more difficult to carry. This challenge varies from parent to parent and child to child based on size, strength, tolerance, and sensitivity; it also depends on the physical and emotional relationship between the parent and the child, as well as on culturally shaped expectations. Eventually, all caregivers encourage locomotor independence in their young. It is a good hypothesis that offspring weight (relative to the offspring's strength as well as mother's size, strength, or age) is highly important and may be the best predictor of the observed shift in chimpanzee mother-infant dyadic behavior compared to simply sex and age of offspring. However, this study did not analyze any metric of any individual's size or strength, which could be due to the obstacles of obtaining those data in the wild.

The sole mention of infant body size by Lonsdorf et al. supports this alternative hypothesis: "A long-term analysis of weight data from Gombe, Tanzania, showed that female chimpanzees are slightly lighter than males up to age 10, when adult dimorphic patterns begin to emerge and eventually result in a male/female body mass ratio of 1.25" (Lonsdorf et al. 2014:2). This implies that male infants and youngsters are likely to be heavier than females at any given age. Body size, not maleness or sex, may explain why males are clinging less and being carried less at earlier chronological ages than females. Body size, not sex, already explains many sex differences in adult primate behavior, including energetic costs, predation risk, food resource accessibility, and substrate use (see, e.g., references in Meredith 2013). Regarding infant chimpanzees, substrate use is especially relevant because mother is a primate's first substrate.

When primate behaviors are related to our conceptions of masculinity and serve as models of our ancestors' behavior (and ours presently), it is tempting to believe that sex differences boil down to sex (Fuentes 2021), rather than some phenomenon that is less conspicuously involved in human conceptions of gender, like gravity.

There is much to learn from chimpanzees about how slight sex differences in physical development can lead to more conspicuous, pronounced, and evolutionarily consequential (sex, life, and death) differences in adult behavior between the sexes. Still, all of that is occurring in a species that develops muscles which are qualitatively different from ours, leading chimpanzees to be significantly stronger than humans (O'Neill et al. 2017). In addition, none of that embodied, social learning requires assumed or hypothetical innate, divergent (or binary) sex differences in psychological contributors to behavior (neither in chimpanzees nor in humans).

Minimal sex differences in average size, mass, and muscle growth and strength, as well as other anatomical and physiological characteristics, may be all that chimpanzees need to develop along some sex-patterned trajectories, with average sex differences that start (metaphorically and literally) millimeters apart and end up, years later, separated by (metaphorical and literal) centimeters. The framework for the study, which seems to be about innate sex-based differences in the psychology of behavior, emanating independently from within each infant and determining the patterned differences we see, is missing this very important, basic embodied and relational (with the mother and playmates) view of development and may be only a part of a much more complex story (as acknowledged by Lonsdorf 2017; Lonsdorf et al. 2014). If the sex differences in chimpanzee infant behavior that we have considered here are determined to be driven by sex differences in the chimpanzee infants' brain and not their body size, then that needs to be linked to the sex differences in adult behavior that are said to be the evolutionary cause of what is going on in the infants. To make the link, it will be necessary to track the development of the infants into adults who exhibit the behaviors under selection theorized to have brought about the infant behaviors.

Big questions remain about how an individual's sex helps to determine their behavior and helps to explain patterned sex differences in behavior. What is more, questions remain about how individuals perceive themselves and one another in such a patterned milieu, and whether such perceptions are involved in their sociosexual behavior. For example, do chimpanzee mothers discriminate between infant sons and daughters, based on sex alone (i.e., genitalia or pheromones), and might that contribute to the patterned sex differences in development that Lonsdorf et al. observed? More broadly: Do nonhuman primates have gender? Do they, for example, have "culturally contextualized social and structural experience as well as expressions of identity" (DuBois and Shattuck-Heidorn 2021)? If they do, would we even be able to recognize it, since it may very well manifest differently?

To our knowledge, Schwartz (2018) is the first researcher in human evolutionary biology to publicly grapple with the existence of gender in nonhuman primates. Focusing on a few key aspects of adult behavior, Schwartz looked at sex biases in the grooming partners of chimps and bonobos and relevant cognition. Here we briefly consider the aspect of cognition.

Is there anything we know about ape cognition that indicates they could form a gender identity or that they could conceive of one in others? Since apes do not have a spoken language, any concept of gender which they may have would be acquired and function differently from ours. Not that humans should lack any capacities apes have, but in addition to any of those possible gender/sex homologies, our ability to reason about abstractions and to transmit that abstract reasoning to others means that our gender is qualitatively different from any that apes possibly have or experience. Because apes lack abstract reasoning (Povinelli 2003), any "knowing," "knowledge," or "understanding," including that of sex (same or different) and gender, should be understood as embodied (Povinelli 2003). This is a challenge to do as a different body, let alone as a different species

It is widely understood that chimpanzees have the ability to know that others are relatively separate and that they have similar capabilities, needs, and desires (Tomasello 2022). Because bonobos are similar to chimps, they too may have this capacity, though neither have it to the same extent as humans. Schwartz (2018) sees this as evidence of "theory of mind" which may confer the potential for chimpanzees and bonobos to recognize their own gender as well as that of others. Yet because traditional wording of this phenomenon has been confusing, it is difficult to interpret what many researchers are actually suggesting about the minds of apes. "Theory of mind" does not refer to concepts or reasoning about the abstract (e.g., minds of others). It describes an important aspect of embodied cognition that is better labeled "body reading" (de Waal 2016). The presence of body reading (i.e., theory of mind) does not demonstrate the holding of a concept about sex or sex differences, which is a key aspect of human gender.

Let us return to the issue of the mother chimpanzee: Based on any experiences with adult males, might she associate her son with those males and push him away physically, thereby contributing to his earlier development of independence compared to her daughters? This seems possible, but it would be difficult to investigate. Many if not all of the studies aimed at determining whether nonhuman primates categorize one another into two distinct sexes, based on conspicuous anatomical cues, involve training them to do so in the first place (de Waal and Pokorny 2008; Schwartz 2018). Such training, however, sounds very much like encouraging a human concept of gender in apes. While there is no compelling evidence that apes gender themselves, if they did, they would have had to build that patterned behavior without sharing it via language.

Although chimpanzees lack a spoken language, there is space in human evolutionary biology for comparing ape gestures to human language (Hobaiter et al. 2022). This would be an area worthy of connecting to the discourse around gender in nonhuman primates, especially as cultural traditions and their transmission are increasingly understood among chimpanzees (Whiten et al. 2021). Language enables humans, including children, to enforce social norms (Tomasello 2019) in ways not detected in other species. Thus, there is a more dynamic and intense, ever-present gender at the culture level in human communities. For example, humans experience and enact patriarchy in ways that chimpanzees do not (despite a tradition of applying the term to both species' social behavior and organization).

The gender/sex framework requires that we put every individual in constant developmental and social context throughout life, and that we probe the causes of their behaviors in that lifetime of developmental and social context. In doing so, we can no longer use "sex," "male," or "female" as placeholders to explain sex differences in behaviors, which will make it easier to apply what we learn from nonhuman primates to understanding ourselves. So, while "overt gender socialization and phenomena such as gender performance seem to be uniquely human" (Meredith 2015), whatever gender might or could be in nonhuman species, and whatever that means for understanding gender in humans, are questions that are best approached utilizing the gender/sex lens.

5 Concluding Remarks

The question whether gender exists in nonhuman primates does not and may never have a straightforward answer. For humans, universal and variable gender/sex entanglement is a powerful cultural phenomenon that rests on conspicuous biological sex differences as well as norms and beliefs associated with them. Those beliefs about gender are increasingly incorporated into science—what primatologists report about sex differences and which evolutionary "causes" scientists accept to explain sex differences in long bone growth. With human evolutionary biology's shift to a gender/sex approach, science will necessarily change, and, as a result, beliefs about gender as it exists in the present world will change as will our relationship with our evolutionary history (for a similar take on gender, see Fuentes 2021).

5.1 Rethinking Causation, Evidence, and Narrative

Historically, in human evolutionary biology, sex differences have been assumed to underlie behavioral differences directly via biology. Traditionally, the quest to comprehend sex differences in behavior has promised to illuminate how evolution works and to reveal something important about human nature. Unfortunately, outcomes have included the confirmation (and enduring perpetuation) of simplistic assumptions about how evolution works, untrue narratives about how it has occurred, and beliefs about human nature that were conceived long before evolutionary science came along. For example, greater male height has served as evidence that men and women are evolutionarily and essentially distinct, and that men are fundamentally more competitive and dominant. This has led to the belief that males, from conception on, evolved to be better built for success, which increases their value and inspires the enduring narrative that men, masculinity, and maleness forged our species' triumph.

When gender and sexual behavior are seen through that biologically determined, adaptationist, and teleological (i.e., for a purpose) lens, then gender roles are seen as being a person's evolutionary purpose. That is the source of human evolutionary biology's power to assist in societal oppression. The belief that biologically based sex differences in behavior have been a sort of "force" that has been responsible for our species' success has supported, even if passively, sexism, misogyny, and patriarchal oppression of people of all genders and sexualities. Instead, gender/sex entanglement lends intellectual legitimacy to evolutionary views that center love, egalitarian norms, and pleasure (e.g., Lindisfarne 2019), which for far too long have, been considered naive or unscientific.

Beliefs about human nature are increasingly built by science. We are misled and biased by our habit of projecting the present onto the past as well as to other species and then applying what we imagined exists in nature back onto ourselves (Hubbard et al. 1979). Gender/sex helps to free us from the blinkered loop in which the field, and especially its interpretation/communication, has been stuck since its inception.

5.2 Applying the Gender/Sex Approach to Research Design

To apply gender/sex approach, human evolutionary biologists should take the following cues from Schellenberg (2019):

1. Identify the specific hypothesized mechanism (instead of simply "sex," "gender," "sex/gender," or "gender/sex").
2. Focus on specific operationalized variables (instead of simply "sex," "gender," "sex/gender," or "gender/sex").
3. If needed, define the variables of "sex," "gender," "sex/gender," or "gender/sex" as specifically as possible; include how data were obtained (e.g., based on genital observations, presence of Y chromosome, self-reporting, museum collection catalog).
4. Whenever possible, use methods that keep researchers blind to those variables.

5. Cling only to testable hypotheses. This will expose alternatives that are either not yet tested or that are untestable, and will both strengthen the study and prevent it from perpetuating status quo, un-evidenced assumptions about sex and gender.

In addition, human evolutionary biologists should consider following Joel and McCarthy (2017), who offer "a framework for defining what is being measured and what it means" and outline how sex differences can be classified on four dimensions: (a) persistent versus transient across the lifespan; (b) context independent versus dependent; (c) dimorphic versus continuous; and (d) a direct versus indirect consequence of sex.

The terms sex, gender, sex/gender, or gender/sex are too big, diverse, variable, and complex to use without explicit definition in scientific research or scholarship. What is more, binary thinking—the assumption that male and female are mutually exclusive categories of whole organisms (as opposed to gametes, or chromosomes)—is incompatible with a more complete understanding of gender/sex. While egg-making individuals are female and sperm-making individuals are male, and (regardless of whether and which gametes they produce) XX individuals are female and XY are male, rarely does one biological aspect of a lifetime, an organism, a sex category (like male or female), or a lineage provide sufficient insight or explanation in human evolutionary biological research—at least not for research involving individual traits and behaviors. What is it specifically about gender/sex that is hypothesized to be the mechanism? Is it height, muscular strength, or estrogen? For further reading and recommendations, see Springer et al. (2012) and DuBois and Shattuck-Heidorn (2021).

Schellenberg's last recommendation involving testable hypotheses is especially challenging because hypotheses about natural history and any alternatives are notoriously difficult to test. In many cases, this should disqualify them as "hypotheses." However, the constant and acute awareness of the hypothesis-testing problem will strengthen gender/sex related science. Schellenberg urges us to interrogate science, asking (not of any animal's trait, but of our own work), "what is the purpose?" As Schellenberg (2019:284) writes, "elucidating why and for which purpose sex/gender is being pursued in the first place could…unveil sex/gender and reveal in its place the actual subjects of interest. And, if given the opportunity, researchers may find ways that allow them…to capture glimpses of these subjects of interest."

5.3 Valuing Scientific Description, Biocultural Approaches, and "Outliers"

So much of human evolutionary biology is a search for causes. Identifying causes is the purpose of the science. What is the value of knowing the primary cause or the percent causation of each contributing factor in an evolving trait, if we are pretending that each contributor is mutually exclusive and measurable, and that each trait is separate? Knowing what contortions and caricatures we make of reality, are these practices worthwhile? Gender/sex frees us from that old atomized, linear, uncluttered, (non)evolutionary way of thinking. Adopting gender/sex means redefining what human evolutionary biology is as a science. Evolution is a vastly interconnected process, so why should we search for measurable causes? It must be, at least partly, because purely descriptive work is less valued. Instead of looking to explain this or that percentage contribution of multiple causes to a phenomenon, why not describe it?

To adopt gender/sex will shift the lens from adults to infants and children, to their development and social contexts. To adopt gender/sex is to embrace the complex context for the development and existence of sex/sexuality and to surrender to a more complicated reality than the adaptationist perspective. To adopt gender/sex will be to rigorously integrate the social sciences and humanities. Gender/sex makes human evolutionary biology fundamentally more anthropological and more biocultural in its theory, methods, and goals. Those who feel more comfortable working with the plasticity, ambiguity, fluidity, and uncertainty of gender/sex entanglement will lead the way. Biocultural anthropologists come to mind. To our knowledge there have been few published papers situated in the field of biocultural anthropology that ponder the similarities of gender/sex entanglement theory and the biocultural approach. The only one we are aware of to date is DuBois and Shattuck-Heidorn (2021).

One of the shifts we must make is to value observational data of the kind that is often disparagingly viewed as "outliers" or as "anecdotes." A good example is the chimpanzee "Donna," whom de Waal (2022) describes as having atypical biological and behavioral characteristics. Accounts of diversity are crucial to evolutionary understanding of variation (for a discussion of "normalcy" and human gender/sex, see DuBois and Shattuck-Heidorn 2021). As observations and analyses accumulate, variability among nonhuman primates like chimpanzees and bonobos will reveal how plastic, context specific, and complicated they are. Gender/sex will require a renewed appreciation for naturalistic description, for its exploration has been overshadowed by the value placed on the collection and analysis of sufficiently large datasets conducive to statistical tests. To date, sex-typical behavior has been determined by averages. However, our focus should include the outliers. Such a shift in perspective will enable us to stop anthropomorphizing our modern-day relatives and conduct, understand, and communicate nonhuman primate research with more nuance and less bias.

Exceptions and those on the margins are not only part of a complete picture, but all biological change in Earth's deep time has occurred *because of the existence of the rare few*. Aligned with constant and necessary biological variation, gender/sex cannot reduce biology to strict binary, distinct, homogenous entities. Gender/sex is perhaps the most powerful lens for addressing variation, the currency of evolution. Gender/sex entanglement applies to our data, samples, and subjects as well as to the scientists who study, interpret, and observe, which is context for bias for everyone but also a formula for scientific strength (Astorino 2019; Meredith and Schmitt 2019; Smith and Archer 2019; Thayer 2019).

5.4 Embracing Epistemic Humility

The gender/sex framework, with its emphasis on embodiment, will help illuminate a path to understanding sex differences in psychology and behavior. Any such phenomenon need not automatically or necessarily be conceived of as stereotyped gender essence, but instead as embodied aspects or tendencies that develop over time and according to experience in a body. Thus, any psychological sex differences that may underpin the infant chimpanzee behavior reported by Lonsdorf et al. (2014) may be better understood by other researchers and the public in a dynamic, gender/sex way. This may lessen the likelihood that such research will be applied too simplistically or too broadly to humans, merely on the basis of a shared genetic code.

Finally, gender/sex bridges the conceptual gap in popular culture between evolutionary history and actual history, by helping to navigate questions that continue to haunt evolutionary thinkers within and beyond academia: Are we wired by, or for, patriarchy? Is matriarchy our ancestral condition, our evolved natural order? If neither apply, then how do we make sense of the diversity of animal and human hierarchies that exist? How are we to understand the violence that upholds hierarchies? What roles do evolution and nonhuman animals play in answering these questions and at what cost—given the popular and scientific traditions of determinism, adaptationism, and teleology that undergird evolution-inspired sexism and racism? What makes human evolutionary biology relevant for asking/answering questions of hierarchy and power relations? Put another way, how is "evolution" more informative or useful an answer to questions about contemporary patterned and organized human behavior than "quantum physics?"

Acknowledgments   We thank the organizers of the Ernst Strüngmann Forum for posing such a stimulating question to us. There are countless answers to the question we were asked to address, and we are grateful to have this opportunity to offer just one. This article benefited from feedback on an early draft from Anne Fausto-Sterling, Charlotte Cornil, Katharina Hoppe, and Gillian Bentley, as well as from editorial advice from Zachary DuBois and Peg McCarthy, and from Dunsworth's experience at the Forum.

References

Ah-King M (2007) Sexual Selection Revisited: Towards a Gender-Neutral Theory and Practice. A Response to Vandermassen's "Sexual Selection: A Tale of Male Bias and Feminist Denial". Eur J Womens Stud 14 (4):341–348. https://doi.org/10.1177/1350506807081883
——— (2022) The History of Sexual Selection Research Provides Insights as to Why Females Are Still Understudied. Nat Commun 13:6976. https://doi.org/10.1038/s41467-022-34770-z
Astorino CM (2019) Beyond Dimorphism: Sexual Polymorphism and Research Bias in Biological Anthropology. Am Anthropol 121 (2):489–490. https://doi.org/10.1111/aman.13224
Auerbach BM, Savell KRR, and Agosto ER (2023) Morphology, Evolution, and the Whole Organism Imperative: Why Evolutionary Questions Need Multi-Trait Evolutionary Quantitative Genetics. Yearb Biol Anthropol 181 (Suppl 76):180–211. https://doi.org/10.1002/ajpa.24733
Bogin B, Varea C, Hermanussen M, and Scheffler C (2018) Human Life Course Biology: A Centennial Perspective of Scholarship on the Human Pattern of Physical Growth and Its Place in Human Biocultural Evolution. Am J Phys Anthropol 165 (4):834–854. https://doi.org/10.1002/ajpa.23357
Cassini M (2020) A Mixed Model of the Evolution of Polygyny and Sexual Size Dimorphism in Mammals. Mamm Rev 50 (1):112–120. https://doi.org/10.1111/mam.12171

Clancy KBH, and Davis JL (2019) Soylent Is People, and WEIRD Is White: Biological Anthropology, Whiteness, and the Limits of the WEIRD. Annu Rev Anthropol 48:169–186. https://doi.org/10.1146/annurev-anthro-102218-011133

Collins PH (1990) Black Feminist Thought: Knowledge Consciousness, and the Politics of Empowerment. Routledge, New York

Cooke L (2022) Bitch: On the Female of the Species. Basic Books, New York

Crenshaw KW (1989) Demarginalizing the Intersection of Race and Sex: A Black Feminist Critique of Antidiscrimination Doctrine, Feminist Theory and Antiracist Politics. Univ Chic Leg Forum 1989 (1):8. https://chicagounbound.uchicago.edu/uclf/vol1989/iss1/8

Darwin C (1871) The Descent of Man and Selection in Relation to Sex. John Murray, London

de Waal F (2016) Are We Smart Enough to Know How Smart Animals Are? W. W. Norton, New York

——— (2022) Different: Gender through the Eyes of a Primatologist. W. W. Norton, New York

de Waal F, and Pokorny J (2008) Faces and Behinds: Chimpanzee Sex Perception. Adv Sci Lett 1 (1):99–103. https://doi.org/10.1166/asl.2008.006

DuBois LZ, and Shattuck-Heidorn H (2021) Challenging the Binary: Gender/Sex and the Bio-Logics of Normalcy. Am J Hum Biol 33 (5):e23623. https://doi.org/10.1002/ajhb.23623

Dunsworth HM (2020) Expanding the Evolutionary Explanations for Sex Differences in the Human Skeleton. Evol Anthropol 29 (3):108–116. https://doi.org/10.1002/evan.21834

Eliot L, Ahmed A, Khan H, and Patel J (2021) Dump the "Dimorphism": Comprehensive Synthesis of Human Brain Studies Reveals Few Male-Female Differences Beyond Size. Neurosci Biobehav Rev 125:667–697. https://doi.org/10.1016/j.neubiorev.2021.02.026

Fausto-Sterling A (2015) How Else Can We Study Sex Differences in Early Infancy? Dev Psychobiol 58 (1):5–16. https://doi.org/10.1002/dev.21345

——— (2019) Gender/Sex, Sexual Orientation, and Identity Are in the Body: How Did They Get There? J Sex Res 56 (4-5):529–555. https://doi.org/10.1080/00224499.2019.1581883

——— (2020) Sexing the Body: Gender Politics and the Construction of Sexuality, Updated edition. Basic Books, New York

Fedigan L (1986) The Changing Roles of Women in Models of Human Evolution. Annu Rev Anthropol 15:25–66

Fisher ML, Burch R, Sokol-Chang R, Wade TJ, and Widman D (2021) Sexuality and Gender in Prehistory. In: Henly T, Rossano M (eds) Psychology and Cognitive Archaeology. Routledge, New York, pp 83–96

Fromonteil S, Marie-Orleach L, Winkler L, and Janicke T (2023) Sexual Selection in Females and the Evolution of Polyandry. PLoS Biol 21 (1):e3001916. https://doi.org/10.1371/journal.pbio.3001916

Fuentes A (2021) Searching for the "Roots" of Masculinity in Primates and the Human Evolutionary Past. Curr Anthropol 62(S23) (S23):15–25. https://doi.org/10.1086/711582

Geary DC (2021) Male, Female: The Evolution of Human Sex Differences, 3rd Edition. APA, Washington, DC

Gowaty P, Yong-Kyu K, and Anderson WW (2012) No Evidence of Sexual Selection in a Repetition of Bateman's Classic Study of Drosophila Melanogaster. PNAS 109 (29):11740–11745. https://doi.org/10.1073/pnas.1207851109

Henrich J, Heine SJ, and Norenzayan A (2010) Most People Are Not WEIRD. Nature 466:29. https://doi.org/10.1038/466029a

Hobaiter C, Graham KE, and Byrne RW (2022) Are Ape Gestures Like Words? Outstanding Issues in Detecting Similarities and Differences between Human Language and Ape Gesture. Philos Trans R Soc Lond B Biol Sci 377 (1860):20210301. https://doi.org/10.1098/rstb.2021.0301

Hoquet T (2020) Bateman's Principles: Why Biology Needs History and Philosophy. Anim Behav 168:e5–e9. https://doi.org/10.1016/j.anbehav.2020.08.010

Hrdy SB (1981/1999) The Woman That Never Evolved. Harvard Univ Press, Cambridge, MA

Hubbard R, Heinifin MS, and Fried B- (1979) Women Look at Biology Looking at Women. Schenkman Publ Co., Cambridge, MA

Joel D (2021) Beyond the Binary: Rethinking Sex and the Brain. Neurosci Biobehav Rev 122:165–175. https://doi.org/10.1016/j.neubiorev.2020.11.018

Joel D, and McCarthy MM (2017) Incorporating Sex as a Biological Variable in Neuropsychiatric Research: Where Are We Now and Where Should We Be? Neuropsychopharmacology 42:379–385. https://doi.org/10.1038/npp.2016.79

Karasik LB, Adolph KE, Fernandes SN, Robinson SR, and Tamis-LeMonda CS (2023) Gahvora Cradling in Tajikistan: Cultural Practices and Associations with Motor Development. Child Dev 94 (4):1049–1067. https://doi.org/10.1111/cdev.13919

Khorasani DG, and Lee S-H (2019) Women in Human Evolution Redux. In: Willermet C, Lee S-H (eds) Evaluating Evidence in Biological Anthropology. Cambridge Univ Press, Cambridge, pp 11–34

Laland K, Sterelny K, Odling-Smee J, Hoppitt W, and Uller T (2011) Cause and Effect in Biology Revisited: Is Mayr's Proximate-Ultimate Dichotomy Still Useful? Science 334 (6062):1512–1516. https://doi.org/10.1126/science.1210879

Lassek WD, and Gaulin SJC (2022) Substantial but Misunderstood Human Sexual Dimorphism Results Mainly from Sexual Selection on Males and Natural Selection on Females. Front Psychol 13:859931. https://doi.org/10.3389/fpsyg.2022.859931

Lindisfarne N (2019) The Roots of Sexual Violence. Masculinities 12:42–58. https://dergipark.org.tr/en/download/article-file/1206114

Lips P (2007) Vitamin D Status and Nutrition in Europe and Asia. J Steroid Biochem Mol Biol 103 (3–5):620–625. https://doi.org/10.1016/j.jsbmb.2006.12.076

Lonsdorf EV (2017) Sex Differences in Nonhuman Primate Behavioral Development. J Neurosci Res 95 (1–2):213–221. https://doi.org/10.1002/jnr.23862

Lonsdorf EV, Markham AC, Heintz MR, et al. (2014) Sex Differences in Wild Chimpanzee Behavior Emerge during Infancy. PLoS One 9 (6):e99099. https://doi.org/10.1371/journal.pone.0099099

Markowitz S (2001) Pelvic Politics: Sexual Dimorphism and Racial Difference. Signs 26 (2):389–414. https://doi.org/10.1515/9780295742595-006

Mayr E (1961) Cause and Effect in Biology. Science 134 (3489):1501.

Meredith SL (2013) Identifying Proximate and Ultimate Causation in the Development of Primate Sex-Typed Social Behavior. In: Clancy KBH, Hinde K, Rutherford JN (eds) Building Babies: Primate Development in Proximate and Ultimate Perspective, vol 37. DIPR. Springer, New York, pp 411–433

——— (2015) Comparative Perspectives on Human Gender Development and Evolution. Yearb Phys Anthropol 156 (S59):72–97. https://doi.org/10.1002/ajpa.22660

Meredith SL, and Schmitt CA (2019) The Outliers Are In: Queer Perspectives on Investigating Variation in Biological Anthropology. Am Anthropol 121 (2):487–489. https://doi.org/10.1111/aman.13223

O'Neill M, Umberger BR, Holowka MB, Larson SG, and Reiser PJ (2017) Chimpanzee Super Strength and Human Skeletal Muscle Evolution. PNAS 114 (28):7343–7348. https://doi.org/10.1073/pnas.1619071114

Ocobock C, and Lacy S (2023) Woman the Hunter: The Physiological Evidence. Am Anthropol 2023 (1):1–12. https://doi.org/10.1111/aman.13915

Povinelli D (2003) Folk Physics for Apes: The Chimpanzee's Theory of How the World Works. Oxford Univ Press, Oxford

Roughgarden J (2004) Evolution's Rainbow: Diversity, Gender, and Sexuality in Nature and People. Univ California Press, Berkely

Savell KRR, Auerbach BM, and Roseman CC (2016) Constraint, Natural Selection, and the Evolution of Human Body Form. PNAS 113 (34):9492–9497. https://doi.org/10.1073/pnas.1603632113

Schellenberg D (2019) Why Does Sex/Gender (Come to) Matter? Researchers' Reasons for Sex/Gender Assessment Illustrate Its Context-Dependencies and Entanglements. Somatechnics 9 (2–3):264–287. https://doi.org/10.3366/soma.2019.0283

Schneider JL, and Iverson JM (2023) Equifinality in Infancy: The Many Paths to Walking. Dev Psychobiol 65 (2):e22370. https://doi.org/10.1002/dev.22370

Schwartz J (2018) Is Gender Unique to Humans? https://www.sapiens.org/culture/gender-identity-nonhuman-animals/

Smith RWA, and Archer SM (2019) Bisexual Science. Am Anthropol 121 (2):491–492. http://doi.org/10.1111/aman.13225

Sobo EJ (2020) Dynamics of Human Biocultural Diversity: A Unified Approach. Routledge, New York

Springer KW, Stellman JM, and Jordan-Young RM (2012) Beyond a Catalogue of Differences: A Theoretical Frame and Good Practice Guidelines for Researching Sex/Gender in Human Health. Soc Sci Med 74 (11):1817–1824. https://doi.org/10.1016/j.socscimed.2011.05.033

Stewart-Williams S (2019) Nurture Alone Can't Explain Male Aggression. https://nautil.us/nurture-alone-cant-explain-male-aggression-237381/

Tang-Martinez Z (2016) Rethinking Bateman's Principles: Challenging Persistent Myths of Sexually Reluctant Females and Promiscuous Males. J Sex Res 53 (4):532–559. https://doi.org/10.1080/00224499.2016.1150938

Thayer ZM (2019) Early Life Adversity and the Value of Diversity. Am Anthropol 121 (2):484–485. https://doi.org/10.1111/aman.13221

Tomasello M (2019) Becoming Human: A Theory of Ontogeny. Belknap/Harvard Univ Press, Cambridge, MA

——— (2022) The Evolution of Agency: Behavioral Organization from Lizards to Humans. MIT Press, Cambridge, MA

Vandermassen G (2004) Sexual Selection: A Tale of Male Bias and Feminist Denial. Eur J Womens Stud 11 (1):9–26. https://doi.org/10.1177/1350506804039812

Whiten A, Harrison RA, McGuigan N, Vale GL, and Watson SK (2021) Collective Knowledge and the Dynamics of Culture in Chimpanzees. Philos Trans R Soc Lond B Biol Sci 377 (1843):20200321. https://doi.org/10.1098/rstb.2020.0321

Wilson ML, Miller CM, and Crouse KN (2017) Humans as a Model Species for Sexual Selection Research. Proc R Soc Lond B Biol Sci 284 (1866):20171320. https://doi.org/10.1098/rspb.2017.1320

Wood W, and Eagly AH (2012) Biosocial Construction of Sex Differences and Similarities in Behavior. In: Olson JM, Zanna MP (eds) Advances in Experimental Social Psychology, vol 46. Academic Press, Burlington, pp 55–123

Zihlman AL (1985) Review: Gathering Stories for Hunting Human Nature. Fem Stud 11 (2):365–377

5

Operationalization, Measurement, and Interpretation of Sex/Gender

Transcending Binaries and Accounting for Context and Entanglement

Stacey A. Ritz, Greta Bauer, Dorte M. Christiansen, Annie Duchesne, Anelis Kaiser Trujillo, and Donna L. Maney

Abstract Given the proliferation of calls to consider sex and gender in biomedicine, it is critical to address how the two concepts, and the relationships between them, are being implemented in a research setting. This chapter considers how we might transcend a simple, binary female–male framing and embrace the idea of the entanglement of sex and gender. The ways that the terms *sex* and *gender* are typically used in biology and health research are considered, with a focus on the relationships between these constructs, and areas of coherence and disagreement in their conceptualization. Problems arise when sex and gender are principally operationalized in terms of a female–male binary, including not only the resulting exclusion of trans, nonbinary, and intersex individuals but also the inadequacy of a binary analytical framework to account for context, overlap, in-group heterogeneity, continuity, and similarity. *Entanglement* and *interaction* are compared and contrasted, three forms of scientific engagement with these ideas are identified, and the implications of intersectionality for the operationalization of sex and gender are considered. In the context of experimentation, an entanglement perspective on sex and gender is explored for what it might enable along with the challenges it presents. As researchers grapple with the incorporation of sex and gender in their work, these frameworks will require ongoing development and refinement, reduced reliance on the dominant binary female–male analytical framing, and a move to a contextual, mechanistic approach that better reflects conceptual complexity, diverse

Stacey A. Ritz (✉)
Dept. of Pathology and Molecular Medicine, McMaster University,
Hamilton, ON L8S 4L8, Canada
Email: ritzsa@mcmaster.ca

Affiliations for the coauthors are available in the List of Contributors

© Frankfurt Institute for Advanced Studies (FIAS) 2025
L. Z. DuBois et al. (eds.), *Sex and Gender*, Strüngmann Forum Reports,
https://doi.org/10.1007/978-3-031-91371-6_5

Group photos (top left to bottom right)   Stacey Ritz, Dorte Christiansen, Greta Bauer, Anelis Kaiser Trujillo, Donna Maney, Annie Duchesne, Greta Bauer, Stacey Ritz, Dorte Christiansen, Donna Maney, Anelis Kaiser Trujillo, Annie Duchesne, Stacey Ritz, Donna Maney, Greta Bauer, Dorte Christiansen, Annie Duchesne, Anelis Kaiser Trujillo. Photos by Norbert Miguletz.

research methods, and intersectional considerations.

Keywords Sex/gender entanglement, binary notions of sex, biomedical science, intersectional research, policy

1 Introduction

In research, operationalization is usually understood as the process through which an unobservable phenomenon gets translated into a set of observable measures so that it can be empirically investigated. The validity of the knowledge being produced from such a process is then contingent on whether the operationalization captures the necessary elements of the represented concept (Haucke et al. 2021). Sex and gender are two such concepts that scholars are increasingly called upon to operationalize in their research, particularly in light of policies, mandates, and guidelines implemented by funders and publishers (Heidari et al. 2024).

Most often, researchers operationalize sex by categorizing individuals into female or male on the basis of either a trait (or set of traits), the researcher's categorization of the individual based on gendered presentation, or explicit self-identification by the research subject. For over fifty years, the validity of this operationalization of sex has been called into question by feminist scholars who distinguish between the concepts of sex and gender as a corrective against the attribution of observed female–male group differences to biological causes, and aim to carve out a conceptual and empirical space within which such differences could also be understood to arise out of social and cultural norms, structures, institutions, and distributions of power and resources (Fausto-Sterling 1987; Keller 1995; Sanz 2017; Unger 1979). At the same time, such a partitioning of gender from sex has the effect of obscuring the ways that the biological and sociocultural are co-constituted, in constant dynamic dialogue, entangled, and difficult to represent empirically. As definitions and boundaries of "sex" and "gender" continue to evolve across scholarly disciplines and research contexts, it is critical to revisit, unpack, and critique the measures and constructs we have relied on in the past in order to incorporate new theoretical insights and to improve our operationalizations.

Here, we describe and extend a discourse on the operationalization and measurement of sex/gender and their entanglement that began in October 2023 at the Ernst Strüngmann Forum. The expertise in our discussion group spanned a wide range of scholarly areas: from women, gender, and sexuality studies to psychology, biology, immunology, and public health as well as the history and philosophy of science. As researchers, several of us work with human subjects or human epidemiological data, while others have more experience with nonhuman animal models and *in vitro* cell culture. Some of us have focused on scholarship of sex and gender throughout our careers, while others have pivoted more recently to engage with these perspectives. Forum participants from other groups who joined our discussions also represented a wide range of expertise, including behavioral neuroendocrinology, anthropology, sociology, and science and technology studies. This broad range of perspectives was helpful to us as we talked, listened, and synthesized.

We organized our group discussions at the Forum around the following key themes, each with a set of questions to help guide and focus discussion:

1.   The use of sex and gender as categories in research: What do categorization schemes (e.g., female, male, woman, man, trans, cis) enable us to achieve with respect to sex/gender? What are the ethical considerations here, what harms may result, and are there contexts in which sex/gender categories should not be applied? What are the manifestations and alternatives to sex/gender categorization?

2.   Entanglement: How can researchers move past shallow "recognition" and "acknowledgment" of sex/gender entanglement and achieve a true pragmatic engagement in their research? What does entanglement demand of researchers?

3.   Intersectionality: What are the implications of doing research that starts from sex/gender without an intersectionality frame?

4.   Nonhuman animals and cells: How can research using animal and *in vitro* models operationalize the concepts of sex and gender, and what kinds of claims can be made about sex/gender in humans on the basis of findings in nonhuman models?

2 Revisiting the Conceptualizations of Sex and Gender

2.1 Distinctions between Sex and Gender

In the 1970s, Western feminist scholarship began to articulate a distinction between the concepts of *sex* and *gender* as a corrective against the essentialist and determinist tendencies to attribute observed female–male differences to biology, and to carve out a discursive and conceptual space in which such differences could also be understood to arise out of social and cultural norms, structures, institutions, and distributions of power and resources (Broverman et al. 1972; Clingman and Fowler 1976). It is important to emphasize, as Purtschert (2022) notes, that there are no simple feminist reference texts from the 1970s that clearly separate sex from gender. The history of the distinction between sex and gender is more complicated than the stories we tell about it today. The systematic separation of these concepts is blurry and happened during conversations at conferences and between colleagues from different fields. It originated in clinical psychology (Money 1955) and traversed through literature (Greer 1970; Millett 1970), sociology (Oakley 1972), anthropology (Rubin 1975), biology (Haraway 1984/1986), and back to sociology (West and Zimmerman 1987). Thus, instead of regarding the split as a clean one that took place between 1950–1970, we should instead acknowledge the complex, ongoing history of the problematization of the relationships between the terms, which has been continuously rewritten. Following this line, we see how the definitions and boundaries of the terms *sex* and *gender* continue to evolve and take different forms in different cultural and disciplinary landscapes. The nature of these definitions and boundaries varies depending on needs, norms, and practices. For instance, nearly every institution with a mandate to address sex and gender in research offers its own definitions to guide its stakeholders. In most of these cases, sex is explicitly associated with "biology" whereas gender is described as a social construct; this distinction has been at least partly motivated by the rampant conflation of the terms in the biological and medical literature (Kaufman et al. 2023).

The terms *sex* and *gender* each have multiple usages and meanings in biomedical research on both humans and nonhuman animals. Both terms can be understood as systems to categorize individuals as well as collections of traits and processes associated with the categories. There is often some conceptual slippage between these and other senses of the terms. In the categorical sense, sex is most often deployed with reference to the categories of female and male, sometimes with a recognition that some individuals do not clearly fit with one of those categories. The traits understood to be associated with sex are typically those directly and indirectly associated with sexual reproduction (e.g., sex chromosomes, reproductive organs and tissues, hormonal profiles). Gender, in the categorical sense, is sometimes used to refer to an individual's *gender identity* (i.e., their own internal sense of being a gendered individual), but it is important to emphasize that gender identity is not equivalent to *gender*. A broader understanding of gender includes a wide range of factors that are part of an individual's cognitive, social, and environmental experience, such as cultural ideologies and norms, gendered expectations and roles, gendered embodiment and performance, institutional structures and power dynamics, and more. As hybrid terminologies, *sex/gender* and *gender/sex* describe the strong interrelatedness of both concepts, although with slightly different emphases (Kaiser 2012; Kaiser et al. 2007, 2009; van Anders 2009, 2015, 2024). Sex/gender refers to the intrinsically inseparable conceptual nature of these two entities, emphasizing their mutual influence and the ongoing construction of *sex*. It simultaneously captures both the identity of an individual and the characteristics or ("biological") processes within them. Gender/sex emphasizes the significance of *gender* over *sex* without disregarding the latter. As noted by van Anders (2024:9), the "/" part of gender/sex is important because without it, "gendersex can imply that gender and sex are inextricable, making for a 'lumpmash' that conflates interconnectedness with inseparability," and thus it can also acknowledge the distinct lived realities of sex and gender for some individuals. For the sake of consistency, we will use sex/gender for the remainder of this chapter, but this decision does not reflect a general preference for one term over the other, and we acknowledge the value of both terms.

We deliberately do not offer our own definitions of sex and gender (or sex/gender) here; conceptual understandings of these concepts shift and evolve with continued scholarship, and the specific contexts in which the terms are employed will demand attention to different aspects of their conceptualization. We believe it makes more sense to build approaches to research that recognize and embrace the evolving understandings of sex and gender. Such approaches demand that investigators themselves explicitly articulate the frameworks of sex and gender they have chosen to employ and to identify the concrete sex/gender-related object(s) of interest (e.g., the categories, hormones, molecular mediators, genes, behaviors, exposures) that are most relevant in the context of their work. We ask investigators themselves to articulate explicitly the frameworks and definitions for what sex/gender means (in the experimental logic of biomedicine, psychology, and anthropology) and recommend that for each study, sex/gender be operationalized and modeled with precise and explicit conceptual understanding. In this way, sex/gender is introduced and used initially only as a first proxy; thereafter it is replaced by actual operationalized and measured variables. This approach allows for a more precise and clear interpretation of results and helps to avoid misunderstandings and biases in research (DuBois and Shattuck-Heidorn 2021; Maney 2016; Odling-Smee et al. 2024; Pape et al. 2024; Richardson 2022; Rippon et al. 2014; Ritz and Greaves 2024; Schellenberg 2019; Schellenberg and Kaiser 2017; Springer et al. 2012).

2.2 Contested Understandings of Gender

Discussions at the Forum revealed a diversity of conceptualizations of gender. There were some areas of general agreement, for example, that gender is created in the social realm, is variable across cultures and time, and is fluid. A common conceptualization used in psychology and much of medical research (as well as in some anthropological and sociological traditions) recognizes gender as having numerous components, including social identity, the psychological or perceived identity, the behavior of the individual in gendered terms, and the socially prescribed and experienced dimensions of femininity and masculinity. These components interact with gendered cultural values, resulting in socially prescribed gender roles that entail gender-specific behaviors, interests, expectations, experiences, and divisions of labor (Johnson et al. 2009). In other words, gender includes but is not limited to the construct of gender identity.

Our conversations at the Forum exposed ways that the concept of gender has been understood and operationalized in different academic settings. For some, the operations of power in social systems, which create gender inequality and oppression, is present and central to their conceptualization of gender; however, not everyone shared this conceptualization. One biomedical researcher indicated that in their field, the word *gender* is used to refer to sex-biased interactions with the environment, and it does not carry implications about power, justice, or oppression. This understanding was resonant for others as well, who tended to think about gendered practices and behavior independently of inequality.

The significance of this disjuncture became particularly apparent during a discussion about sex-biased maternal care in laboratory rats. In the late 1970s and early 1980s, researchers observed that mother rats spend more time, on average, licking male pups than female pups[1] to initiate defecation and urination. Time spent licking the male pups, which was found to depend on pheromone signals, was correlated with aspects of their behavior as adults (Moore 1982; Moore and Morelli 1979; Richmond and Sachs 1984). For some, this example was seen as analogous to the "pink hat–blue hat" phenomenon in humans—the finding that adults engage in different types of play and talk with babies perceived to be female or male (Araujo et al. 2022; Burnham and Harris 1992; Cahill 1989; Leone and Robertson 1989). Others felt strongly that these were not similar phenomena. Initially, these positions proved

---

[1] This behavior is not dimorphic with respect to the sex category of the pups, although it is often portrayed to be. The data reported by Richmond and Sachs (1984), for example, have a 67% overlap in the distribution of anogenital licking time between female and male pups at day 7, the point of greatest disparity.

to be quite polarizing, but through further discussion, the reasons for this discrepancy emerged. Those who understood these two examples as analogous were focused principally on the notion that both examples illustrate how sex/gender-biased early-life experiences can shape the future trajectories of the individuals. In contrast, those who did not see them as similar were more focused on the origins and meanings of the experiences themselves, noting that in rats they were triggered by chemical signals (pheromones) rather than sociocultural norms (hat color and associated assumptions about gender and gender-appropriate behavior); in other words, the differential treatment in humans is connected to gender ideologies, whereas anogenital licking in rats is (presumably) not a matter of gender ideology or power-related social and cultural norms of gender. This issue was debated for some time without a resolution or consensus, yet it proved valuable in that it highlighted a divergence in the ways that the construct of gender is understood and employed, and its implications for biomedical research with nonhuman animals. We believe that continuing to unpack and explore the implications of this disjuncture will be fruitful and useful to the ongoing efforts to address sex and gender in research.

The use of the term *gender* when discussing findings in nonhuman animals proved to be a particularly crucial question. As noted above, researchers have typically been instructed by funding agencies and research policies that (a) sex is biological and gender is social, and (b) that nonhuman animals do not have gender. If gender is understood solely to mean *gender identity* (an understanding with which we disagree), then indeed it would make little sense to use the term *gender* for nonhuman animals since we do not know whether they have any sensibility of gender identity for themselves (or if they do, whether it resembles anything that we mean by gender identity in humans). At the same time, it is clear that in many species there are aspects of social and environmental experience that vary among individuals in relation to sex, such as the example of anogenital licking of neonates by maternal rats. We found it difficult to resolve the question of how we should talk about the social experiences of animals without invoking the term *gender*. If such findings can be described only in terms of sex (because scientists have been told not to use the term *gender*), then there is a risk of inadvertently reinforcing a biologically essentialist and determinist understanding of the phenomenon; if we discuss them as *gender*, we run the risk of anthropomorphizing the experiences of nonhuman animals.

With gender understood as rooted in social and cultural contexts, questions arise about whether the meanings and implications of many types of gendered environmental and social experiences in humans can be reproduced in meaningful ways in nonhuman animals. Some Forum participants voiced strong caution around attempts to model aspects of human gender in animals, contending that the potential for misinterpretation, misattribution, and the dissemination of harm is very high. For example, although the experiences of forced copulation in nonhuman animals have a number of elements in common with rape in humans, it would be highly inappropriate to understand these as equivalent, because the impact of rape in humans includes a range of psychological and emotional aspects that arise not only from the physical act itself and that do not have meaningful correlates in animals. These include the ways that sexual violence is structurally constructed and treated across different societies, how sexual violence is treated within the legal system, how beliefs about gender and sexuality perpetuate a sense of responsibility of the victims, the expectations related to trust and consent that people bring to these interactions, and the meaning attributed to such experiences by society and by the victims themselves. Similarly, experimental animal models of human mental health conditions, such as depression or addiction, can reproduce some but not all important contributing factors in these conditions. For example, animal models can probably shed some light on how hormonal fluctuations can affect neurotransmitter expression, but it is also clear that gender socialization, norms, beliefs, and biases influence the human manifestations of these types of disorders which likely cannot be modeled in animals (and hormones themselves are also influenced by social experience). These points should not be taken as a general critique of nonhuman animal research; however, when the research is probing phenomena or pathways that are likely to be influenced by social experience and structures, researchers need to be highly attentive to a model's limitations, give careful consideration to the potential influence of social stereotypes and bias on their interpretations, and recognize that the transferability and generalizability of such findings to humans demands heightened critical scrutiny, as some feminist scholars have rightly pointed out (Fedigan 1992; Gungor et al. 2019; Pape 2021).

Several biomedical scientists at the Forum described recent shifts in language used in their fields to refer to nonhuman animals. They noted that it is relatively rare in contemporary practice for a researcher to describe rats as "depressed" or "anxious"; instead, terms such as "depression-*like* behaviors" are used. Similarly, instead of describing a mouse as "afraid" if it froze during a behavioral evaluation, a researcher is now more likely to say simply that the mouse "froze," which avoids attributing emotional valence to the mouse. Although this way to limit anthropomorphism is not specific to considerations of sex/gender per se, it exemplifies how we might shift our descriptions of observations related to sex/gender, bringing a more critical and rigorous approach to the reporting of relevant findings (Annandale and Hammarström 2011; Madsen et al. 2017; Ritz and Greaves 2022; Sánchez 2007).

3 Problems with the Operationalization of Sex and Gender as Binary in Research

In recent decades, several major biomedical funders and journals have implemented policies requiring researchers to attend to sex and/or gender considerations. These policies include the *Sex as a Biological Variable* (SABV) policy of the US National Institutes of Health (Arnegard et al. 2020; NIH 2015; see also Pape, this volume), the *Sex and Gender-Based Analysis* policy of the Canadian Institutes of Health Research (Government of Canada 2018; Health Canada 2017), the *Horizon Europe Guidance on Gender Equality Plans* of the European Research Foundation (European Commission 2021) and the *Sex and Gender Equity in Research* (SAGER) guidelines (Heidari et al. 2016, 2024; Peters et al. 2021). Although such policies vary considerably in their specific requirements, most ask researchers to include females/women and males/men and disaggregate findings by sex or gender category. Statistical comparison of groups of females and males is typically not explicitly required, but is often done regardless (Garcia-Sifuentes and Maney 2021).

These types of policies were motivated by a history of strong male bias in clinical and preclinical biomedical research, and by instances in which the lack of consideration of sex/gender influences is perceived to have created and perpetuated gendered health inequities (Criado-Perez 2019). Their broad implementation by high-level funding bodies signals a commitment by the academic research enterprise to take these considerations seriously. At the same time, we believe that some serious and compelling problems have inadvertently been created by the disproportionate focus on female–male difference and the suggestion that the categories themselves are the most important factors affecting human health.

The uncritical consideration of sex/gender as a binary categorical comparison of female versus male calls on and amplifies cultural beliefs that sex/gender variations are manifestations of biologically innate differences between females and males. When left unchecked, these types of cultural beliefs can influence and constrain the kinds of hypotheses that are generated, how analyses are undertaken, and the interpretations derived from the data. The systematic disaggregation of data into binary sex/gender categories can invite comparison where none is warranted, generating an overemphasis on differences between group means at the expense of appreciating diversity and heterogeneity within and between those categories and the overlap between them (Bauer 2023; Joel et al. 2015; Pape et al. 2024; Patsopoulos et al. 2007; Rippon et al. 2014; Sanchis-Segura and Wilcox 2024); this, in turn, may have major implications on the translation of that work into practice and policy. Although we certainly do not deny that the use of sex/gender categories can be useful for some purposes, there are significant problems that can arise by overfocusing on binaries and difference; emphasizing the categories per se diverts attention from the actual mechanistic factors that mediate sex/gender-related variation (DuBois and Shattuck-Heidorn 2021; Pape et al. 2024).

Of greatest concern for us is the tendency to conduct a binary female–male comparison, find a statistically significant difference between the two groups, and then make a recommendation that men and women function "differently," or should receive different treatments, interventions, or policy recommendations. There are several crucial problems with this logic (Bryant et al. 2019; Fausto-Sterling 2000; Fine and Fidler 2014; Joel 2016; Maney 2016). First, it neglects to account for heterogeneity within and overlap between categories. Even if the means statistically are significantly different from one another, it is relatively rare for the female and male distributions to be so disparate and distinct that it would warrant differential treatment, and there is rarely homogeneity within the categories. Since findings of difference between categories are based on groupings, these average group measures and differences may not apply to all or even to most individuals within those groups. By overgeneralizing to the entire category based on the mean for that category and making recommendations for treatment or intervention on that basis, individuals who are further from the mean for their group will be increasingly likely to be misclassified and treated inappropriately. Second, binary comparisons across sex/gender category can lead to false positive findings of difference and misguided attempts to attribute those findings to the usual suspects (e.g., hormones) in the absence of a broader consideration of possible contributors (Fine 2012; Maney and Rich-Edwards 2023; Rippon et al.

2014; Williams et al. 2023). Third, binary comparisons can lead to the categorization of individuals as "typical" and "atypical" which can perpetuate stereotypes, stigma, and discrimination, exacerbated further by the noninclusive nature of binary categorization around sex/gender.

As mandates for the consideration of sex/gender evolve, it will be essential to avoid reliance on simplistic female–male binary group comparisons. Several participants at the Forum suggested that a better strategy would be to take a hypothesis-free approach to data analysis; that is, to let the data speak independently of a priori hypotheses (to the extent that is possible, given that the data are collected by humans). Indeed, several participants are already doing these types of analyses by looking for clusters that emerge from the data rather than dividing the sample into sex categories a priori (e.g., Sanchis-Segura et al. 2022). In the implementation of this approach, multiple variables are considered in one model: those unrelated to sex/gender (e.g., age, place of living) as well as sex/gender-related variables (e.g., sex chromosome complement, sex/gender assigned at birth, sex/gender identity, sex/gender-associated hormones). With no a priori hypothesis and a "letting the data speak" approach, analyses will show which of those variables have the highest explanatory value. Encouraging investigators to measure plausibly relevant variables will enable analyses that are less reliant on binary categorization, and research policies mandating consideration of sex/gender could evolve to foster more mechanism-oriented approaches. We must consider, however, that if each of these variables is nonetheless implemented in a binary way—sex chromosome complement (XX and XY), sex/gender identity (women, men)—the problem of binarity will not automatically be solved (Bryant et al. 2019) and issues of inclusivity may remain. Moving away from a reliance on binary categorization will be an important mechanism to achieve a more multidimensional representation and understanding of sex/gender.

The policy discourse mandating the inclusion of sex/gender considerations and the inclusion of female and male subjects (including the rhetoric that women are underrepresented as subjects in research) has seeded a belief that most research done on males is completely irrelevant to women. This notion that every aspect of female and male physiologies is profoundly different, not comparable, and routinely require different treatments is clearly not true. Although there are some elements that are fairly dimorphic in structure (like certain reproductive organs), most aspects of human functioning are reasonably similar, with shifts in distributions rather than dimorphism (see Chapter 4). Indeed, understanding the relevance of sex and gender is valuable to help refine knowledge and intervention, and for addressing those instances in which female–male difference is consequential in some way. For the most part, however, men and women are similar: when we see statistical differences, they are almost always characterized by relatively small shifts in distributions, not dimorphisms in fundamental mechanisms that would warrant dichotomous treatment on the basis of sex category (see Chapter 2; Hyde 2005; Hyde et al. 2019; Richardson 2010; Zell et al. 2015).

Rhetoric surrounding consideration of sex and gender in research often cites their "impact," "influence," or "role" in health and disease. We find this language problematic, as sex and gender are categories, not causes. This criticism came up often at the Forum: categories are organizational schemas devised and applied by humans for specific purposes. With respect to sex/gender, the categories of female, male, man, or woman are not only inadequate to capture the diversity of human form and experience, they are also imperfect proxies for understanding the actual mechanisms that underlie sex/gender-related variability. While sex/gender categories can sometimes be useful, even when binary, the vast existing canon of interdisciplinary knowledge shows that the complexity of sex/gender is not adequately captured by two groups (Eliot et al. 2021; Joel 2021), which often have cis-heteronormative interpretations (Ashley et al. 2024; Ciccia 2024). Expanding our conceptual toolbox beyond categories will serve to broaden our potential to understand the mechanisms through which sex/gender operates, reflect on how we operationalize sex/gender in experimentation, and mitigate the potential for harm that can arise when we do not appropriately account for heterogeneity, overlap, and the limitations of our categorization schemes.

4 Embracing Entanglement to Enhance our Conceptualization of Sex/Gender

4.1 Distinguishing Entanglement from Interaction

Beyond issues of definitions and operationalization, one of the most important challenges to considering sex and gender categories is their near inseparability in a research context—akin to the inseparability of nature and nurture. It is uncontroversial that all living organisms interact with, respond to, and adapt to environment and experience, and are materially shaped by it. The physical material and arrangement of our bodies is generated in active, ongoing dialogue with the physical and social world we find ourselves in. This view, that biological and environmental factors act on one another to produce developmental and health outcomes, is referred to as *interactionism*. The term "interactionist consensus" understands interactionism not as a stance or approach that can be contrasted with a non-interactionist one, but as a baseline position, acknowledged across scientific fields, that does not entail commitment to any particular scientific methodology (Ferreira Ruiz and Umerez 2021; Kitcher 2003).

In the context of sex/gender science, interactionist approaches are those that strongly distinguish sex-related from gender-related factors; a given factor is framed to be *either* sex-related *or* gender-related. In a common genre of publications in this field, lists of such factors are produced under the headings of "sex" and "gender" (including in some of our own publications; e.g., Ritz et al. 2014). Under interactionism, the task of the researcher is taken to be to parcel out the additive or multiplicative contribution of various discrete factors, such as chromosomal sex (deemed "biology") and primary household income earner status (deemed "gender"). Keller (1985/1995) has called this representation a "bucket" or "particulate" and "oppositional" model for parsing nature and nurture. An example of an interactionist frame would be an approach that primarily focuses on how "biological" factors produce sex-related disparities in a condition like Alzheimer disease while acknowledging that social and environmental factors also likely influence outcomes. Such approaches do not typically consider how such interactions may impact the so-called sex-related biological effects. Often, this approach assumes symmetry, separability, and orthogonality between these factors; for example, that it is possible to calculate the relative magnitude of their importance for the phenotype of interest. Such assumptions prevent an understanding of how the factors may be confounded and co-constituted.

Interactionism is thus most productively contrasted not with a crude non-interactionism—a position rarely held—but with various co-constitutive, dialogical approaches to conceptualizing development, which are aligned with what this Forum referred to as entanglement. Our group spent considerable time contemplating the distinction between entanglement and interaction, and what the implications of such a distinction might be for research. For us, a useful point of distinction was that interactionism sees biological and environmental/cultural/social factors as discrete and distinguishable from one another, whereas entanglement sees them as co-constituted and in dynamic, looping relationships, with every factor understood as being both sex-and-gender at the same time (Figure 5.1). Examples of entanglement approaches include biosocial or biocultural research in the field of anthropology (DuBois et al. 2021), developmental systems theory (Griffiths and Gray 2005), constructivist interactionism (Oyama 2006), dialectical biology (Lewontin and Levins 2007), dynamic systems theory (Fausto-Sterling 2021), process ontologies (Dupré 2020), agentic realism (Barad 2007), Haraway's concepts of naturecultures and material semiotics (Haraway 1990, 1992), and feminist new materialisms (Coole and Frost 2010; Hird 2002, 2003; Kaiser 2016; Sheridan 2002). Inherently, these approaches understand the phenotypic implications of developmental factors as plastic, and hence modifiable by these co-constitutive processes. In sex/gender science, methodologies for engaging these complexity-affirming, co-constitutive approaches can also be enacted in the context of considerations of intersectionality (discussed below).

There are a few different senses in which we can recognize the entanglement of sex and gender. In the most basic sense, our social world is entangled with our objects of study because we are human beings who are a part of that social world, who know the world through language and cognitive frameworks deeply conditioned by our culture and sociality. It is vital to acknowledge that our knowledge systems are affected by power dynamics, that we are gendered humans in a gendered world looking at sex, and that our claims about gender and sex in turn alter the social world in which we live. In this epistemological sense, recognition of sex/gender entanglement means that we acknowledge that our thinking about sex/gender is already preconditioned by the ideologies and experiences we have been exposed to about the meanings of sex and gender, and a recognition that we always construct our definitions and operationalizations of sex in the lab guided by our knowledge of gender and awareness of our own history with it.

In contrast, an ontological recognition of entanglement has a different focus; that is, on the nature of our objects of investigation themselves. This aspect of entanglement is focused on how sex- and gender-related factors are interrelated with one another at the level of material, objective reality. In other words, the ontological claim of sex/gender entanglement is that sex and gender can only be understood as co-constituted, as opposed to the alternative, interactionist claim of interdependence, wherein "biological" and "environmental" variables are best modeled discretely

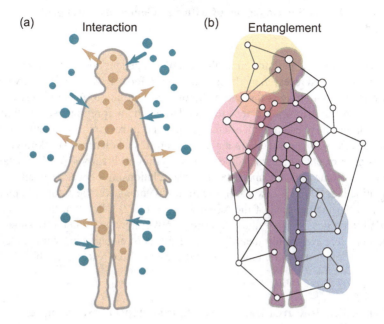

Figure 5.1 Sex, gender, and interactionism versus entanglement. An interactionist view of sex and gender tends to draw a sharp line between them as different in kind (sex as biological and internal to the body and gender as social and external to the body). Under this view, although sex and gender can impact one another, they are seen as independent and treated as such (e.g., gonadal hormones and gender roles may each influence behavior, but little attention is given to how they influence each other). In contrast, an entanglement view of sex and gender does not typically draw a sharp distinction between them. Rather, entanglement envisions sex- and gender-related factors as not able to be completely isolated from each other—all such factors are both sex-and-gender at the same time. Research from an entanglement perspective may examine relationships between clusters of variables (represented by the colored blobs), while recognizing that these are not neatly separable into "sex" and "gender" and remain part of larger networks of interregulation even when these are beyond the scope of the study to attempt to account for directly.

and are properly conceptualized as isolated factors that interact with one another.

Functionally speaking, in thinking about the operationalization of an entanglement approach, we arrived at the view that it is possible to utilize an entanglement approach as a framework for the study of sex/gender that allows the clarification and questioning of one's assumptions in the development of research questions, planning of research design, and operationalizing the variables. For example, a research team might ask themselves: What are the assumptions and possible range of results and interpretations of this research setup if we assume that our variables are entangled and co-constituted, compared with a situation for which we assume that they are distinct and can be studied as distinct? Through critical reflection, the answers to these questions can then guide researchers in the decisions they will make about study design, analysis, and interpretation. Additionally, researchers should keep in mind that, epistemologically speaking, they are to a small or large extent always putting their own cultural ideas of gender into their understanding of sex.

4.2 Types of Scientific Engagement with Sex/Gender as Entangled and Interacting

In our view, engaging with the entanglement of sex/gender does not entail a particular set of prescribed research methods, but rather offers a framework for how we ask questions, the assumptions that we make or are unwilling to make in research design, and what constraints we need to attend to when interpreting research findings. In considering how entanglement can be engaged in research, we identified three main ways that some researchers have taken up this challenge: (a) acknowledging the existence of entanglement without interrogating it directly, (b) using entanglement as the framework for understanding the relationship between sex and gender but treating it as a "black box," and (c) aiming to understand the nature of entanglement as the principal goal of the research (Figure 5.2). Note that this typology is intended to be descriptive rather than prescriptive; none are inherently superior to the others. Here we simply identify several forms of engagement with sex/gender entanglement that depend on the goals and context of the research itself.

4.3 Biomedical Research from an Entangled Perspectives: Examples

A detailed example of work falling under the "processes and mechanisms" category of Figure 5.2 comes from feminist scientist Gillian Einstein, who has developed a situated approach to neuroscience. Her group proposes an epistemology holding

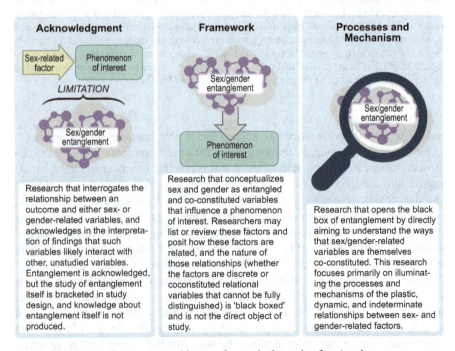

Figure 5.2   Forms of engagement with entanglement in the study of sex/gender.

that knowledge about the nervous system is situated within the multiple hierarchical and socially constructed interactions that involve participants' experiences, experimenter's positionality, and technological constraints. Einstein has deployed this situated, entangled approach to neuroscience to produce critical sex/gender analysis at the interface of intersecting social identities and varying biologies. For instance, Brown et al. (2022) utilized this approach to explore entanglement within the context of sport-related mild traumatic brain injuries (TBIs) and generate new investigative avenues to be tested experimentally. The study examined the apparent paradox that women, who tend to be exposed to fewer sport-related concussions than men, nevertheless carry a greater symptom burden, show more deficits across certain cognitive domains, and take longer to recover. While this disparity had been associated with men reporting fewer symptoms, Brown et al. highlight that women in sports, perhaps due to gender-related assumptions and biases around strength, musculature, and femininity, undergo less extensive strengthening and endurance training compared to their male counterparts. These differences in training could lead to disproportionately greater strength of neck musculature in men. Also absent in the characterization of difference is the extent to which women suffer TBI through intimate partner violence. TBI through intimate partner violence is a substantially different type of TBI, with stigma, fear of judgment, fear of the perpetrator, and traumatic stress complicating recovery and generating additional cognitive disruptions. Crucially, Brown et al. critically analyzed how gendered experiences may shape how sex-related variables, such as estrogen levels, interact with TBI. By treating sex and gender as entangled, Brown et al. opened new avenues of study, such as how chronic unpredictable stress influences the potential effects of ovarian hormones on recovery of the blood brain barrier, while also highlighting how, more generally, gendered experiences tend to be overlooked in clinical neuroscience.

The situated neuroscience approach has also led to the development of methods that can allow for an entangled understanding of sex/gender (Einstein et al. 2012). Einstein proposed a very mixed methods approach (Einstein 2024; Hankivsky et al. 2017), combining qualitative, quantitative behavioral, and quantitative neurophysiological methodologies to investigate the experiences of pain in Somali-Canadian women who had been subjected to genital cutting (Perović et al. 2021). By combining in-depth interviews about women's experiences of pain with standardized questionnaires and physiological assessments, Perović et al. were able to expose inconsistencies between different domains of measurement (physiological pain and pain reporting) using qualitative data to contextualize gendered experiences of pain within experiences of immigration and cultural acceptance. Their work exposed entanglement between contextualized, intersectional gender-related variables and physiological pathways and measures of pain in the brain. This study also revealed limitations in standardized pain questionnaires arising from cultural variation in recounting pain experience and showed that immigration and cultural conformity intersect to inform women's conceptualization of relevant sources of pain.

Approaching sex and gender as entangled can generate novel inquiries and methods as well as mitigate some of the harms embedded within essentialist assumptions. Nonetheless, the implementation and systematization of entanglement approaches into all types and programs of research poses challenges. If entanglement demands scientific practices that prevent researchers from using their available tools and expertise, heeding its demands becomes a practical impossibility for many. How can we support investigators to bridge their existing expertise and methodology to meaningfully engage with questions of entanglement? This question informed our articulation of engagement with entanglement in Figure 5.2, recognizing that some forms of experimentation and investigation may allow for direct interrogation of entanglement, while others may use entanglement principally as a conceptual tool.

If the framework of entanglement is taken as a set of guiding questions and frameworks, the task before the researcher is not to "disentangle" and apportion causality, but rather to distinguish between different uses and operationalizations of sex- and gender- related variables and understand their limitations for drawing inferences about causality. For complex, multidimensional, socially embedded categories such as gender and sex, it is often better to not obliterate heterogeneity and complexity, but to find ways to consider them, inviting multiple, alternative theoretical models to the table. Complexity-affirming approaches that embrace entanglement need not mean including all possible variables, but rather call for wisdom and a critical lens in framing, research design, and interpretation to enable feasible and implementable study designs while remaining attentive to the implications of complexity and entanglement.

5 Intersectionality, Entanglement, and Sex/Gender

Intersectionality is a framework through which we recognize that bodies, physiologies, and behaviors can be shaped by structural factors beyond the individual. Thus, in health and biological research, intersectionality as a concept generates robust, situated, contextualized theories of human sociostructural systems in which power is embedded. We emphasize that intersectionality is not a theory about identities but is better conceived of in terms of social status or social location. For example, one does not have to identify as poor to be thought of as or be poor; one does not have to identify as Black to be racialized as Black. So although intersectionality is not about identity categories per se, it is one example of a group-based approach, as opposed to a process- or institution-based approach (Choo and Ferree 2010).

As one of the many sociostructural systems of power, sex/gender is frequently understood to be one important element of an intersectional framework. If we take sex/gender as our starting point and central focus (as some sex/gender scholarship, particularly in biomedicine, currently does), intersectionality theory reminds us that we will end up in a place different from where we would be if we were to begin instead with a construct that acknowledges multiple oppressions and positionalities. One important theoretical conceptual tool for sex/gender research is relationality (Collins 2019), which serves as a series of thinking tools and approaches to understanding interlocking systems of power. Relationality helps us think about how sex/gender will be shaped by and "move" differently in conjunction, for example, with socioeconomic status and classism, race and racism, sexual orientation and heterosexism. There are three major approaches to relational thinking (Collins 2019). The first is *addition*: What happens when we add something like race/ethnicity to research on sex/gender? The example used by Collins is heteropatriarchy, where the addition of heterosexism to sexism creates a new understanding. The second approach is *articulation*; we can study how two separate things that are joined move together, like a joint. The third approach is *co-formation*, which is mutual co-constitution; sometimes two things come together to form a new thing that cannot be separated. One metaphor we can use to understand the concept of co-formation is a cake: We cannot unblend the cake, yet we can say I study chocolate, you study sugar, and somebody else studies eggs, and we are all trying to understand cake. Similarly, can we separate sex/gender from other things once it is blended together and embodied in human beings? All three of these relationality approaches can be used in work about sex/gender and its entanglement. For further discussion about the relationships between intersectionality, sex, gender, and entanglement, see Chapter 7 and 9.

One significant question for many researchers is how to apply intersectionality across different types of research. A common notion is that to produce knowledge about sex/gender, we first start with cells, build upward to the organismic level, then to social influences, and only at that point is intersectionality relevant. An alternative view is that there is no scaling "up" or "down" here, but that intersectional sex/gender entanglement works in ways that are not captured by a hierarchical or reductionist conceptualization: we do not simply conceptualize knowledge as going from the petri dish "up," but instead recognize that the perspectives and concepts we brought to the experiment in the petri dish were shaped by the social context in which it was carried out. Similarly, if one takes only the perspective of whole humans within a social context and then moves straight to the dish, one could end up testing the wrong hypothesis. Moving back and forth between levels of analysis has the potential to produce a more complete picture; Fausto-Sterling's (2005, 2008) exploration of sex/gender and bone density is an insightful example of this.

Much biomedical research starts with observations of the population-level distributions of health outcomes, an approach that permits hypotheses to be tested in a dish. For the laboratory question and model to be suitable, it is essential that the population component is intersectional. An intersectional analysis that includes sex/gender could and often does generate a set of questions different from what an analysis that includes sex/gender alone would generate. At the Forum, some of our discussion pondered questions of how cells in a dish model whole people who are also, in fact, intersectionally embodied, also contemplating the differences between primary, immortalized, or transgenically modified cells (see also Ritz 2017). In addition, we discussed whether one could record not just sex, but also age and race of the people from whom cells were taken, and the feeling was that doing so would be both naïve and overly complex. Indeed, it is important that we understand the strengths and limitations of any research approach, and address sex/gender and intersectional considerations in ways that are contextually appropriate.

The benefits of an intersectionality approach will vary for different types of work with humans, namely clinical trials and observational research. In clinical trials, we have random assignment to exposure groups, which is a strategy to minimize the effect of confounding by other measured or unmeasured variables (including race or socioeconomic status); however, from an intersectionality perspective, we are not only concerned about confounding (which will be controlled by randomization), but about the potential for effect measure modification. The ways that power structures become embodied in physiology and behavior play out in ways that may cause a treatment to have different effects in different contexts. This potential context dependence is relevant, for example, in cases of predisposing factors or comorbidities that may modify effects on outcome, which we can understand as ways that social power potentially becomes embodied in our research participants. In both clinical trials and observational research in whole humans, many factors can make our results discrepant from those of other studies, and thus specific to a population and not generalizable. Note that a discrepancy is not necessarily a problem, but rather an important point of potential variability that may reflect different social contexts and their embodiment. Moreover, an understanding of intersectional social power helps us understand who might and might not participate in our research.

In population research, we often cannot deal with confounding variables through random assignment because of risk (e.g., to assign participants to smoking) or pragmatics (e.g., we cannot feasibly or ethically assign participants to pregnancy, or to live in a certain location). We test for relationships between exposures and outcomes, designing our studies and analyzing our data in ways that attempt to isolate sex/gender effects. However, we must keep in mind that who is exposed to our variables of interest depends on many factors that are embedded in social power, which intersectionality reminds us is not uniform across sex/gender categories.

In studies of humans, it is important to carefully consider that multiple sources of discrimination (e.g., sexism, racism, homophobia) are interconnected and could tempt us to account for the same experience multiple times. As a potential solution, some researchers have turned to attribution-free measures of discrimination such as the Intersectional Discrimination Index (InDI) (Scheim and Bauer 2019; Bastos et al. 2025). This tool is explicitly designed to measure discrimination as it may be experienced across a wide range of intersections. At the Forum, we noted lack of agreement on the risks or benefits of attribution-free measures of discrimination such as the InDI, and research into their usefulness is ongoing. It is important to note, however, the use of the InDI does not make a study inherently intersectional; it only provides a measure designed to compare across intersectional groups defined by social, power-related categories.

6 Experimental Sciences Under an Entanglement Framework: Paths Forward

For experimental researchers working in the field of biomedicine—from human clinical trials to *in vitro* cell culture—the knowledge-building power of experimentation lies in the ability to control variables, establishing experimental conditions that allow causal inferences. Thus, to some extent there is an inherent tension between the experimental requirement to isolate and control variables with entanglement's recognition of the complexity and dynamism of sex/gender. It will be challenging to incorporate concepts of entanglement into experimental research design focused on the isolation, control, and manipulation of individual variables that entanglement tells us cannot actually be controlled, manipulated in isolation, or, in some cases, even defined.

Indeed, some of the sensibilities we bring to experimentation and models can misdirect our attention. Experiments that reveal a molecular pathway, for example mediating the relationship between a stimulus and an outcome, are sometimes perceived as revealing the "root causes" of difference. But "molecular mechanisms" should not be equated with the idea of "root causes" in this way. As one Forum participant noted during our discussions, "just because something can be shown in a lab and is molecular, does not mean it is a cause...an entanglement approach would recognize that both systems [sex and gender] interact throughout development to produce these phenotypes." For example, the relationship between acute epinephrine levels and PTSD appears to be moderated by sex; the associations were found to be positive in men but negative in women (D. M. Christiansen, unpublished observation). Such a result could serve as a starting point for further explorations of mechanism, including considering which aspects related to sex and gender contribute to the observed group differences. Identifying this molecular link to epinephrine expression does not necessarily mean that psychosocial and cultural experiences have no role to play, or that epinephrine-based treatment would be beneficial to one sex and detrimental in another.

Sex/gender inclusion policies ask researchers to bring sex/gender into their work, not to fundamentally change the nature of the work they are doing or to make sex/gender the focus of their work. Yet even that small request has profound implications that are not often considered. One point of discussion, both within our group and among other participants, was the question whether any single variable, in this case sex or gender category,[2] ought to be institutionally mandated for incorporation into every study. Recognizing the importance of intersectionality, why has sex/gender been selected as being of such high priority that every study has to include it? This may seem a strange question for us—a group of scholars who are deeply invested in the consideration of sex/gender in research—to be raising. Of course, we do believe that it is important for all scientists to bring well-informed considerations regarding sex and gender to their programs of research. The question we raise is about whether that means that sex and gender need to be examined as a variable in every study. It is particularly problematic that the incorporation of sex/gender categories into every study sometimes seems to come often at the expense of rigorous analytical methods (Garcia-Sifuentes and Maney 2021; Joel et al. 2015; Maney et al. 2023; Pape et al. 2024), which flies in the face of the stated goals of the policies—to increase rigor. We contend that instead of mandating the inclusion of sex/gender into all research in any specific way, it would be more fruitful to require researchers to consider thoughtfully and provide justification about whether and how to account for sex/gender in the context of their research, taking into account the existing knowledge about sex/gender in the discipline, the assets and limitations of the methods and models being employed, and the methods necessary to ask the questions rigorously. We do not believe that this suggestion lets researchers off the hook for the consideration of sex/gender; rather than calling for a specific approach or framework universally, it instead calls for a more nuanced consideration of when and how it is most appropriate to operationalize sex/gender in a given research

---

2 A number of participants at the Forum have questioned whether sex or gender category should really be considered a "variable" at all, arguing that they would be better considered as constructs consisting of multiple variables (e.g., Pape et al. 2024; Schellenberg and Kaiser 2017).

context. We hope that such considerations would foster a deeper engagement with how sex/gender might be best represented in their research.

We recommend the following steps to enhance effective communication, collaboration, and practice among scientists, policy makers, and gender and feminist scholars:

- Recognize the practical, conceptual, and logical limitations of attempts to separate sex from gender. Doing so will require careful attention to terminology; that is, whether to use "sex," "gender," or "sex/gender." These terms are situated within linguistic and cultural settings that should be considered.
- Take opportunities to revisit, refine, and advance the conceptual frameworks used to guide the consideration of sex/gender in research.
- Encourage researchers to reduce reliance on binary female–male comparisons by embracing a more mechanism-informed scientific paradigm in which sex- and gender-related *factors* drive research design rather than sex and gender *categories*. Importantly, "sex-related" and "gender-related," as terms, should themselves be treated as temporary categories—at intermediate stages of our research—as they risk perpetuating the sex-gender divide. We believe there is great value in removing the focus from sex/gender category and placing it onto more concrete, measurable variables that are closer to causal mechanisms. Such sex/gender-transformative approaches will allow us to probe more deeply the specific sex/gender-related factors that mediate observed sex/gender differences and enable more targeted interventions. Where sex/gender category is the only information available to the researcher, particular care must be taken in interpretation to avoid unwarranted dichotomization by examining the nature of the data distributions so that the extent of overlap and heterogeneity between and within the groups is apparent.
- Refine guidance and policy to reflect conceptual complexity, diverse research methods, and intersectional considerations. We question whether it is necessary or valid to ask *every* piece of research to address sex/gender and acknowledge contexts in which other factors may be of higher priority for investigation. Some otherwise powerful methods, models, and tools may not always be well suited to addressing questions related to sex/gender.
- Recognize the potential harms that come with such a sharp focus on sex and gender as the most important source of variation.

Our conversations at the Forum stimulated much reflection on, and critique of, how sex and gender are currently incorporated into research. Of course, many questions were raised but not answered. We look forward to participating in more conversations on these topics; for example, we are particularly excited about the growing availability of novel, powerful statistical approaches that will move biomedicine past the practice of comparing female/male means. We envision a future in which proxy categories such as "sex" and "gender" are replaced with more meaningful, informative explanatory variables. These variables will be revealed by research approaches that consider and engage sex/gender entanglement in psychological, clinical, and public health research.

Acknowledgments   The authors are grateful to Sarah Richardson for her insightful contributions to our discussions and her valuable input on an early draft of this manuscript. We also want to thank the other participants at the Forum who joined and contributed to our group's discussion sessions, the Forum co-chairs and Program Advisory Committee, and the Ernst Strüngmann Foundation for supporting this event.

References

Annandale E, and Hammarström A (2011) Constructing the "Gender-Specific Body": A Critical Discourse Analysis of Publications in the Field of Gender-Specific Medicine. Health 15 (6):571–587. https://doi.org/10.1177/1363459310364157

Araujo V, Schaffer D, Costa AB, and Musse SR (2022) Towards Virtual Humans without Gender Stereotyped Visual Features. ACM, New York. doi:10.1145/3550340.3564232

Arnegard ME, Whitten LA, Hunter C, and Clayton JA (2020) Sex as a Biological Variable: A 5-Year Progress Report and Call to Action. J Womens Health 29 (6):858–864. https://doi.org/10.1089/jwh.2019.8247

Ashley F, Brightly-Brown S, and Rider GN (2024) Beyond the Trans/Cis Binary: Introducing New Terms Will Enrich Gender Research. Nature 630 (8016):293–295. https://doi.org/10.1038/d41586-024-01719-9

Barad K (2007) Meeting the Universe Halfway: Quantum Physics and the Entanglement of Matter and Meaning. Duke Univ Press, Durham

Bastos JL, Gebrekristos LT, Dale SK, del Rio-González AM, Bauer GR, Scheim AI (2025) The inner workings of the Intersectional Discrimination Index: (Re)assessing the internal validity of the anticipated, day-to-day, and major discrimination measures. Stigma Health https://doi.org/10.1037/sah0000611

Bauer GR (2023) Sex and Gender Multidimensionality in Epidemiologic Research. Am J Epidemiol 192 (1):122–132. https://doi.org/10.1093/aje/kwac173

Broverman IK, Vogel SR, Broverman DM, Clarkson FE, and Rosenkrantz PS (1972) Sex-Role Stereotypes: A Current Appraisal. J Soc Issues 28 (2):59–78. https://doi.org/10.1111/j.1540-4560.1972.tb00018.x

Brown A, Karkaby L, Perovic M, Shafi R, and Einstein G (2022) Sex and Gender Science: The World Writes on the Body. In: Gibson C, Galea LAM (eds) Sex Differences in Brain Function and Dysfunction, vol 62. Springer, Cham, pp 3–25

Bryant KL, Grossi G, and Kaiser A (2019) Feminist Interventions on the Sex/Gender Question in Neuroimaging Research. https://sfonline.barnard.edu/feminist-interventions-on-the-sex-gender-question-in-neuroimaging-research/

Burnham DK, and Harris MB (1992) Effects of Real Gender and Labeled Gender on Adults' Perceptions of Infants. J Genet Psychol 153 (2):165–183. https://doi.org/10.1080/00221325.1992.10753711

Cahill SE (1989) Fashioning Males and Females: Appearance Management and the Social Reproduction of Gender. Symb Interact 12 (2):281–298. https://doi.org/10.1525/si.1989.12.2.281

Choo HY, and Ferree MM (2010) Practicing Intersectionality in Sociological Research: A Critical Analysis of Inclusions, Interactions, and Institutions in the Study of Inequalities. Sociol Theory 28 (2):129–149. https://doi.org/10.1111/j.1467-9558.2010.01370.x

Ciccia L (2024) ¿Por Qué Es Necesario Eliminar la Categoría Sexo del Ámbito Biomédico? Hacia la Noción de Bioprocesos en la Era Posgenómica" Dossier "Estudios Trans". Interdisciplina 12 (32):105–129. https://www.revistas.unam.mx/index.php/inter/article/view/86922

Clingman J, and Fowler MG (1976) Gender Roles and Human Sexuality. J Pers Assess 40 (3):276–284. https://doi.org/10.1207/s15327752jpa4003_7

Collins PH, (2019) Intersectionality as Critical Social Theory. Duke Univ Press, Durham

Coole D, and Frost S (eds) (2010) New Materialisms: Ontology, Agency, and Politics Duke Univ Press, Durham

Criado-Perez C (2019) Invisible Women: Data Bias in a World Designed for Men. Abrams, New York

DuBois LZ, Gibb JK, Juster RP, and Powers SI (2021) Biocultural Approaches to Transgender and Gender Diverse Experience and Health: Integrating Biomarkers and Advancing Gender/Sex Research. Am J Hum Biol 33 (1):e23555. https://doi.org/10.1002/ajhb.23555

DuBois LZ, and Shattuck-Heidorn H (2021) Challenging the Binary: Gender/Sex and the Bio-Logics of Normalcy. Am J Hum Biol 33 (5):e23623. https://doi.org/10.1002/ajhb.23623

Dupré J (2020) Life as Process. Epistemol Philos Sci 57 (2):96–113. https://doi.org/10.5840/eps202057224

Einstein G (2024) Key Research Concepts. https://einsteinlab.ca/about-us/our-research-focus/key-research-concepts/

Einstein G, Au AS, Klemensberg J, Shin EM, and Pun N (2012) The Gendered Ovary: Whole Body Effects of Oophorectomy. Can J Nurs Res 44 (3):7–17. https://pubmed.ncbi.nlm.nih.gov/23156189/

Eliot L, Ahmed A, Khan H, and Patel J (2021) Dump the "Dimorphism": Comprehensive Synthesis of Human Brain Studies Reveals Few Male-Female Differences Beyond Size. Neurosci Biobehav Rev 125:667–697. https://doi.org/10.1016/j.neubiorev.2021.02.026

European Commission (2021) Directorate General for Research and Innovation. Horizon Europe Guidance on Gender Equality Plans. https://data.europa.eu/doi/10.2777/876509

Fausto-Sterling A (1987) Society Writes Biology / Biology Constructs Gender. Daedalus 116 (4):61–76. https://www.jstor.org/stable/20025124

——— (2000) The Five Sexes, Revisited. Sciences (New York) 40 (4):18–25. https://doi.org/10.1002/j.2326-1951.2000.tb03504.x

——— (2008) The Bare Bones of Race. Soc Stud Sci 38 (5):657–694. https://doi.org/10.1177/0306312708091925

——— (2021) A Dynamic Systems Framework for Gender/Sex Development: From Sensory Input in Infancy to Subjective Certainty in Toddlerhood. Front Hum Neurosci 15:613789. https://doi.org/10.3389/fnhum.2021.613789

Fausto-Sterling A (2005) The Bare Bones of Sex: Part 1—Sex and Gender. Signs 302 (2):1491–1527. https://doi.org/10.1086/424932

Fedigan LM (1992) Primate Paradigms: Sex Roles and Social Bonds. Univ Chicago Press, Chicago

Ferreira Ruiz M, and Umerez J (2021) Interactionism, Post-Interactionism, and Causal Complexity: Lessons from the Philosophy of Causation. Front Psychol 12:590533. https://doi.org/10.3389/fpsyg.2021.590533

Fine C (2012) Explaining, or Sustaining, the Status Quo? The Potentially Self-Fulfilling Effects of "Hardwired" Accounts of Sex Differences. Neuroethics 5 (3):285–294. https://doi.org/10.1007/s12152-011-9118-4

Fine C, and Fidler F (2014) Sex and Power: Why Sex/Gender Neuroscience Should Motivate Statistical Reform. In: Clausen J, Levy N (eds) Handbook of Neuroethics. Springer, Dordrecht, pp 1447–1462

Garcia-Sifuentes Y, and Maney DL (2021) Reporting and Misreporting of Sex Differences in the Biological Sciences. eLife 10:e70817. https://doi.org/10.7554/eLife.70817

Government of Canada (2018) Sex and Gender in Health Research – CIHR. https://cihr-irsc.gc.ca/e/50833.html

Greer G (1970) The Female Eunuch. 1st Farrar, Straus and Giroux ed edn. Farrar, Straus and Giroux, New York

Griffiths PE, and Gray RD (2005) Discussion: Three Ways to Misunderstand Developmental Systems Theory. Biol Philos 20 (2–3):417–425. https://doi.org/10.1007/s10539-004-0758-1

Gungor NZ, Duchesne A, and Bluhm R (2019) A Conversation around the Integration of Sex and Gender When Modeling Aspects of Fear, Anxiety, and PTSD in Animals. https://sfonline.barnard.edu/a-conversation-around-the-integration-of-sex-and-gender-when-modeling-aspects-of-fear-anxiety-and-ptsd-in-animals/

Hankivsky O, Doyal L, Einstein G, et al. (2017) The Odd Couple: Using Biomedical and Intersectional Approaches to Address Health Inequities. Glob Health Action 10 (Suppl 2):1326686. https://doi.org/10.1080/16549716.2017.1326686

Haraway DJ (1984/1986) Primatology Is Politics by Other Means. In: PSA: Proceedings of the Biennial Meeting of the Philosophy of Science Association, vol 1984. Philosophy of Science Assoc., Lansing, pp 489–524

——— (1990) Simians, Cyborgs, and Women: The Reinvention of Nature. Routledge, New York

——— (1992) Otherworldly Conversations; Terran Topics; Local Terms. Sci Cult (Lond) 3 (1):64–98. https://doi.org/10.1080/09505439209526336

Haucke M, Hoekstra R, and Van Ravenzwaaij D (2021) When Numbers Fail: Do Researchers Agree on Operationalization of Published Research? Royal Society Open Science 8 (9):191354. https://doi.org/10.1098/rsos.191354

Health Canada (2017) Health Portfolio Sex and Gender-Based Analysis Policy. https://www.canada.ca/en/health-canada/corporate/transparency/heath-portfolio-sex-gender-based-analysis-policy.html

Heidari S, Babor TF, De Castro P, Tort S, and Curno M (2016) Sex and Gender Equity in Research: Rationale for the SAGER Guidelines and Recommended Use. Res Integr Peer Rev 1:2. https://doi.org/10.1186/s41073-016-0007-6

Heidari S, Fernandez DGE, Coates A, et al. (2024) WHO's Adoption of SAGER Guidelines and GATHER: Setting Standards for Better Science with Sex and Gender in Mind. Lancet 403 (10423):226–228. https://doi.org/10.1016/S0140-6736(23)02807-6

Hird MJ (2002) Re(Pro)Ducing Sexual Difference. Parallax 8 (4):94–107. https://doi.org/10.1080/1353464022000027993

——— (2003) From the Culture of Matter to the Matter of Culture: Feminist Explorations of Nature and Science. Sociol Res Online 8 (1):92–103. https://doi.org/10.5153/sro.780

Hyde JS (2005) The Gender Similarities Hypothesis. Am Psychol 60 (6):581–592. https://doi.org/10.1037/0003-066X.60.6.581

Hyde JS, Bigler RS, Joel D, Tate CC, and van Anders SM (2019) The Future of Sex and Gender in Psychology: Five Challenges to the Gender Binary. Am Psychol 74 (2):171. https://psycnet.apa.org/doi/10.1037/amp0000307

Joel D (2016) Captured in Terminology: Sex, Sex Categories, and Sex Differences. Fem Psychol 26 (3):335–345. http://doi.org/10.1177/0959353516645367

——— (2021) Beyond the Binary: Rethinking Sex and the Brain. Neurosci Biobehav Rev 122:165–175. https://doi.org/10.1016/j.neubiorev.2020.11.018

Joel D, Kaiser A, Richardson SS, et al. (2015) A Discussion on Experiments and Experimentation: NIH to Balance Sex in Cell and Animal Studies. Catalyst 1 (1):1–13. https://doi.org/10.28968/cftt.v1i1.28821

Johnson JL, Greaves L, and Repta R (2009) Better Science with Sex and Gender: Facilitating the Use of a Sex and Gender-Based Analysis in Health Research. Int J Equity Health 8 (14):1–11. https://doi.org/10.1186/1475-9276-8-14

Kaiser A (2012) Re-Conceptualizing "Sex" and "Gender" in the Human Brain. Z Psychol 220 (2):130–136. https://doi.org/10.1027/2151-2604/a000104

——— (2016) Sex/Gender Matters and Sex/Gender Materialities in the Brain. In: Pitts-Taylor V (ed) Mattering: Feminism, Science, and Materialism. Biopolitics. New York Univ Press, New York, pp 122–139

Kaiser A, Haller S, Schmitz S, and Nitsch C (2009) On Sex/Gender Related Similarities and Differences in fMRI Language Research. Brain Res Rev 61 (2):49–59. https://doi.org/10.1016/j.brainresrev.2009.03.005

Kaiser A, Kuenzli E, Zappatore D, and Nitsch C (2007) On Females' Lateral and Males' Bilateral Activation during Language Production: A fMRI Study. Int J Psychophysiol 63 (2):192–198. https://doi.org/10.1016/j.ijpsycho.2006.03.008

Kaufman M, Eschliman E, and Sanchez Karver T (2023) Differentiating Sex and Gender in Health Research to Achieve Gender Equity. Bull World Health Org 101 (10):666–671. https://doi.org/10.2471/BLT.22.289310

Keller EF (1985/1995) Reflections on Gender and Science. Yale Univ Press, New Haven
——— (1995) Gender and Science: Origin, History, and Politics. Osiris 10 (1):26–38. https://doi.org/10.1086/368741

Kitcher P (2003) Battling the Undead: How (and How Not) to Resist Genetic Determinism. In: In Mendel's Mirror: Philosophical Reflections on Biology. Oxford Univ Press, Oxford, pp 283–300

Leone C, and Robertson K (1989) Some Effects of Sex-Linked Clothing and Gender Schema on the Stereotyping of Infants. J Soc Psychol 129 (5):609–619. https://doi.org/10.1080/0 0224545.1989.9713779

Lewontin R, and Levins R (2007) Biology under the Influence: Dialectical Essays on the Coevolution of Nature and Society. New York Univ Press, New York

Madsen TE, Bourjeily G, Hasnain M, et al. (2017) Http://Dx.Doi.Org/10.1089/Gg.2017.0005. Gend Genome 1 (3):122–128. https://doi.org/10.1089/gg.2017.0005

Maney DL (2016) Perils and Pitfalls of Reporting Sex Differences. Philos Trans R Soc Lond B Biol Sci 371 (1688):20150119. https://doi.org/10.1098/rstb.2015.0119

Maney DL, Karkazis K, and Hagen KBS (2023) Considering Sex as a Variable at a Research University: Knowledge, Attitudes, and Practices. J Womens Health 32 (8):843–851. https://doi.org/10.1089/jwh.2022.0522

Maney DL, and Rich-Edwards JW (2023) Sex-Inclusive Biomedicine: Are New Policies Increasing Rigor and Reproducibility? Womens Health Issues 33 (5):461–464. https://doi.org/10.1016/j.whi.2023.03.004

Millett K (1970) Sexual Politics. Doubleday, New York

Money J (1955) Hermaphroditism, Gender and Precocity in Hyperadrenocorticism: Psychologic Findings. Bull Johns Hopkins Hosp 96 (6):253–264. https://pubmed.ncbi.nlm.nih.gov/14378807/

Moore CL (1982) Maternal Behavior of Rats Is Affected by Hormonal Condition of Pups. J Comp Physiol Psychol 96 (1):123–129. https://doi.org/10.1037/h0077866

Moore CL, and Morelli GA (1979) Mother Rats Interact Differently with Male and Female Offspring. J Comp Physiol Psychol 93 (4):677–684. https://doi.org/10.1037/h0077599

NIH (2015) NOT-OD-15-102: Consideration of Sex as a Biological Variable in NIH-Funded Research. https://grants.nih.gov/grants/guide/notice-files/not-od-15-102.html

Oakley A (1972) Sex, Gender, and Society. [7th print.] edn. Harper and Row, New York

Odling-Smee L, Ashley F, Ritz SA, McCarthy MM, and Baker N (2024) Sex and Gender Discussions Don't Need to Be Toxic. https://www.nature.com/articles/d41586-024-01311-1

Oyama S (2006) Speaking of Nature. In: Haila Y, Dyke C (eds) How Nature Speaks: The Dynamics of the Human Ecological Condition. Duke Univ Press, Durham, pp 257–258

Pape M (2021) Lost in Translation? Beyond Sex as a Biological Variable in Animal Research. Health Sociol Rev 30 (3):275–291. https://doi.org/10.1080/14461242.2021.1969981

Pape M, Miyagi M, Ritz SA, et al. (2024) Sex Contextualism in Laboratory Research: Enhancing Rigor and Precision in the Study of Sex-Related Variables. Cell 187 (6):1316–1326. https://doi.org/10.1016/j.cell.2024.02.008

Patsopoulos NA, Tatsioni A, and Ioannidis JPA (2007) Claims of Sex Differences: An Empirical Assessment in Genetic Associations. JAMA 298 (8):880. https://doi.org/10.1001/jama.298.8.880

Perović M, Jacobson D, Glazer E, Pukall C, and Einstein G (2021) Are You in Pain If You Say You Are Not? Accounts of Pain in Somali–Canadian Women with Female Genital Cutting. Pain 162 (4):1144–1152. https://doi.org/10.1097/j.pain.0000000000002121

Peters SAE, Babor TF, Norton RN, et al. (2021) Fifth Anniversary of the Sex and Gender Equity in Research (SAGER) Guidelines: Taking Stock and Looking Ahead. BMJ Glob Health 6 (11):e007853. https://doi.org/10.1136/bmjgh-2021-007853

Purtschert P (2022) Staying with the Gender Trouble. Feministisch Stud 40 (2):350–359. https://doi.org/10.1515/fs-2022-0048

Richardson SS (2010) Sexes, Species, and Genomes: Why Males and Females Are Not Like Humans and Chimpanzees. Biol Philos 25 (5):823–841. https://doi.org/10.1007/s10539-010-9207-5

——— (2022) Sex Contextualism. PTPBio 14:2. https://doi.org/10.3998/ptpbio.2096

Richmond G, and Sachs BD (1984) Maternal Discrimination of Pup Sex in Rats. Dev Psychobiol 17 (1):87–89. https://doi.org/10.1002/dev.420170108

Rippon G, Jordan-Young R, Kaiser A, and Fine C (2014) Recommendations for Sex/Gender Neuroimaging Research: Key Principles and Implications for Research Design, Analysis, and Interpretation. Front Hum Neurosci 8:650. https://doi.org/10.3389/fnhum.2014.00650

Ritz SA (2017) Complexities of Addressing Sex in Cell Culture Research. Signs 42 (2):307–327. https://doi.org/10.1086/688181

Ritz SA, Antle DM, Côté J, et al. (2014) First Steps for Integrating Sex and Gender Considerations into Basic Experimental Biomedical Research. FASEB J 28 (1):4–13. https://doi.org/10.1096/fj.13-233395

Ritz SA, and Greaves L (2022) Transcending the Male–Female Binary in Biomedical Research: Constellations, Heterogeneity, and Mechanism When Considering Sex and Gender. Int J Environ Res Public Health 19 (7):4083. https://doi.org/10.3390/ijerph19074083

——— (2024) We Need More-Nuanced Approaches to Exploring Sex and Gender in Research. Nature 629 (8010):34–36. https://doi.org/10.1038/d41586-024-01204-3

Rubin G (1975) The Traffic in Women: Notes on the Political Economy of Sex. In: Reiter RR (ed) Toward an Anthropology of Women. Monthly Review Press, New York, pp 157–210

Sánchez D (2007) The Truth about Sexual Difference: Scientific Discourse and Cultural Transfer. The Translator 13 (2):171–194. https://doi.org/10.1080/13556509.2007.10799237

Sanchis-Segura C, Aguirre N, Cruz-Gómez ÁJ, Félix S, and Forn C (2022) Beyond "Sex Prediction": Estimating and Interpreting Multivariate Sex Differences and Similarities in the Brain. Neuroimage 257:119343. https://doi.org/10.1016/j.neuroimage.2022.119343

Sanchis-Segura C, and Wilcox RR (2024) From Means to Meaning in the Study of Sex/Gender Differences and Similarities. Front Neuroendocrin 73:101133. https://doi.org/10.1016/j.yfrne.2024.101133

Sanz V (2017) No Way out of the Binary: A Critical History of the Scientific Production of Sex. Signs 43 (1):1–27. https://doi.org/10.1086/692517

Scheim AI, and Bauer GR (2019) The Intersectional Discrimination Index: Development and Validation of Measures of Self-Reported Enacted and Anticipated Discrimination for Intercategorical Analysis. Soc Sci Med 226:225–235. https://doi.org/10.1016/j.socscimed.2018.12.016

Schellenberg D (2019) Why Does Sex/Gender (Come to) Matter? Researchers' Reasons for Sex/Gender Assessment Illustrate Its Context-Dependencies and Entanglements. Somatechnics 9 (2–3):264–287. https://doi.org/10.3366/soma.2019.0283

Schellenberg D, and Kaiser A (2017) The Sex/Gender Distinction: Beyond F and M. In: Travis C, Whyte JW (eds) Apa Handbook of the Psychology of Women. APA, Washington, DC, pp 165–187

Sheridan S (2002) Words and Things: Some Feminist Debates on Culture and Materialism. Aust Fem Stud 17 (37):23–30. https://doi.org/10.1080/08164640220123425

Springer KW, Stellman JM, and Jordan-Young RM (2012) Beyond a Catalogue of Differences: A Theoretical Frame and Good Practice Guidelines for Researching Sex/Gender in Human Health. Soc Sci Med 74 (11):1817–1824. https://doi.org/10.1016/j.socscimed.2011.05.033

Unger RK (1979) Toward a Redefinition of Sex and Gender. Am Psychol 34 (11):1085–1094. https://doi.org/10.1037/0003-066X.34.11.1085

van Anders SM (2009) Androgens and Diversity in Adult Human Partnering. In: Endocrinology of Social Relationships. Harvard Univ Press, Boston, pp 340–363

———— (2015) Beyond Sexual Orientation: Integrating Gender/Sex and Diverse Sexualities via Sexual Configurations Theory. Arch Sex Behav 44 (5):1177–1213. https://doi.org/10.1007/s10508-015-0490-8

———— (2024) Gender/sex/ual Diversity and Biobehavioral Research. Psychol Sex Orientat Gend Divers 11 (3):471–487. https://doi.org/10.1037/sgd0000609

West C, and Zimmerman DH (1987) Doing Gender. Gend Soc 1 (2):125–151. https://doi.org/10.1177/0891243287001002002

Williams JS, Fattori MR, Honeyborne IR, and Ritz SA (2023) Considering Hormones as Sex- and Gender-Related Factors in Biomedical Research: Challenging False Dichotomies and Embracing Complexity. Horm Behav 156:105442. https://doi.org/10.1016/j.yhbeh.2023.105442

Zell E, Krizan Z, and Teeter SR (2015) Evaluating Gender Similarities and Differences Using Metasynthesis. Am Psychol 70 (1):10–20. https://doi.org/10.1037/a0038208

6

Gender and Sex Entanglement in Neuroscience

A Neurofeminist Perspective

Annie Duchesne

Abstract From the discovery of the gonadal neuroendocrine axis to projects mapping gender/sex differences in brain and behavior, research in neuroscience has laid the foundation in biosciences for the investigation of sex and, to a lesser degree, gender. Given its role in sex and gender research, neuroscience has also been a central site of critical engagement by feminist science scholars, giving rise to neurofeminism, a subfield where neuroscience and feminist perspectives on science intersect. To date, neurofeminism has produced critiques, research frameworks, methodologies, epistemologies, and neuroscientific knowledge that coalesce to advance complex and emancipatory understandings of brain, body, and mind. This chapter aims to demonstrate the instrumental role of neurofeminist research and perspectives in producing alternative operationalizations of sex and gender, particularly with respect to their interrelation. First, a critical overview of dominant and emerging models for investigating sex and gender in neuroscience is provided to highlight benefits of approaching sex and gender as biosocially entangled. Second, the neurofeminist perspective on sex and gender entanglement is further characterized through a series of examples from human and nonhuman animal research. Consideration is then given to potential challenges associated with the neurofeminist approach to entanglement. Finally, the generative potential of neurofeminist science scholarship to improve the science of sex and gender is demonstrated.

Keywords Neurofeminist research, sex/gender entanglement, operationalization of sex and gender, intersectionality human and nonhuman animal research

Annie Duchesne (✉)
Psychology Department, University of Northern British Columbia, Prince George, BC Canada;
Psychology Department, University of Quebec in Trois Riviere, QC, Canada
Email: annie.duchesne@unbc.ca

© Frankfurt Institute for Advanced Studies (FIAS) 2025
L. Z. DuBois et al. (eds.), *Sex and Gender*, Strüngmann Forum Reports,
https://doi.org/10.1007/978-3-031-91371-6_6

1 Operationalizing Sex and Gender: Compounding Complexities

Sex and gender are complex constructs[1] that give rise to a diversity of material, psychological, and social existences. The inherent complexity of these constructs poses significant and unavoidable challenges to their scientific investigation, among the foremost of which is the central problem of dimension reduction (Mitchell 2009). Research that attempts to operationalize complex constructs faces questions of how to best reduce their dimensionality to develop viable constructs for measuring and interpreting observations while still producing generalizable knowledge. Today, sex and gender constructs are commonly articulated in humans as the respective biological and psychosocial domains of male and femaleness (White et al. 2021). This operationalization has oriented the development of models, methods, and measures that emphasize sex and gender as independent constructs (Eliot et al. 2023). While this articulation has permitted the production of scientific evidence demonstrating the unique contribution of gendered experiences to gender/sex[2] differences (e.g., gender stereotype and discrimination; Ellemers 2018), it has also limited our ability to articulate a crucial dimension of the complexities of sex and gender: their interrelation.

In addition to the problem of dimension reduction, the operationalization of complex constructs presents challenges related to the availability of current scientific knowledge to inform their articulation. Current evidence is often insufficient to provide a comprehensive description of complex constructs. Therefore, we rely more heavily on our common sense and beliefs to speculate about which features of a complex phenomenon are important for constituting the best description of an empirically viable construct (Mitchell 2009). Consequently, complex multidimensional phenomena are at much greater risk of being misrepresented through empirical reduction compared to unidimensional phenomena (e.g., temperature). However, unlike many complex constructs, our beliefs and values about sex and gender are deeply embedded in power differentials that structure our society, from legal, educational, and health systems to how we communicate, experience, and therefore think and represent gender/sex realities in science (Schiebinger 2004). Beliefs about the innate intellectual inferiority of women have grounded scientific articulations and interpretations of gender/sex differences in the biological sciences (Bleier 1976).While those oppressive beliefs no longer prevail in contemporary science, biases regarding, for example, the innately different nature of men and women's brains and behavior remains prevalent, leaving open the possibility for discriminatory scientific discourse (Fine 2013). Given the risks that gender/sex complexity poses for the perpetuation of oppressive ideologies through scientific inquiry, the operationalization of sex and gender must critically contend with current discriminatory beliefs about gender/sex realities.

---

[1] A complex construct provides a description of a complex phenomenon having many components that interact with one another and with the environment in a nonlinear fashion. Complex phenomena emerge from complex systems, such as physiological systems or social institutions (Mitchell 2009).

[2] A hybrid term describing the embeddedness of sex and gender, commonly employed to highlight the indissociable contribution and interrelation when referring to broad group differences, such as between men and women, which cannot be reduced to either of the constructs. Gender/sex does not signify that they can be reduced to one another but that they intersect (Kaiser 2012; van Anders 2024).

Initiatives to transform the science of sex and gender must allow for better representation of their interrelation to accommodate their inherent sociopolitical complexities. Biosocial entanglement is a conceptual approach centered on the interrelation of complex constructs that explicitly articulates biological realities as socially situated. This conceptual approach has been employed for many decades in developmental science (Gottlieb 2007; Lerner 1978; Oyama 2000) and is currently on the rise in fields such as behavioral genetics (Uchiyama et al. 2022) and affective science (Boiger and Mesquita 2012; Lindquist et al. 2022) to advance biosocial understandings of human phenomena. Although marginal in sex and gender science, biosocial entanglement has been employed for decades by feminist science scholars who, for instance, proposed frameworks of women's health centered on the dynamic interrelation between sex-related components and gendered experiences (Fausto-Sterling 2005). Unlike most approaches of biosocial entanglement, a feminist science perspective on gender/sex considers multiple levels of entanglement including that which occurs between (a) sex and gender dimensions of an individual, (b) gender/sex and power dynamics, and (c) science and society. This approach has the unique potential to advance the neuroscience of sex and gender while minimizing the risk of reinforcing discriminatory ideologies.

2 The Critical Importance of Neuroscience in the Operationalization of Sex and Gender

Neuroscience is a field of scientific inquiry dedicated to understanding the development, structure, and function of the nervous system. Though neuroscience was initially classified as a subfield of biology, neuroscientific phenomena are now being investigated from within many natural (e.g., neurochemistry, neuroengineering, computational neuroscience), health (psychiatry, neurology, neuropharmacology), and social sciences (e.g., neuropsychology, neuroeducation, neuromarketing), highlighting the complexity and far-reaching impact of neuroscientific knowledge in how we understand ourselves and others, and shape the environment around us (Altimus et al. 2020). Neuroscientific findings play a central role in shaping public opinion and discourse (Racine et al. 2010), through, for instance, increasing the public credibility of psychological explanation, even when the specific neuroscientific findings discussed are unrelated to the psychological phenomenon at hand (Bennett and McLaughlin 2024). Neuroscientific research also plays an increasingly large role in supporting political agendas related to education and mental health (Brun et al. 2024). Critically analyzing the operationalization of sex and gender in neuroscience, therefore, has the potential to inform diverse scientific and other cultural understandings of human existence.

Neuroscience has shaped and continues to shape the ways sex and, to a lesser degree, gender are operationalized. From the discovery of the gonadal neuroendocrine axis (Plant 2015), to the development of animal models dissociating gonadal from chromosomal effects (Arnold and Chen 2009), to projects mapping gender/sex differences in brain development (Kaczkurkin et al. 2019), structure (Ingalhalikar et al. 2014), and function (Ryali et al. 2024), research in neuroscience has laid the foundation for the investigation of sex-related variables in the biomedical sciences (Becker et al. 2005; Miller et al. 2017), and it continues to be a leading field in the development of models and methods for the investigation of sex and gender (Dalla et al. 2024; Reale et al. 2023; Wierenga et al. 2024). Despite the many contributions of neuroscience to the study of sex and gender, androcentric bias[3] within the field has been widely criticized in recent years for invisibilizing female biological realities and, therefore, compromising our ability to generalize experimental research findings to a broader population, including, in particular, women (de Lange et al. 2021; Shansky and Woolley 2016). Today, the systematic inclusion of women participants, female organisms, or female subindividual components (e.g., cells or organoids) is becoming standard practice alongside the characterization of binary sex differences (Beltz et al. 2019).

Neurofeminism—a subfield where neuroscience and feminist science intersect (Bluhm et al. 2012; Friedrichs and Kellmeyer 2022)—has contributed to empirical and conceptual research, while contending with sexist interpretations of sex differences in the brain and the relation of these neural differences to behavior. Neurofeminism challenges assumptions regarding the existence (Eliot et al. 2021; Joel 2021), origin (Fine 2010; Joel 2012; Jordan-Young and Karkazis 2019; Saguy et al. 2021), and operationalization (Joel and Fausto-Sterling 2016; Maney 2016; Sanchis-Segura and Wilcox 2024) of sex differences in the brain and interrogates the validity of translating sex differences observed in nonhuman animal models into explanations for human gender/sex realities (Eliot and Richardson 2016; Gungor et al. 2019). To date, neurofeminism has produced critiques (Kuria 2014; Llaveria Caselles 2021; Walsh and Einstein 2020), research frameworks (Hyde et al. 2019; Joel 2021; Rippon et al. 2014; Springer et al. 2012), methodologies (Brown et al. 2022), epistemologies (Roy 2018), and neuroscientific knowledge (Joel et al. 2015; Perović et al. 2021; van Anders 2024) that coalesce to advance a sociopolitically located understanding of gender/sex realities within the brain, body, and mind. The neurofeminist perspective also recognizes that the social location of a researcher and discipline inform the characterization of complex phenomena, which in turn informs the development or use of scientific methods and practices. Recognizing the potential impacts of positionality promotes critical interrogation of how power dynamics within the enterprise of science may marginalize, erase, or hegemonize some perspectives over others (Duchesne and Kaiser Trujillo 2021; Einstein 2012; Roy 2012). By grounding models and methods for representing gender/sex entanglement in a neurofeminist perspective, we will be better able to characterize the compounding complexity within both the phenomena of study and the enterprise of science itself.

In this chapter, I aim to demonstrate the generative potential of neurofeminist approaches to gender/sex entanglement by first highlighting biases that exist within and arise from current and emerging operationalizations of sex and gender in mainstream neuroscience. This overview focuses on dominant and emerging models, methods, and interpretations which make apparent the biases that neurofeminism addresses, and is not intended to be an exhaustive account. Thereafter, I detail the neurofeminist perspective of sex and gender entanglement through a series of examples of current and potential avenues for research and reflection. I conclude with a brief discussion of some of the challenges related to using a framework of entanglement for the empirical

---

[3] Androcentric bias occurs when male body and experience are treated as the norm, thus restricting the study of other bodies, which are treated as marginal and/or abnormal (Schiebinger 1999).

investigation of complex constructs for the neuroscience of gender/sex and beyond.

3 Operationalization of Sex and Gender in Neuroscience: Compounding Binaries

While both sex- and gender-related constructs are generally thought to influence the development and functioning of the nervous system in humans, neuroscience, like most biosciences, is primarily concerned with the contribution of sex as a biological construct (Cahill 2006; Eliot et al. 2021). In human neuroscience, a majority of research into sex-related effects is derived from systematic comparison between groups of cis men and cis women (Rechlin et al. 2022). Neurobiological differences observed between male and female organisms are often *assumed* (unfalsifiably) to be the product of sexually driven evolutionary pressures caused by innate differences in sex-associated traits (Cahill 2006; Spets and Slotnick 2022). Under this biologically essentialist assumption, sex differences tend to be interpreted as the direct manifestation of sex-related biological mechanisms. Biologically essentialist assumptions of sex differences in the nervous system have also informed conceptualizations of gender identity, gender roles, and sexual orientation as directly resulting from the effect of gonadal hormones on brain development (Hines 2011; Roselli 2018). Decades of critical feminist research has challenged not only essentialist assumptions about biologically innate sex differences in the nervous system, but also the validity of a binary conceptualization as adequate operationalization of sex-related effects. Most importantly, critical neurofeminist analysis has been instrumental in exposing the harmful consequences for women, gender-diverse, and sexually marginalized populations of proposing unfalsifiable biologically essentialist neuroscientific explanations rooted in binary operationalization of sex (Fausto-Sterling 2019; Jordan-Young 2012; Llaveria Caselles 2021).

Novel articulations of sex and gender that focus on variation in sex-related components across sexed animals, rather than binary comparison, place a greater emphasis on the neurobiological correlates of gendered experiences (Bölte et al. 2023; Eliot et al. 2023; Wierenga et al. 2024). These new approaches to sex and gender allow for investigation of their distinct and interactive effects that move past the male/female binary. However, both dominant and emerging approaches to operationalizing sex and gender in neuroscience tend to converge in dichotomizing gender/sex realities through a nature/nurture binary, continuing to ground articulations of sex in biologically essentialist beliefs and treating sex and gender as independent constructs.

3.1 Dichotomizing and Essentializing Gender/Sex Effects in the Nervous System

Sex-related variation is often described in terms of sexual dimorphism or differences (Joel and McCarthy 2017). Sexual dimorphism is commonly understood as the physiological processes that orient a morphologically undifferentiated embryo down a particular path of sexual development, giving rise to two sexually distinct phenotypes or dimorphs (e.g., genitalia and gonads) generally, although not consistently, associated with sex-related chromosomes (Joel 2012). Sexual dimorphism can also be observed in terms of size, but such a size difference must be large enough (i.e., nonoverlapping between sexes) to be considered morphologically distinct. For instance, the robust nucleus of the arcopallium, a key brain region of the vocal motor system in zebra finches, is initially sexually monomorphic: it atrophies as females develop but enlarges in males, resulting in a five-fold larger structure in mature male finches (Nixdorf-Bergweiler 1996). To date, we have no evidence of a phenotype within the human brain that would be considered sexually dimorphic (Joel et al. 2020). In fact, sexual dimorphism in the mammalian brain is considered more of an exception than a rule. One such exception, documented in rats, is the sexually dimorphic nucleus of the preoptic area, which is disproportionately larger in males (McCarthy 2023). While the cues that trigger sexually dimorphic developmental trajectories can be environmental and/or genetic, sexually dimorphic phenotypes that remain stable across time and context are generally considered sex-determined or innate (Joel and McCarthy 2017). Consequently, labeling an observed sex difference as dimorphic (sometimes inadvertently) implies that the difference is of such magnitude that it is considered morphologically distinct, biologically determined, and minimally impacted by variation in context.

"Sex differences" on the other hand, refers to monomorphic traits across individuals that present overlapping yet distinct distributions between male and female organisms (e.g., differences in circulating levels of testosterone). Sex differences can be classified as divergent (same processes leading to different outcomes) or convergent (different processes leading to same outcome), and may be documented as quantitative (e.g., differences in brain volume) or qualitative (e.g., differences in the nature of a quantitatively similar relationship between two variables as a function of sex; Beltz et al. 2019). The mapping of sex differences onto a male versus female sex binary constitutes the most common experimental approach to studying sex effects in neuroscience, in both humans and nonhuman animal models (Lafta et al. 2024). While this approach to characterizing sex differences offers insight into the functioning of the nervous system, the imposition of a male/female binary operationalization of biological sex as a construct through which group differences in brain and behavior are examined often results in the assumed existence of a biologically driven explanation in neuroscience *whether or not differences are observed* (Garcia-Sifuentes and Maney 2021).

Sex-related effects in neuroscience are mainly quantified by computing differences between males and females. However, knowledge about the potential biological mechanisms underlying these effects must be produced through experimental manipulation. Manipulative approaches include the use of genetic, pharmacological, and surgical procedures to systemically or locally modulate sex-related components, procedures that are mostly carried out in nonhuman animal models (Sagoshi et al. 2020). For instance, genetically modified mouse models allow chromosomal sex-associated traits to be investigated independently of gonadal effects (reviewed in Arnold and Chen 2009). In humans, however, the most common approach to evaluating sex-associated neuroendocrine variables is by indirectly (i.e., not manipulatively) examining reproductive phenomena such as puberty and the menstrual cycle, which are used as proxies for phasic variation in gonadal hormones (Dubol et al. 2021). Sex effects in humans are also characterized by studying groups of individuals with conditions associated with nonnormative development and/or activity of sex-associated traits (e.g., people with congenital adrenal hyperplasia; Khalifeh et al. 2022), creating pseudo-experimental opportunities to isolate the effects of sex-associated factors.

To date, quantitative gender/sex differences in brain structure and function in humans are often small and difficult to replicate (Eliot 2024; Eliot et al. 2021), an observation that also extends to the neurobehavioral correlates of gonadal hormones (Buskbjerg et al. 2019; Dubol et al. 2021; Pletzer et al. 2023). These inconsistencies can partly be attributed to methodological limitations associated with sample size (Wierenga et al. 2024) and measurement quality of sex-associated variables[4] (Celec et al. 2015; Schmalenberger et al. 2021; Wierenga et al. 2024); however, biases related to the biologically essentialist conceptualization and operationalization of sex-related effects remain (Joel and McCarthy 2017; Maney 2016; Wierenga et al. 2024). Feminist science scholars have been instrumental in describing the ramifications of these biases in the production of neuroscientific knowledge about gender/sex realities. For instance, critical feminist analysis has demonstrated that despite minimal consistent empirical evidence for binary gender/sex differences in the human brain, observed differences tend to be interpreted as indicative of two "kinds" of brains specialized for different things, a framing that is then reinforced in the popularization of the research findings (Maney 2015; Rippon et al. 2021). Moreover, these differences are often discussed as sexually dimorphic, and therefore biologically determined, providing "neuroscientific grounds" for strengthening existing gender stereotypes and discrimination (Jordan-Young and Rumiati 2012; Saguy et al. 2021). Critical review of analytical strategies has also revealed that sex differences are often reported from neuroscientific studies in which analyses of males and females were conducted separately, an analytic approach that not only requires an a priori assumption that these groups should be treated differently, but also tends to statistically inflate differences (Garcia-Sifuentes and Maney 2021). Analytical bias in sex difference findings has further been observed wherein variation in the selection of parameters for data preprocessing and other methods (e.g., machine-learning algorithms) has been shown to drastically influence the presence or absence of sex differences (Sanchis-Segura et al. 2022, 2023). Finally, recent feminist critique of the sexual selection theory, which is commonly referred to as an evolutionary justification for the presence of sex difference in the human brain (Cahill 2006), underscores its androcentric bias and minimal consideration of the phenotypic diversity that exists within sexes, particularly in female animals (Ah-King 2022).

One way to address inconsistencies and biases within literature on gender/sex is to consider the role of context, in particular gendered experiences, in relation to sex-related effects. Failure to do so when conducting and interpreting neuroscientific research in sex differences has been identified as a central source of bias, with

---

[4] For instance, in a recent review of 77 neuroimaging studies that assessed neural correlates of the menstrual cycle, Dubol et al. (2021) rated the findings, in over 75% of the studies, as low or very low confidence, due to lack of appropriate controls and ambiguity in the assessment of menstrual cycle phase, among other issues.

harmful consequences for women and gender diverse people (Eliot and Richardson 2016). For instance, a common formulation in neuroendocrinology is that variation in ovarian hormones renders females more vulnerable to stress-related disorders, often with minimal or no consideration of social or environmental factors that may influence experiences of stress (e.g., gendered differences in caring responsibilities), therefore reinforcing sexist beliefs that women are innately "more emotionally vulnerable" (Kundakovic and Rocks 2022; Li and Graham 2017). Relatedly, neuroscientific investigation conducted with "nontypical" biological sex presentations (e.g., individuals with congenital adrenal hyperplasia) tends to focus on aligning people neurobiologically as either male or female while ignoring their medically and socially marginalizing experiences (Jordan-Young and Rumiati 2012). For instance, gender/sex differences observed in brain morphometry vary as a function of individuals' chronic stress (Shang et al. 2024). The context contingency of sex differences is also increasingly documented in neuroscientific research conducted in nonhuman animal models. Recently, Mitchell et al. (2022) demonstrated that cage size, shock intensity, and number of training trials all moderated sex differences observed in fear-conditioned darting behaviors in rats (Mitchell et al. 2022). Darting behaviors were initially considered a female-specific fear response, but these behaviors can also be observed in males under certain conditions, and early-life stress can eliminate darting behaviors in females (Manzano Nieves et al. 2023). Early-life environment has also been shown to alter the development and function of sexually dimorphic brain regions (preoptic area) in laboratory rats, suggesting that the degree of sexual dimorphism in those regions could be context contingent (Eck et al. 2022; Halladay and Herron 2023). Emerging literature demonstrating context contingency of sex differences calls into question the existence of "pure" (i.e., translatable) sex differences across species and further challenges the legitimacy of generalizing decontextualized research on sex differences (Pape et al. 2024).

Conducting more rigorous, context-informed investigation of gender/sex differences will undoubtedly improve data quality and reduce interpretive bias. However, no methodological enhancements can overcome the conceptually and empirically limiting operationalization of sex as a male versus female binary. Further, continuing to ground research in the essentialist assumption that sex differences in the nervous system result from innately "programmed" differences between males and females will continue to produce and reproduce unstable and harmful knowledge (Maney 2016). The suggestion that there is more variability within sexes than can be accounted for under sexual selection theory renders this conceptualization of sexual dimorphism cryptic. Altogether, the male versus female sex binary is considered by many an inadvisable operationalization of sex-based realities in research (Eliot et al. 2023; Gungor et al. 2019; Hyde et al. 2019; Joel 2021; Ritz and Greaves 2022).

3.2 Novel Operationalizations of Sex and Gender: Sex and Gender Dichotomies

Novel frameworks for sex and gender in neuroscience have recently been proposed that address some of the critiques advanced by feminist science scholars regarding the limits associated with a binary and "sex-centric" operationalization of sex and gender (Bölte et al. 2023; Eliot et al. 2023; Wierenga et al. 2024). These frameworks are organized around definitions of sex and gender as multidimensional continuous constructs. *Sex* is defined as a biological phenomenon resulting in a series of components that are differently expressed across time and levels of organization (external genitalia, secondary sex characteristics, gonads, chromosomes, and hormones), rooting the construct in clearly stated and measurable biological processes rather than as a "natural kind" inferred from differences between individuals. Increasing consideration has also been given to nonbinary distribution of sex-related variables (Wierenga et al. 2024). This approach centers on the characterization of sex-related components across organisms (e.g., promoting the study androgens in both females and estrogens in males), welcoming rather than excluding diverse sex-related manifestation, and thereby mitigating bias in our understanding of sex-related components across individuals. When considering operationalizations of sex across studies, limitations to this approach are apparent. Review of this literature demonstrates that although no single operationalization comprehensively captures sex as a construct (e.g., sex can be operationalized as chromosomal, or by proxy of neuroendocrine activity, such as levels of circulating estrogens), researchers (including myself) who use a particular operationalization tend to continue using it over time. In other words, we may become "blinded" by our operationalization, developing methods and measures that are inadvertently designed to reflect the *construct* back to us. This bias is also apparent in how animal models remain largely used to characterize natural or "pure" biological phenomena, which by extension tends to articulate nonhuman animals as acultural beings. Finally, while this operationalization of sex emphasizes a continuous, rather than categorical, distribution of sex-related effects on the nervous system, it does not directly challenge biologically essentialist assumptions related to the construct of sex.

Gender in neuroscience has traditionally been characterized as an irreducible phenomenon that reflects a person's identity and expressions (e.g., gender presentation, adherence to or rejection of cultural expectations associated with sex), and has therefore been considered intractable in nonhuman animal research (McCarthy 2023). Until recently, the neurobiology of gender in humans has received little attention (Rauch and Eliot 2022). A small number of studies investigate neural correlates of gender roles, attitudes, and behaviors in relation to the male versus female sex binary (reviewed in Rauch and Eliot 2022). However, two main tools to assess gender roles via self-report were validated in student populations more than fifty years ago (see Oertelt-Prigione 2023); more broadly, operationalizations of gender as a measure of sociostructurally decontextualized masculine and feminine attitudes limit our understanding by masking the multidimensionality and power-laden nature of gender.

Novel approaches favor measures such as the Stanford gender-related variables for health research (GVHR) scale that avoid binarization of gender by providing measures of gender norms, related traits, and relations (Nielsen et al. 2021). These approaches also encourage the integration of gendered experiences, such as caretaking duties and career development (Wierenga et al. 2024). Greater consideration of environmental domains, such as stress exposure, rearing environment, and maternal care, has been recommended when interrogating the role of sex-related component in nonhuman animals. Recommendations have also been made regarding the dissemination of neuroscientific research on sex and gender on the basis of the discipline's epistemological authority (Eliot et al. 2023) and the potentially harmful impact that biased interpretation can have, particularly for people in marginalized populations. Specifically, caution is advised against speculative interpretations, the use of dichotomizing language when disseminating gender/sex differences research, and producing oversimplified translational interpretations that risk essentializing gender/sex differences. To represent more transparently the degrees of variation and similarity that exist between differently sexed and gendered individuals, researchers are encouraged to report negative findings and contextualize their interpretations appropriately with respect to effect sizes and applied relevance. Finally, studying more diverse populations, so that knowledge produced about sex and gender becomes more representative, is also recommended.

These novel approaches are grounded in a more inclusive, contextually sensitive, and humble articulation of sex and gender. They highlight that both "biological" and "social" aspects of maleness and femaleness are equally relevant to understanding gender/sex differences and are, therefore, undoubtedly better situated to grapple with the compounding complexities of these constructs than the more common binary operationalization. For a recent example of the value of this work in capturing variation and similarity across gender/sex realities, see Dhamala et al. (2024). However, despite recognizing the biopsychosocial contributions associated with sex and gender variations, newer operationalizations still treat sex and gender as distinct and independent constructs, implicitly reinforcing another dichotomy between nature versus nurture, and in doing so, continuing to perpetuate certain biases. One such bias regards a lack of consideration for the role that gendered experiences may have in the development and functioning of sex-associated traits. Sex-associated phenomena such as the menstrual cycle and pregnancy tend to be operationalized and validated solely with respect to their hormonal dimensions (Eliot et al. 2023), with no consideration of how the biology of reproductive phenomena can be modulated by social and cultural practices such as gender-related experience of stress (e.g., menstruation-related stress and discomfort). Further, the dichotomizing of sex and gender into nature versus nurture tends to reinforce rather than mitigate essentialist assumptions about sex-related effects, which are conceptualized as sources of variation arising purely from nature.

Moving beyond the compounding of categorical frameworks additionally requires iterative conceptual and theoretical revisions to the ways in which we select the most appropriate (or least limiting) dimensions to reduce. Gender/sex entanglement constitutes one avenue to move beyond compounding binaries and minimize undue and harmful essentialist assumptions and interpretations. By articulating the construct of sex as socially situated and therefore interrelated with gender, biological sex cannot be operationalized as acontextual and essential in origin. By characterizing sex and gender as entangled, we commit to the idea that these constructs, alongside other biosocial dimensions of human identity, cannot be studied independently. This approach requires concepts and theories in which explanatory models of sex and gender are conceived as interconnected systems that dynamically inform the development and functioning of the nervous system. To contend with the multiplicity of gender constructions and experiences, and to represent these constructions and experiences in a manner that empowers and emancipates, critical analysis regarding the current practice of neuroscience needs to take place, and new methods that consider people's lived experiences must be developed. Broader consideration not only of who is being studied, but also who is conducting the study, will help to correct the perpetuation of androcentric bias in neuroscience and reconnect us to the question of diversity. Who decides what "knowledge" is, how it is disseminated, and for whom it will have positive or negative impacts?

4 A Neurofeminist Perspective on Gender/Sex Entanglement

Attending to the complexities of the interrelations between gender and sex is a hallmark of feminist science. Societies create frameworks such as gender to determine what is relevant, normal, and essential to sexed organisms. By assigning meaning and value to sex-associated traits, phenomena, and consequently to individuals, gendered constructions inform how we behave, the technologies we develop, and our built environment. By extension, gendered constructions inform our biologies. Approaching sex and gender as entangled constructs is critical to characterize biological ramifications of gender/sex at different levels of individual and societal functioning. Importantly, gendered constructions, such as the societal tendency to conceptualize all non-male and non-White bodies and behaviors as "deviations from the norm," also inform what aspects of a biobehavioral phenomenon are considered relevant for neuroscientific study. Accordingly, while there is clear justification to operationalize sex and gender as biosocially entangled, how we go about characterizing this entanglement as a discipline must also be critically analyzed.

Gender/sex entanglement signifies that gendered experiences inform the development and functioning of biological systems, including sex-associated traits (Rippon et al. 2014; Springer et al. 2012). For instance, stressful experiences, parental care, and interpersonal closeness are all domains of experience that display gender-related differences, and all have also been shown to alter circulating levels of gonadal hormones (Brown et al. 2009; Pletzer et al. 2021; van Anders et al. 2012). Conversely, the administration of gonadal hormones has been shown to influence biobehavioral responses to stressful situations (Roca et al. 2003; Rubinow et al. 2005). While the existence of bidirectional relationships between gonadal hormones and the environment is uncontroversial in behavioral neuroendocrinology, entanglement between gendered experiences and the endocrine system is rarely considered in neuroscience.

4.1 Gender/Sex Entanglement in Human Research: Reconceptualizing the Menstrual Cycle

Feminist scholars have long argued in favor of considering the constitutive role of cultural beliefs, social status, and power dynamics in the study of reproductive phenomena (Bobel et al. 2020). Lessons from their work can inform the neurofeminist characterization of biosocially entangled phenomena. For instance, the menstrual cycle is a biosocial phenomenon yet in neuroscience it is operationalized uniquely as a hormonal process decontextualized from experience. The widespread belief that non-menstruating people are considered "the norm" and menstruating people are considered the "other" (i.e., menstruation stigma) is embedded in our relations and institutions (Chrisler 2011), and the othering of menstruating people is reflected through external and internalized experiences of shame, disgust, concealment, and inferiority with psychological consequences such as hypervigilance and self-consciousness (Johnston-Robledo and Chrisler 2020). Menstruating people often change and/or are asked to change their behavior to minimize stigma and comply to sociocultural norms. For example, during menstruation many experience greater self-consciousness during exercise such that they avoid physical activities or change their usual exercise routines to minimize these experiences (Kolić et al. 2021). The negative social construction of the menstrual cycle extends beyond menstruation into the premenstrual phase, where people report an increase in body shame and body dissatisfaction (Kaczmarek and Trambacz-Oleszak 2016). In addition, distinct influences of power dynamics, cis-normativity, and menstrunormativity interact to further disadvantage and stigmatize transgender people who menstruate (Rydström 2020). Experiences of "period poverty" (i.e., lack of material and educational resources to manage one's period) also significantly influence people's experiences during menstruation (Cardoso et al. 2021).

Such gendered constructions of menstruating bodies translate into gendered, embodied sources of stress. Physiological stress responses are observed in people subjected to stigma (Schvey et al. 2014). Similarly, exposure to stressful situations can lead to the release of estrogens and progesterone, with the progesterone response being particularly pronounced in the context of social rejection (Pletzer et al. 2021; Wirth 2011). Further, changes in behavior, such as reductions in physical exercise and social activities observed during menstruation, can contribute directly to changes in the activity of gonadal systems or indirectly through the effects of behavior change on physiological stress systems (Jasienska et al. 2006). By providing meaning and value to menstrual physiology, gendered experiences shape neuroendocrine activity and the functions of ovarian hormones. Therefore, brain and behavior differences that were initially interpreted as the "sole effect" of ovarian hormones can be reconsidered as possible consequences of gendered experience that *manifest* as changes in ovarian hormone levels. These gendered experiences can inform how we investigate the effects of ovarian hormones across the menstrual cycle in people of different gender and marginalized identities, where greater differences in the occurrence and types of gendered experience and discrimination could be expected to have still further impact on stress physiology and ovarian hormone activity. Conceptualizing sex and gender as entangled can advance our understanding of neuroendocrine regulation of ovarian hormones and promote questioning of how a diversity of gendered experiences may impact neurophysiology at many levels.

4.2 Gender/Sex Entanglement in Nonhuman Animal Research: Reconceptualizing Ovarian Hormones

Animal research can also benefit from operationalizing gender/sex as biosocially entangled. For instance, considering phenomena that regulate the functioning of ovarian hormones (e.g., stress exposure) can provide insight into the socially learned aspects of gender without requiring gendered experience to be modeled as subjective. Investigating the interplay between the gonadal and stress systems, proposed by DuBois and Shattuck-Heidorn (2021), is one way to capture biosocial entanglement between sex and gender. Systematically attending to context contingencies and conceptualizing sex-associated traits as biosocial constructs will orient investigations away from essentialization of sex differences toward modeling a more complex, diverse, and dynamic gender/sex entanglement, regardless of species. In the same vein, Richardson (2022) proposed a framework wherein the context for anticipated sex differences in biomedical research should be made explicit, and experiments should be designed to test the proposed model rather than systematically computing sex differences in the absence of a guiding explanatory model. Importantly, adopting a contextualized and entangled approach to gender/sex safeguards against possible translational missteps (Gungor et al. 2019), which tend to occur when human conditions or phenomena are modeled in nonhuman animals mainly according to their phenotypical resemblance or face validity (e.g., animal models of sexual violence; de M. Oliveira et al. 2022). An entanglement approach is more concerned with etiological and ethological validity, focusing on interrogating the mechanisms at play in a biosocial phenomenon (e.g., how the experience of stress may change the impact of ovarian hormones on brain and behavior; Pape et al. 2024).

The characterization of gender/sex as biosocially entangled aligns with an increasing number of theories in behavioral genetics (Uchiyama et al. 2022), affective neuroscience (Satpute and Lindquist 2019), and psychopharmacology (Branchi and Giuliani 2021) that consider context, and in particular culture, as constitutive in the evolution, development, and functioning of various biobehavioral processes. A compelling example of the benefits of taking a biosocial approach to understanding brain function can be observed in Branchi and Giuliani's framework for investigating the therapeutic efficacy of psychotropes in the wake of documented inconsistencies (Branchi and Giuliani 2021). They introduce a distinction between instructive causality, in which an action determines a specific effect, and permissive causality, which allows an action to promote many effects, and argue that pharmacological treatments such as selective serotonin reuptake inhibitors (SSRIs) are best understood as permissive. By promoting brain plasticity, SSRIs place the brain in a state of *potential change* for an instructive effect (e.g., therapy) to take place. Such a conceptualization is supported by a study that investigated relationships between citalopram dosage (a widely prescribed SSRI), quality of environment (based on education, employment, and income), and reported mood in over 4,000 patients in the United States registered to a clinical trial. Findings demonstrated that higher dosage of citalopram predicted a stronger correlation between environment quality and reported depressive symptoms (Viglione et al. 2019). In other words, cumulative citalopram dose appeared to amplify the effect of the environment on depressive symptoms (i.e., high quality environment + high dose of citalopram = reduction in depressive symptoms; low quality environment + high dose of citalopram = worsening of depressive symptoms). Using an animal model of chronic stress and environmental enrichment, Branchi et al. (2013) observed a similar permissive role of SSRIs. Specifically, mice administered SSRIs while being chronically exposed to mild stressors demonstrated worse depressive-like behavior (e.g., sucrose preference test) and physiology (e.g., corticoid levels) compared to animals receiving the control solution; the opposite effects were observed when mice receive SSRIs in an enriched environment. This model demonstrates that the environment creates unique contexts for the physiological effects of SSRIs on brain and behavior. In other words, the efficacy of SSRIs must be understood as entangled with environmental context.

Although speculative at this point, this framework can be extended and applied to gender/sex research to help test hypotheses that aim to reconcile inconsistencies in the effects of gonadal hormones on brain and behavior, which, like antidepressant efficacy, vary widely. Like antidepressants, one central neuroendocrine effect of gonadal hormones is potentiation of neuroplasticity (Been et al. 2022), therefore arguably placing the organism in a "permissive state." Such a framework can generate novel biosocial conceptualization of conditions such as premenstrual syndrome (PMS) and premenstrual dysphoric disorder, conditions in which hormonal variation differentially impacts individual affective responses. Interestingly, research documenting the sociostructural correlates of PMS demonstrates that people are more likely to report PMS if they have an unequal share of household or child-rearing responsibilities (Coughlin 1990; Ussher 2003) or are experiencing relationship strain (Kuczmierczyk et al. 1992). By differentiating biobehavioral effects in terms of causality, this approach offers a generative framework to devise new neurobiological models of gender/sex entanglement.

The continued development of encultured and embodied models in neuroscience creates unique opportunities to bring the wealth of knowledge that has been produced independently about sex and gender into an entangled perspective (Fausto-Sterling 2021). Models of entanglement can be tested empirically or via data simulation, where alternative hypothesized versions of entanglement (e.g., assuming ovarian hormones as permissive vs. instructive) could be simulated to further inform the development of experiments (for an example, see Cross et al. 2023). As we recognize that environment and social learning shape organisms' biological and behavioral repertoires, the richness and complexity of the animal's environment become central to developing a biosocial understanding of gender/sex differences. Several approaches to increase behavioral and biological diversity in the lab, also known as rewilding, would be particularly relevant (Zipple et al. 2023). Finally, while the main focus of biosocial entanglement described here centers on socially induced gendered experiences, entanglement can be characterized through many points of entry, such as through gendered differences in environmental exposure to endocrine-disrupting chemicals (Ritz and Greaves 2022).

4.3 Gender/Sex as Entangled with Other Social Categories

Gender is neither fixed nor homogenous. Although *gendered experience* is generally operationalized as the domain of experiences and behaviors socially ascribed to men and women (e.g., intimate partner violence, childcare), articulating these experiences *only* through a binary masks the multiplicity of sociocultural interactions from which gender emerges.

Neurofeminists have emphasized the need for neuroscience to contend with the multiplicity of gender/sex experiences (Kuria 2014; Rippon et al. 2014). Intersectionality theory has been proposed by many feminist scholars as a generative framework to ground the characterization of gender/sex as biosocially entangled (Duchesne and Kaiser Trujillo 2021; Hankivsky et al. 2017; Shattuck-Heidorn and Richardson 2019). Intersectionality conceptualizes social group memberships as interdependent, resulting from sociohistorically inherited and structurally embedded power dynamics (Bowleg 2008). Intersectionality, grounded in Black feminist activism, was developed to critically analyze the experience of African American women within the legal system and to counter the underrepresentation of Black women's experiences in gender and critical race studies (Crenshaw 1989). Today, intersectionality is employed across many disciplines to critically inform the development of research aimed at understanding the sociostructural axes of oppression related to group memberships and engage in social justice actions and goals that contribute to dismantling social injustice (Moradi and Grzanka 2017). However, the consideration of intersectionality in gender/sex neuroscience research remains marginal (Duchesne and Kaiser Trujillo 2021).

Recently, Carter et al. (2022) critically analyzed neuroscientific research on stress using a Black feminist intersectional lens. Their analyses demonstrated that the embodiment of racism has been mostly articulated through a "minority stress" framework, which characterizes experiences of stigma and discrimination related to social oppression as chronic sources of stress. Traditional stress assessments have operationalized chronic stress as a series of cumulative events, without accounting for potential interactions with gender/sex or other dimensions of context. To better model the experiences of Black women, Carter and colleagues recommend using stress models, such as the proliferation model of stress (Pearlin et al. 2005), which connect stressful conditions or situations at different degrees of social organization and characterize the severity of a stressor by the degree to which it sets other stressful experiences in motion. Highly proliferating sources of stress such as overpolicing are often structural in nature. Further, Carter et al. (2022) suggest that research attempting to understand the embodiment of Black women's experiences of racism and sexism will benefit from employing conceptualizations that consider stress-related outcomes as the manifest "cost" of adapting to oppressive environmental contexts, rather than solely as dysfunction. These findings highlight how implicit biases about dysfunction and the fixedness of the brain enter into the neuroscientific interpretation of individual differences, leaving space for deterministic reiteration of inequalities. Such critical analyses can provide alternative operationalizations upon which more representative biobehavioral models can be developed.

4.4 Gender/Sex as Entangled with the Practice of Neuroscience

Neurofeminists have proposed frameworks wherein the nervous system is understood as situated within a multilevel, biosocially entangled context, in which the relationships between knowers and the sociohistorical contextualization of the phenomenon are considered constitutive of neuroscientific knowledge. Einstein's "very mixed methods" approach is a strong example of how situating the nervous system biosocially can facilitate an integration of traditional quantitative modes of neuroscientific inquiry with qualitative interview data. This approach has already successfully generated novel insights into reconciling inconsistencies in different domains of pain experience (physiological and self-report) and methodological limitations related to cultural variation in pain experiences/reporting in Somali-Canadian women who had experienced female genital cutting (Brown et al. 2022; Perović et al. 2021). Similarly, Roy (2018) proposed a transformative approach to conducting experimental research rooted in feminist theory and activism. In a landmark transdisciplinary project, Roy brought together neuroendocrinologists and reproductive rights activists to discuss women's reproductive health and related policies, integrating perspectives to co-produce new insights. Such an approach creates spaces where sociohistorical and empirical views regarding gender/sex can be integrated to generate hypotheses related to biosocial engagement that can inform the future of what, how, and from what position to practice science.

Although these approaches consider biosocial entanglement as critical in knowledge production, they do not allow for an equally entangled interrogation of how context contributes to neural representation. Representational similarity analysis (RSA) is an analytical tool that shows promise for studying the brain in a context-sensitive manner (Popal et al. 2019). Rather than measuring the brain's response to stimuli and making inferences about "how the brain works," RSA involves creating hypothesized multidimensional spaces for a particular set of stimuli (i.e., dissimilarity matrices) and comparing these against observed patterns of brain activity to understand how complex information is neurally represented. In social and affective neuroscience, RSA has demonstrated that social categories (e.g., gender/sex, race, social status) associated with a particular stimulus (e.g., a face) are inextricably embedded together in the neural process of face recognition rather than represented as discrete "features" (i.e., are represented in an entangled manner) (Popal et al. 2019). To date, the vast majority of RSA studies have focused on mapping stimulus dimensions, but this technique could be adapted to explore variation in participants' gendered experiences. This application of RSA constitutes a methodological innovation that can help advance an understanding of gender/sex as entangled while attempting to preserve the sociocultural complexity of gendered experiences.

An intersectional and inclusive perspective can profoundly transform the production of neuroscientific knowledge, leading to better resolution and awareness of currently underrepresented populations. However, for this enterprise to produce emancipatory and empowering knowledge, the practice of neuroscience must also be transformed. Edminston and Juster (2022) describe how current practices (e.g., recruitment, screening) in neuroscientific research conducted in transgender populations limit the representation of transgender communities to more privileged and socially affluent groups (Edmiston and Juster 2022). Further, the threat of transition care disclosure, participation in studies with little to no direct benefit to participants, and research fatigue all contribute to an unwelcoming, extractive research environment. Edminston and Juster propose adopting a community-based participatory framework wherein neuroscientific research questions and practices are developed in concert with members of gender-diverse communities. Community-based neuroscience has already been employed to a limited degree in adult (Weng et al. 2020) and youth populations through citizen science projects (Green et al. 2022). Finally, developing inclusive scientific practice must also extend to inclusivity in who is conducting the research, as questions of diversity and inclusion are inextricably connected to advancing a neuroscience of gender/sex that moves beyond the reiteration of discriminatory biases.

4.5 Challenges Ahead

Adopting a neurofeminist perspective is not without challenges. For pragmatic reasons, the neurofeminist perspective was described herein by referring to sex- and gender-associated variables. However, the use of these terms in the context of entanglement might benefit from reevaluation. Sex, gender, and gender/sex are categories that refer to an assemblage of assumptions and traits embedded in fixedness and homogeneity. Should novel terminology be developed that captures the dynamic interdependence of these constructs, or should we simply avoid the use of the terms sex and gender in favor of more precise descriptions of the variables at play?

A neurofeminist perspective also demands conceptual and methodological transdisciplinarity, which poses challenges for training and the pace of knowledge production. Transdisciplinarity necessarily means that knowledge produced will not have the same resolution or disciplinary location as that produced via disciplinary research. The transdisciplinary researcher is, in some ways, at arm's length from the latest research developments, which may lead to the persistence of some outdated ideas and practices. However, abandoning all disciplinary expertise is not the goal. Rather, while consensus should be reached related to the characterization and assumptions of sex and gender, both disciplinary and transdisciplinary approaches will help build a more diverse research ecosystem, where innovations and practices from one perspective inform the reflection and progression of another.

Adopting a neurofeminist perspective also poses some risk for scholars. Neurofeminism is currently a marginal approach to the study of sex and gender, which is already a small subfield of neuroscience. Further, feminism has a long history of being epistemologically silenced and discredited (Fricker 2007). Mitigating strategies related to epistemic exclusion that arise from adopting a neurofeminist perspective must be considered, particularly for early-career trainees. Strategies can include joining academic network groups (e.g., the Neurogendering Network) or organizations engaged in similar work and approaches, or by providing and supporting resources produced for and by feminist scholars (e.g., Richmond et al. 2022).

Sex and gender research owes its development to many feminist/queer and critical race scholars, and needs this work to continue transforming scientific practice. However, the widespread defunding and closure of departments in humanities, particularly critical studies, significantly challenge further advancements in sex and gender research. Neuroscientists (including myself) have benefited from transdisciplinary research and should in turn consider how to wield the epistemic authority of the discipline to reciprocally support the important critical and theoretical work of the social sciences and humanities.

5 Conclusion

Sex and gender are inarguably significant constructs in advancing our understanding of the nervous system. Similarly uncontroversial is that the characterization and operationalization of sex and gender must account for the context-contingent, dynamic, and power-laden nature of these complex constructs. Failure to do so inevitably perpetuates problems such as androcentric bias. Enhancing resolution on sex and gender can be partly accomplished through frameworks that interrogate both constructs. However, this approach leaves space for potentially harmful reiteration of biased assumptions. Approaching sex and gender from a neurofeminist perspective, where sex and gender are considered biosocially entangled, offers one means of addressing these limitations. Numerous theoretical and methodological avenues of neurofeminist engagement can be implemented at all levels of research, including a critical reanalysis of past findings to inform the interpretation of current findings and proposing new neuroscientific theories, methods, or analytical strategies that balance the dimension reduction inherent to scientific study against the need for context sensitivity. Although neurofeminism is not immune to biases and challenges, it constitutes a promising alternative to contend with the complex nature of sex and gender, and to transform neuroscientific practices broadly.

Acknowledgments I would like to acknowledge the Ernst Strüngmann Forum and the organizing committee for inviting me to share some of my views on sex and gender research; my lab members, members of the neurogenderings network, and the Canadian Organization for the study of Gender and Sex Research for the many constructive discussions and debates. Finally, I would like to acknowledge the thorough editing and so very insightful comments by Nicole White.

References

Ah-King M (2022) The History of Sexual Selection Research Provides Insights as to Why Females Are Still Understudied. Nat Commun 13:6976. https://doi.org/10.1038/s41467-022-34770-z

Altimus CM, Marlin BJ, Charalambakis NE, et al. (2020) The Next 50 Years of Neuroscience. J Neurosci 40 (1):101–106. https://www.jneurosci.org/content/40/1/101.abstract

Arnold AP, and Chen X (2009) What Does the "Four Core Genotypes" Mouse Model Tell Us about Sex Differences in the Brain and Other Tissues? Front Neuroendocrin 30 (1):1–9. https://doi.org/10.1016/j.yfrne.2008.11.001

Becker JB, Arnold AP, Berkley KJ, et al. (2005) Strategies and Methods for Research on Sex Differences in Brain and Behavior. Endocrinology 146 (4):1650–1673. https://doi.org/0.1210/en.2004-1142

Been LE, Sheppard PAS, Galea LAM, and Glasper ER (2022) Hormones and Neuroplasticity: A Lifetime of Adaptive Responses. Neurosci Biobehav Rev 132:679–690. https://doi.org/10.1016/j.neubiorev.2021.11.029

Beltz AM, Beery AK, and Becker JB (2019) Analysis of Sex Differences in Pre-Clinical and Clinical Data Sets. Neuropsychopharmacology 44 (13):2155–2158. https://doi.org/10.1038/s41386-019-0524-3

Bennett EM, and McLaughlin PJ (2024) Neuroscience Explanations Really Do Satisfy: A Systematic Review and Meta-Analysis of the Seductive Allure of Neuroscience. Public Underst Sci 33 (3):290–307. https://doi.org/10.1177/09636625231205005

Bleier R (1976) Myths of the Biological Inferiority of Women: An Exploration of the Sociology of Biological Research. Univ Mich Papers Womens Stud 2 (2):39–63. https://quod.lib.umich.edu/m/mfs/acp0359.0002.002/00000047

Bluhm R, Maibom HL, and Jacobson AJ (2012) Neurofeminism: Issues at the Intersection of Feminist Theory and Cognitive Science. Palgrave Macmillan, New York

Bobel C, Winkler IT, Fahs B, et al. (eds) (2020) The Palgrave Handbook of Critical Menstruation Studies. Palgrave Macmillan, London

Boiger M, and Mesquita B (2012) The Construction of Emotion in Interactions, Relationships, and Cultures. Emot Rev 4 (3):221–229. https://doi.org/10.1177/1754073912439765

Bölte S, Neufeld J, Marschik PB, et al. (2023) Sex and Gender in Neurodevelopmental Conditions. Nat Rev Neurol 19 (3):136–159. https://doi.org/10.1038/s41582-023-00774-6

Bowleg L (2008) When Black + Lesbian + Woman ≠ Black Lesbian Woman: The Methodological Challenges of Qualitative and Quantitative Intersectionality Research. Sex Roles 59 (5):312–325. https://doi.org/10.1007/s11199-008-9400-z

Branchi I, and Giuliani A (2021) Shaping Therapeutic Trajectories in Mental Health: Instructive vs. Permissive Causality. Eur Neuropsychopharmacol 43:1–9. https://doi.org/10.1016/j.euroneuro.2020.12.001

Branchi I, Santarelli S, Capoccia S, et al. (2013) Antidepressant Treatment Outcome Depends on the Quality of the Living Environment: A Pre-Clinical Investigation in Mice. PLoS One 8 (4):e62226. https://doi.org/10.1371/journal.pone.0062226

Brown A, Karkaby L, Perovic M, Shafi R, and Einstein G (2022) Sex and Gender Science: The World Writes on the Body. In: Gibson C, Galea LAM (eds) Sex Differences in Brain Function and Dysfunction, vol 62. Springer, Cham, pp 3–25

Brown SL, Fredrickson BL, Wirth MM, et al. (2009) Social Closeness Increases Salivary Progesterone in Humans. Horm Behav 56 (1):108–111. https://doi.org/10.1016/j.yhbeh.2009.03.022

Brun C, Boraud T, and Gonon F (2024) The Neoliberal Leaning of the Neuroscience Discourse When It Deals with Mental Health and Learning Disorders. Neurobiol Dis:106544. https://doi.org/10.1016/j.nbd.2024.106544

Buskbjerg CR, Gravholt CH, Dalby HR, Amidi A, and Zachariae R (2019) Testosterone Supplementation and Cognitive Functioning in Men: A Systematic Review and Meta-Analysis. J Endocr Soc 3 (8):1465–1484. https://doi.org/10.1210/js.2019-00119

Cahill L (2006) Why Sex Matters for Neuroscience. Nature Reviews Neuroscience 7 (6):477–484. https://doi.org/10.1038/nrn1909

Cardoso LF, Scolese AM, Hamidaddin A, and Gupta J (2021) Period Poverty and Mental Health Implications among College-Aged Women in the United States. BMC Womens Health 21 (1):1–7. https://doi.org/10.1186/s12905-020-01149-5

Carter S, Mekawi Y, Sheikh I, et al. (2022) Approaching Mental Health Equity in Neuroscience for Black Women across the Lifespan: Biological Embedding of Racism from Black Feminist Conceptual Frameworks. Biol Psychiatry Cogn Neurosci Neuroimaging 7 (12):1235–1241. https://doi.org/10.1016/j.bpsc.2022.08.007

Celec P, Ostatníková D, and Hodosy J (2015) On the Effects of Testosterone on Brain Behavioral Functions. Front Neurosci 9:116423. https://doi.org/10.3389/fnins.2015.00012

Chrisler JC (2011) Leaks, Lumps, and Lines: Stigma and Women's Bodies. Psychol Women Q 35 (2):202–214. https://psycnet.apa.org/doi/10.1177/0361684310397698

Coughlin PC (1990) Premenstrual Syndrome: How Marital Satisfaction and Role Choice Affect Symptom Severity. Social Work 35 (4):351–355. https://pubmed.ncbi.nlm.nih.gov/2392713/

Crenshaw KW (1989) Demarginalizing the Intersection of Race and Sex: A Black Feminist Critique of Antidiscrimination Doctrine, Feminist Theory and Antiracist Politics. Univ Chic Leg Forum 1989 (1):8. https://chicagounbound.uchicago.edu/uclf/vol1989/iss1/8

Cross C, Boothroyd L, and Jefferson C (2023) Agent-Based Models of the Cultural Evolution of Occupational Gender Roles. Royal Society Open Science 10 (6):221346. https://doi.org/10.1098/rsos.221346

Dalla C, Jaric I, Pavlidi P, et al. (2024) Practical Solutions for Including Sex as a Biological Variable (SABV) in Preclinical Neuropsychopharmacological Research. J Neurosci Methods 401:110003. https://doi.org/10.1016/j.jneumeth.2023.110003

de Lange AMG, Jacobs EG, and Galea LA (2021) The Scientific Body of Knowledge: Whose Body Does It Serve? A Spotlight on Women's Brain Health. Front Endocrinol 60:100898. https://doi.org/10.1016/j.yfrne.2020.100898

de M. Oliveira VE, de Jong TR, and Neumann ID (2022) Modelling Sexual Violence in Male Rats: The Sexual Aggression Test (SxAT). Transl Psychiatry 12 (1):207. https://doi.org/10.1038/s41398-022-01973-3

Dhamala E, Bassett DS, Yeo BT, and Holmes AJ (2024) Functional Brain Networks Are Associated with Both Sex and Gender in Children. Sci Adv 10 (28):eadn4202. https://doi.org/10.1126/sciadv.adn4202

DuBois LZ, and Shattuck-Heidorn H (2021) Challenging the Binary: Gender/Sex and the Bio-Logics of Normalcy. Am J Hum Biol 33 (5):e23623. https://doi.org/10.1002/ajhb.23623

Dubol M, Epperson CN, Sacher J, et al. (2021) Neuroimaging the Menstrual Cycle: A Multimodal Systematic Review. Front Neuroendocrin 60:100878. https://doi.org/10.1016/j.yfrne.2020.100878

Duchesne A, and Kaiser Trujillo A (2021) Reflections on Neurofeminism and Intersectionality Using Insights from Psychology. Front Hum Neurosci 15:684412. https://doi.org/10.3389/fnhum.2021.684412

Eck SR, Palmer JL, Bavley CC, et al. (2022) Effects of Early Life Adversity on Male Reproductive Behavior and the Medial Preoptic Area Transcriptome. Neuropsychopharmacology 47 (6):1231–1239. https://doi.org/10.1038/s41386-022-01282-9

Edmiston EK, and Juster R-P (2022) Refining Research and Representation of Sexual and Gender Diversity in Neuroscience. Biol Psychiatry Cogn Neurosci Neuroimaging 7 (12):1251–1257. https://doi.org/10.1016/j.bpsc.2022.07.007

Einstein G (2012) Situated Neuroscience: Exploring Biologies of Diversity. In: Bluhm R, Jacobson AJ, Maibom HL (eds) Neurofeminism. Palgrave Macmillan, London, pp 145–174

Eliot L (2024) Remembering the Null Hypothesis When Searching for Brain Sex Differences. Biol Sex Differ 15 (1):1–7. https://doi.org/10.1186/s13293-024-00585-4

Eliot L, Ahmed A, Khan H, and Patel J (2021) Dump the "Dimorphism": Comprehensive Synthesis of Human Brain Studies Reveals Few Male-Female Differences Beyond Size. Neurosci Biobehav Rev 125:667–697. https://doi.org/10.1016/j.neubiorev.2021.02.026

Eliot L, Beery AK, Jacobs EG, et al. (2023) Why and How to Account for Sex and Gender in Brain and Behavioral Research. J Neurosci 43 (37):6344–6356. https://doi.org/10.1523/jneurosci.0020-23.2023

Eliot L, and Richardson SS (2016) Sex in Context: Limitations of Animal Studies for Addressing Human Sex/Gender Neurobehavioral Health Disparities. J Neurosci 36 (47):11823–11830. https://doi.org/10.1523/jneurosci.1391-16.2016

Ellemers N (2018) Gender Stereotypes. Annu Rev Psychol 69 (1):275–298. https://doi.org/10.1146/annurev-psych-122216-011719

Fausto-Sterling A (2019) Gender/Sex, Sexual Orientation, and Identity Are in the Body: How Did They Get There? J Sex Res 56 (4-5):529–555. https://doi.org/10.1080/00224499.2019.1581883

————— (2021) A Dynamic Systems Framework for Gender/Sex Development: From Sensory Input in Infancy to Subjective Certainty in Toddlerhood. Front Hum Neurosci 15:613789. https://doi.org/10.3389/fnhum.2021.613789

Fausto-Sterling A (2005) The Bare Bones of Sex: Part 1—Sex and Gender. Signs 302 (2):1491–1527. https://doi.org/10.1086/424932

Fine C (2010) Delusions of Gender: How Our Minds, Society, and Neurosexism Create Difference. W. W. Norton, New York

————— (2013) Is There Neurosexism in Functional Neuroimaging Investigations of Sex Differences? Neuroethics 6 (2):369–340. https://doi.org/10.1007/s12152-012-9169-1

Fricker M (2007) Epistemic Injustice: Power and the Ethics of Knowing. Oxford Univ Press, New York

Friedrichs K, and Kellmeyer P (2022) Neurofeminism: Feminist Critiques of Research on Sex/Gender Differences in the Neurosciences. Eur J Neurosci 56 (11):5987–6002. https://doi.org/10.1111/ejn.15834

Garcia-Sifuentes Y, and Maney DL (2021) Reporting and Misreporting of Sex Differences in the Biological Sciences. eLife 10:e70817. https://doi.org/10.7554/eLife.70817

Gottlieb G (2007) Probabilistic Epigenesis. Dev Sci 10 (1):1–11. https://doi.org/10.1111/j.1467-7687.2007.00556.x

Green KH, Van De Groep IH, Te Brinke LW, et al. (2022) A Perspective on Enhancing Representative Samples in Developmental Human Neuroscience: Connecting Science to Society. Front Integr Neurosci 16:981657. https://doi.org/10.3389/fnint.2022.981657

Gungor NZ, Duchesne A, and Bluhm R (2019) A Conversation around the Integration of Sex and Gender When Modeling Aspects of Fear, Anxiety, and PTSD in Animals. https://sfonline.barnard.edu/a-conversation-around-the-integration-of-sex-and-gender-when-modeling-aspects-of-fear-anxiety-and-ptsd-in-animals/

Halladay LR, and Herron SM (2023) Lasting Impact of Postnatal Maternal Separation on the Developing BNST: Lifelong Socioemotional Consequences. Neuropharmacology 225:109404. https://doi.org/10.1016/j.neuropharm.2022.109404

Hankivsky O, Doyal L, Einstein G, et al. (2017) The Odd Couple: Using Biomedical and Intersectional Approaches to Address Health Inequities. Glob Health Action 10 (Suppl 2):1326686. https://doi.org/10.1080/16549716.2017.1326686

Hines M (2011) Prenatal Endocrine Influences on Sexual Orientation and on Sexually Differentiated Childhood Behavior. Front Neuroendocrin 32 (2):170–182. https://doi.org/10.1016/j.yfrne.2011.02.006

Hyde JS, Bigler RS, Joel D, Tate CC, and van Anders SM (2019) The Future of Sex and Gender in Psychology: Five Challenges to the Gender Binary. Am Psychol 74 (2):171. https://psycnet.apa.org/doi/10.1037/amp0000307

Ingalhikar M, Smith A, Parker D, et al. (2014) Sex Differences in the Structural Connectome of the Human Brain. PNAS 111 (2):823–828. https://doi.org/10.1073/pnas.1316909110

Jasienska G, Ziomkiewicz A, Thune I, Lipson SF, and Ellison PT (2006) Habitual Physical Activity and Estradiol Levels in Women of Reproductive Age. Eur J Cancer Prevent 15 (5):439–445. https://doi.org/10.1097/00008469-200610000-00009

Joel D (2012) Genetic-Gonadal-Genitals Sex (3G-Sex) and the Misconception of Brain and Gender, or, Why 3G-Males and 3G-Females Have Intersex Brain and Intersex Gender. Biol Sex Differ 3 (27):1–6. https://doi.org/10.1186/2042-6410-3-27

————— (2021) Beyond the Binary: Rethinking Sex and the Brain. Neurosci Biobehav Rev 122:165–175. https://doi.org/10.1016/j.neubiorev.2020.11.018

Joel D, Berman Z, Tavor I, et al. (2015) Sex Beyond the Genitalia: The Human Brain Mosaic. PNAS 112 (50):15468–15473. https://doi.org/10.1073/pnas.1509654112

Joel D, and Fausto-Sterling A (2016) Beyond Sex Differences: New Approaches for Thinking about Variation in Brain Structure and Function. Philos Trans R Soc Lond B Biol Sci 371 (1688):20150451. https://doi.org/10.1098/rstb.2015.0451

Joel D, Garcia-Falgueras A, and Swaab DF (2020) The Complex Relationships between Sex and the Brain. Neuroscientist 26 (2):156–169. https://doi.org/10.1177/1073858419867298

Joel D, and McCarthy MM (2017) Incorporating Sex as a Biological Variable in Neuropsychiatric Research: Where Are We Now and Where Should We Be? Neuropsychopharmacology 42:379–385. https://doi.org/10.1038/npp.2016.79

Johnston-Robledo I, and Chrisler JC (2020) The Menstrual Mark: Menstruation as Social Stigma. In: Bobel C, Winkler IT, Fahs B, Hasson KA, Kissling EA, Roberts T-A (eds) The Palgrave Handbook of Critical Menstruation Studies. Palgrave Macmillan, London, pp 181–199

Jordan-Young R, and Rumiati RI (2012) Hardwired for Sexism? Approaches to Sex/Gender in Neuroscience. Neuroethics 5 (3):305–315. https://doi.org/10.1007/s12152-011-9134-4

Jordan-Young RM (2012) Hormones, Context, and "Brain Gender": A Review of Evidence from Congenital Adrenal Hyperplasia. Soc Sci Med 74 (11):1738–1744. https://doi.org/10.1016/j.socscimed.2011.08.026

Jordan-Young RM, and Karkazis K (2019) Testosterone: An Unauthorized Biography. Harvard Univ Press, Cambridge, MA

Kaczkurkin AN, Raznahan A, and Satterthwaite TD (2019) Sex Differences in the Developing Brain: Insights from Multimodal Neuroimaging. Neuropsychopharmacology 44 (1):71–85. https://doi.org/10.1038/s41386-018-0111-z

Kaczmarek M, and Trambacz-Oleszak S (2016) The Association between Menstrual Cycle Characteristics and Perceived Body Image: A Cross-Sectional Survey of Polish Female Adolescents. J Biosoc Sci 48 (3):374–390. https://doi.org/10.1017/s0021932015000292

Kaiser A (2012) Re-Conceptualizing "Sex" and "Gender" in the Human Brain. Z Psychol 220 (2):130–136. https://doi.org/10.1027/2151-2604/a000104

Khalifeh N, Omary A, Cotter DL, et al. (2022) Congenital Adrenal Hyperplasia and Brain Health: A Systematic Review of Structural, Functional, and Diffusion Magnetic Resonance Imaging (MRI) Investigations. J Child Neurol 37 (8–9):758–783. https://doi.org/10.1177/08830738221100886

Kolić PV, Sims DT, Hicks K, Thomas L, and Morse CI (2021) Physical Activity and the Menstrual Cycle: A Mixed-Methods Study of Women's Experiences. Women Sport Phys Act J 29 (1):47–58. https://doi.org/10.1123/wspaj.2020-0050

Kuczmierczyk AR, Labrum AH, and Johnson CC (1992) Perception of Family and Work Environments in Women with Premenstrual Syndrome. J Psychosom Res 36 (8):787–795. https://doi.org/10.1016/0022-3999(92)90137-q

Kundakovic M, and Rocks D (2022) Sex Hormone Fluctuation and Increased Female Risk for Depression and Anxiety Disorders: From Clinical Evidence to Molecular Mechanisms. Front Neuroendocrin 66:101010. https://doi.org/10.1016/j.yfrne.2022.101010

Kuria EN (2014) Theorizing Race(Ism) While Neurogendering. In: Schmitz S, Höppner G (eds) Gendered Neurocultures: Feminist and Queer Perspectives on Current Brain Discourses. Zaglossus, Vienna, pp 109–123

Lafta MS, Mwinyi J, Affatato O, et al. (2024) Exploring Sex Differences: Insights into Gene Expression, Neuroanatomy, Neurochemistry, Cognition, and Pathology. Front Neurosci 18:1340108. https://doi.org/10.3389/fnins.2024.1340108

Lerner RM (1978) Nature, Nurture, and Dynamic Interactionism. Hum Dev 21 (1):1–20

Li SH, and Graham BM (2017) Why Are Women So Vulnerable to Anxiety, Trauma-Related and Stress-Related Disorders? The Potential Role of Sex Hormones. Lancet Psychiatry 4 (1):73–82. https://doi.org/10.1016/s2215-0366(16)30358-3

Lindquist KA, Jackson JC, Leshin J, Satpute AB, and Gendron M (2022) The Cultural Evolution of Emotion. Nat Rev Psychol 1 (11):669–681. https://doi.org/10.1038/s44159-022-00105-4

Llaveria Caselles E (2021) Epistemic Injustice in Brain Studies of (Trans) Gender Identity. Front Sociol 6:608328. https://doi.org/10.3389/fsoc.2021.608328

Maney DL (2015) Just Like a Circus: The Public Consumption of Sex Differences. Curr Top Behav Neurosci 19:279–296. https://doi.org/10.1007/7854_2014_339

——— (2016) Perils and Pitfalls of Reporting Sex Differences. Philos Trans R Soc Lond B Biol Sci 371 (1688):20150119. https://doi.org/10.1098/rstb.2015.0119

Manzano Nieves G, Bravo M, and Bath KG (2023) Early Life Adversity Ablates Sex Differences in Active versus Passive Threat Responding in Mice. Stress 26 (1):1–32. https://doi.org/10.1080/10253890.2023.2244598

McCarthy MM (2023) Sex Differences in the Brain: Focus on Developmental Mechanisms. In: Legato MJ (ed) Principles of Gender-Specific Medicine. Academic Press, Cambridge, MA, pp 159–180

Miller LR, Marks C, Becker JB, et al. (2017) Considering Sex as a Biological Variable in Preclinical Research. FASEB J 31 (1):29–34. https://doi.org/10.1096/fj.201600781R

Mitchell JR, Trettel SG, Li AJ, et al. (2022) Darting across Space and Time: Parametric Modulators of Sex-Biased Conditioned Fear Responses. Learn Mem 29 (7):171–180. https://doi.org/10.1101/lm.053587.122

Mitchell SD (2009) Unsimple Truths: Science, Complexity, and Policy. Univ Chicago Press, Chicago

Moradi B, and Grzanka PR (2017) Using Intersectionality Responsibly: Toward Critical Epistemology, Structural Analysis, and Social Justice Activism. J Couns Psychol 64 (5):500. https://doi.org/10.1037/cou0000203

Nielsen MW, Stefanick ML, Peragine D, et al. (2021) Gender-Related Variables for Health Research. Biol Sex Differ 12 (1):23. https://doi.org/10.1186/s13293-021-00366-3

Nixdorf-Bergweiler BE (1996) Divergent and Parallel Development in Volume Sizes of Telencephalic Song Nuclei in and Female Zebra Finches. J Comp Neurol 375 (3):445–456. https://doi.org/10.1002/(sici)1096-9861(19961118)375:3%3C445::aid-cne7%3E3.0.co;2-2

Oertelt-Prigione S (2023) The Operationalization of Gender in Medicine. In: Legato MJ (ed) Principles of Gender-Specific Medicine. Academic Press, Cambridge, MA, pp 503–512

Oyama S (2000) The Ontogeny of Information: Developmental Systems and Evolution. Duke Univ Press, Durham

Pape M, Miyagi M, Ritz SA, et al. (2024) Sex Contextualism in Laboratory Research: Enhancing Rigor and Precision in the Study of Sex-Related Variables. Cell 187 (6):1316–1326. https://doi.org/10.1016/j.cell.2024.02.008

Pearlin LI, Schieman S, Fazio EM, and Meersman SC (2005) Stress, Health, and the Life Course: Some Conceptual Perspectives. J Health Soc Behav 46 (2):205–219. https://doi.org/10.1177/002214650504600206

Perović M, Jacobson D, Glazer E, Pukall C, and Einstein G (2021) Are You in Pain If You Say You Are Not? Accounts of Pain in Somali–Canadian Women with Female Genital Cutting. Pain 162 (4):1144–1152. https://doi.org/10.1097/j.pain.0000000000002121

Plant TM (2015) 60 Years of Neuroendocrinology: The Hypothalamo-Pituitary–Gonadal Axis. J Endocrinol 226 (2):T41–T54. https://doi.org/10.1530/joe-15-0113

Pletzer B, Poppelaars ES, Klackl J, and Jonas E (2021) The Gonadal Response to Social Stress and Its Relationship to Cortisol. Stress 24 (6):866–875. https://doi.org/10.1080/10253890.2021.1891220

Pletzer B, Winkler-Crepaz K, and Hillerer KM (2023) Progesterone and Contraceptive Progestin Actions on the Brain: A Systematic Review of Animal Studies and Comparison to Human Neuroimaging Studies. Front Neuroendocrin 69:101060. https://doi.org/10.1016/j.yfrne.2023.101060

Popal H, Wang Y, and Olson IR (2019) A Guide to Representational Similarity Analysis for Social Neuroscience. Soc Cogn Affect Neurosci 14 (11):1243–1253. https://doi.org/10.1093/scan/nsz099

Racine E, Waldman S, Rosenberg J, and Illes J (2010) Contemporary Neuroscience in the Media. Soc Sci Med 71 (4):725–733. https://doi.org/10.1016/j.socscimed.2010.05.017

Rauch JM, and Eliot L (2022) Breaking the Binary: Gender versus Sex Analysis in Human Brain Imaging. Neuroimage 264:119732. https://doi.org/10.1016/j.neuroimage.2022.119732

Reale C, Invernizzi F, Panteghini C, and Garavaglia B (2023) Genetics, Sex, and Gender. J Neurosci Res 101 (5):553–562. https://doi.org/10.1002/jnr.24945

Rechlin RK, Splinter TFL, Hodges TE, Albert AY, and Galea LAM (2022) An Analysis of Neuroscience and Psychiatry Papers Published from 2009 and 2019 Outlines Opportunities for Increasing Discovery of Sex Differences. Nat Commun 13 (1):2137. https://doi.org/10.1038/s41467-022-29903-3

Richardson SS (2022) Sex Contextualism. PTPBio 14, no. 2. https://doi.org/10.3998/ptpbio.2096

Richmond K, Settles IH, and Shields SA (2022) Feminist Scholars on the Road to Tenure: The Personal Is Professional. Cognella Academic Publ, Solana Beach

Rippon G, Eliot L, Genon S, and Joel D (2021) How Hype and Hyperbole Distort the Neuroscience of Sex Differences. PLoS Biol 19 (5):e3001253. https://doi.org/10.1371/journal.pbio.3001253

Rippon G, Jordan-Young R, Kaiser A, and Fine C (2014) Recommendations for Sex/Gender Neuroimaging Research: Key Principles and Implications for Research Design, Analysis, and Interpretation. Front Hum Neurosci 8:650. https://doi.org/10.3389/fnhum.2014.00650

Ritz SA, and Greaves L (2022) Transcending the Male–Female Binary in Biomedical Research: Constellations, Heterogeneity, and Mechanism When Considering Sex and Gender. Int J Environ Res Public Health 19 (7):4083. https://doi.org/10.3390/ijerph19074083

Roca CA, Schmidt PJ, Altemus M, et al. (2003) Differential Menstrual Cycle Regulation of Hypothalamic-Pituitary-Adrenal Axis in Women with Premenstrual Syndrome and Controls. J Clin Endocrinol Metab 88 (7):3057–3063. https://doi.org/10.1210/jc.2002-021570

Roselli CE (2018) Neurobiology of Gender Identity and Sexual Orientation. J Neuroendocrinol 30 (7):e12562. https://doi.org/10.1111/jne.12562

Roy D (2012) Cosmopolitics and the Brain: The Co-Becoming of Practices in Feminism and Neuroscience. In: Bluhm R, Jacobson AJ, Maibom HL (eds) Neurofeminism. Palgrave Macmillan, London, pp 175–192

———— (2018) Molecular Feminisms: Biology, Becomings, and Life in the Lab. Univ Washington Press, Seattle

Rubinow DR, Roca CA, Schmidt PJ, et al. (2005) Testosterone Suppression of CRH-Stimulated Cortisol in Men. Neuropsychopharmacology 30 (10):1906–1912. https://doi.org/10.1038/sj.npp.1300742

Ryali S, Zhang Y, De Los Angeles C, Supekar K, and Menon V (2024) Deep Learning Models Reveal Replicable, Generalizable, and Behaviorally Relevant Sex Differences in Human Functional Brain Organization. PNAS 121 (9):e2310012121. https://doi.org/10.1073/pnas.2310012121

Rydström K (2020) Degendering Menstruation: Making Trans Menstruators Matter. In: Bobel C, Winkler IT, Fahs B, Hasson KA, Kissling EA, Roberts T-A (eds) The Palgrave Handbook of Critical Menstruation Studies. Palgrave Macmillan, London, pp 945–959

Sagoshi S, Maejima S, Morishita M, et al. (2020) Detection and Characterization of Estrogen Receptor Beta Expression in the Brain with Newly Developed Transgenic Mice. Neuroscience 438:182–197. https://doi.org/10.1016/j.neuroscience.2020.04.047

Saguy T, Reifen-Tagar M, and Joel D (2021) The Gender-Binary Cycle: The Perpetual Relations between a Biological-Essentialist View of Gender, Gender Ideology, and Gender-Labelling and Sorting. Philos Trans R Soc Lond B Biol Sci 376 (1822):20200141. https://doi.org/10.1098/rstb.2020.0141

Sanchis-Segura C, Aguirre N, Cruz-Gómez ÁJ, Félix S, and Forn C (2022) Beyond "Sex Prediction": Estimating and Interpreting Multivariate Sex Differences and Similarities in the Brain. Neuroimage 257:119343. https://doi.org/10.1016/j.neuroimage.2022.119343

Sanchis-Segura C, and Wilcox RR (2024) From Means to Meaning in the Study of Sex/Gender Differences and Similarities. Front Neuroendocrin 73:101133. https://doi.org/10.1016/j.yfrne.2024.101133

Sanchis-Segura C, Wilcox RR, Cruz-Gómez AJ, et al. (2023) Univariate and Multivariate Sex Differences and Similarities in Gray Matter Volume within Essential Language-Processing Areas. Biol Sex Differ 14 (1):90. https://doi.org/10.1186/s13293-023-00575-y

Satpute AB, and Lindquist KA (2019) The Default Mode Network's Role in Discrete Emotion. Trends Cogn Sci 23 (10):851–864. https://doi.org/10.1016/j.tics.2019.07.003

Schiebinger LL (1999) Has Feminism Changed Science? vol 0. Harvard Univ Press, Cambridge, MA

——— (2004) Nature's Body: Gender in the Making of Modern Science. Rutgers Univ Press, New Brunswick

Schmalenberger KM, Tauseef HA, Barone JC, et al. (2021) How to Study the Menstrual Cycle: Practical Tools and Recommendations. Psychoneuroendocrinology 123:104895. https://doi.org/10.1016/j.psyneuen.2020.104895

Schvey NA, Puhl RM, and Brownell KD (2014) The Stress of Stigma: Exploring the Effect of Weight Stigma on Cortisol Reactivity. Psychosom Med 76 (2):156–162. https://doi.org/10.1097/psy.0000000000000031

Shang Z, Liu N, Ouyang H, et al. (2024) Sex-Based Differences in Brain Morphometry under Chronic Stress: A Pilot MRI Study. Heliyon 10 (9):e30354. https://doi.org/10.1016/j.heliyon.2024.e30354

Shansky RM, and Woolley CS (2016) Considering Sex as a Biological Variable Will Be Valuable for Neuroscience Research. J Neurosci 36 (47):11817–11822. https://doi.org/10.1523/JNEUROSCI.1390-16.2016

Shattuck-Heidorn H, and Richardson SS (2019) Sex/Gender and the Biosocial Turn. https://sfonline.barnard.edu/sex-gender-and-the-biosocial-turn/

Spets DS, and Slotnick SD (2022) It's Time for Sex in Cognitive Neuroscience. Cogn Neurosci 13 (1):1–9. https://doi.org/10.1080/17588928.2021.1996343

Springer KW, Stellman JM, and Jordan-Young RM (2012) Beyond a Catalogue of Differences: A Theoretical Frame and Good Practice Guidelines for Researching Sex/Gender in Human Health. Soc Sci Med 74 (11):1817–1824. https://doi.org/10.1016/j.socscimed.2011.05.033

Uchiyama R, Spicer R, and Muthukrishna M (2022) Cultural Evolution of Genetic Heritability. Behav Brain Sci 45:e152. https://doi.org/10.1017/s0140525x21000893

Ussher JM (2003) The Ongoing Silencing of Women in Families: An Analysis and Rethinking of Premenstrual Syndrome and Therapy. J Fam Ther 25 (4):388–405. https://doi.org/10.1111/1467-6427.00257

van Anders SM (2024) Gender/sex/ual Diversity and Biobehavioral Research. Psychol Sex Orientat Gend Divers 11 (3):471–487. https://doi.org/10.1037/sgd0000609

van Anders SM, Tolman RM, and Volling BL (2012) Baby Cries and Nurturance Affect Testosterone in Men. Horm Behav 61 (1):31–36. https://doi.org/10.1016/j.yhbeh.2011.09.012

Viglione A, Chiarotti F, Poggini S, Giuliani A, and Branchi I (2019) Predicting Antidepressant Treatment Outcome Based on Socioeconomic Status and Citalopram Dose. Pharmacogenomics J 19 (6):538–546. https://doi.org/10.1038/s41397-019-0080-6

Walsh R, and Einstein G (2020) Transgender Embodiment: A Feminist, Situated Neuroscience Perspective. INSEP–J Int Netw Sex Ethics Polit 8 (SI):9–10. https://doi.org/10.3224/insep.si2020.04

Weng HY, Ikeda MP, Lewis-Peacock JA, et al. (2020) Toward a Compassionate Intersectional Neuroscience: Increasing Diversity and Equity in Contemplative Neuroscience. Front Psychol 11:573134. https://doi.org/10.3389/fpsyg.2020.573134

White J, Tannenbaum C, Klinge I, Schiebinger L, and Clayton JA (2021) The Integration of Sex and Gender Considerations into Biomedical Research: Lessons from International Funding Agencies. J Clin Endocrinol Metab 106 (10):3034–3048. https://doi.org/10.1210/clinem/dgab434

Wierenga LM, Ruigrok A, Aksnes ER, et al. (2024) Recommendations for a Better Understanding of Sex and Gender in Neuroscience of Mental Health. Biol Psychiatry Glob Open Sci 4 (2):100283. https://doi.org/10.1016/j.bpsgos.2023.100283

Wirth MM (2011) Beyond the HPA Axis: Progesterone-Derived Neuroactive Steroids in Human Stress and Emotion. Front Endocrinol 2:19. https://doi.org/10.3389/fendo.2011.00019

Zipple MN, Vogt CC, and Sheehan MJ (2023) Re-Wilding Model Organisms: Opportunities to Test Causal Mechanisms in Social Determinants of Health and Aging. Neurosci Biobehav Rev 152:105238. https://doi.org/10.1016/j.neubiorev.2023.105238

7

Intersectionality, Sex/Gender Entanglement, and Research Design

Greta Bauer

Abstract   Intersectionality is a theoretical framework emerging from US Black, Chicana, and Indigenous feminisms that considers the interlocking nature of processes of oppression across sex/gender, race/ethnicity, sexual orientation, social class, and other social positions. While not originating in applications to research methodology, its core ideas are apparent in Black feminist sociological, legal, and other forms of scholarship even prior to the emergence of the word intersectionality nearly 35 years ago. Since that time, social scientists have provided thinking tools with which to better incorporate intersectional thinking. These include Patricia Hill Collins's ideas of relationality through the addition of categories of difference, such as adding race and ethnicity to sex and gender, and Leslie McCall's differentiation between anti-, intra-, and intercategorical approaches to intersectional complexity. Given intersectionality's core foci on social power, inequity, and social context, what happens when we add intersectionality into the field of sex/gender complexity and entanglement is explored, along with its meaning for research methods. In doing so, a new concept of intersectional entanglement is developed and explored, rooted in Collins's ideas of relationality and embodied through a wide range of biopsychosocial processes. Some research design considerations and questions for sex and gender scholars in the context of intersectional entanglement are outlined.

Keywords   Sex, gender, research methods, theory, entanglement, intersectionality

Greta Bauer (✉)
Eli Coleman Institute for Sexual and Gender Health, University of Minnesota Medical School, Minneapolis, MN 55455, USA
Email: gbauer@umn.edu

© Frankfurt Institute for Advanced Studies (FIAS) 2025
L. Z. DuBois et al. (eds.), *Sex and Gender*, Strüngmann Forum Reports,
https://doi.org/10.1007/978-3-031-91371-6_7

1 A Brief Introduction to Intersectionality

Intersectionality—a Black feminist theoretical framework with roots in Chicana and Native American feminism—did not come out of an academic setting and was not designed for research methods applications (Collins and Bilge 2020). Its roots are in community organizing, equity, and civil rights, such as in the 1970s work of the Combahee River Collective, which explicitly approached the entwinement of oppression related to sex, gender, race, ethnicity, sexual orientation, and social class in their foundational statement (Combahee River Collective 1977). In the 1980s and 1990s, intersectionality emerged into academic work primarily through the germinal works of two scholars: Kimberlé Crenshaw (1989, 1991) in legal studies, who gave it the name intersectionality, and Patricia Hill Collins (1990) in sociology, who theorized the intersectional *matrix of domination*. Other Black feminist scholars used language of *multiple jeopardy* (e.g., King 1988). Regardless of conceptual language used, these works shared an understanding of what Collins calls relationality. One of intersectionality's core ideas, relationality is the complexity through which oppression is structured across social groups such that they can be understood to function differently when joined together, and the ways in which their confluences can co-form something new and potentially indivisible that is different than the sum of its parts (Collins 2019). This formulation is seen as a core understanding of intersectionality in later scholarly work that includes biological sex and/or social gender, as in Lisa Bowleg's paper's title "When Black + lesbian + woman ≠ Black lesbian woman" (Bowleg 2008).

From a research perspective, intersectionality thus tells us that we cannot understand human experience or human health at a particular social intersection as the sum of its parts. While this has obvious implications for some quantitative methods that literally sum average experiences across single categories to describe experiences at an intersection, it has profound implications far beyond this (Bauer 2014). It raises questions on how social power, resources, and decision-making power are structured in research teams; who decides which questions are important and how they will be framed; how research participants are identified and brought into a study; and how results are produced, interpreted, and shared, including with whom and in what form (Agénor 2020; Bauer 2014; Bowleg 2012).

As a traveling theory, intersectionality has moved through communities, across geographies beyond the United States, and through academic disciplines (Cho et al. 2013). While rooted in the particularities of American sociopolitical experience and community advocacy, intersectionality's core ideas of nonadditivity of the effects of social power and marginalization have created a lens for examining assumptions inherent in academic work and community practice across fields and countries. At the same time, understandings of intersectionality itself have evolved differently in different communities and disciplines, and concerns have been raised about what it means for intersectionality to travel into academic research—in particular into quantitative research (Bauer 2014; Bowleg 2008; Cho et al. 2013). Fields that have a history of shallow engagement with rich theoretical constructs, such as epidemiology and public health (Krieger and Zierler 1997), risk oversimplifying or even shunting aside intersectionality's core ideas, including its foundational focus on social power and inequity.

2 Adding in Intersectional Approaches to Sex and Gender Research

Much of the academic work in sex, gender, and research methods has not explicitly engaged with intersectionality theoretical frameworks. While intersectionality does not emerge from sex or gender research, it is intricately linked. In a systematic review of 681 academic papers, in which the authors took an explicitly intersectional approach to an original quantitative analysis, 76.7% included at least one dimension of sex or gender in that analysis (Bauer et al. 2021). Nearly as many included at least one dimension of race or ethnicity, reflecting intersectionality's embeddedness in Black feminism.

What would it mean then for sex and gender researchers to more explicitly incorporate intersectionality into this body of work? Does it even make sense to "add in" intersectionality to a field, as if applying a corrective to white feminist work or to nonfeminist sex differences research? This approach is evident in some policy settings. For example, in Canada, where the government implemented a gender-based analysis (GBA) approach to policy following the Beijing Women's Conference in 1995, intersectionality was explicitly added to this approach in 2011 by Status of Women Canada. The Canadian government agency then adopted the acronym GBA+, with the "plus" specifying that an intersectional approach and understanding was required for future GBA analyses of policy impacts. This type of language may suggest, however, that intersectionality is just an optional add-on to sex and gender work, or something that somehow emerges from it. "Adding in" intersectionality must mean integrating it at a deep level, rather than a superficial add-on component that leaves the core of the field unchallenged. As we will see, a deeper engagement using the idea of entanglement may help us better understand how social power structures and processes entwine biological sex and social gender with other social positions or groups in interesting and important ways.

Collins writes that introducing intersectionality into established fields has the potential to generate debate on accepted frameworks, potentially producing paradigm shifts (Collins 2019). Thus, there is the potential for transformation of sex and gender research through intersectional approaches that ground us in understandings of social power while introducing complexity into "master categories" such as sex/gender. Collins argues for a desegregation of knowledges that unites the humanities and sciences, and unites the study of sexism with that of colonialism, racism, and other similar systems of power that are siloed in ways that limit our thinking.

3 Sex/Gender Multidimensionality and Entanglement

Sex and gender have been brought more explicitly into health research methodology through both academic arguments on methodology (Krieger 2003) and administrative requirements tied to funding mechanisms. In Canada, for example, all proposals to the Canadian Institutes for Health Research (CIHR) are evaluated based on required statements on how the research addresses biological sex and (separately) social gender (CIHR 2018). Multidimensionality of sex and gender goes well beyond a split between the biological and social to incorporate many primary, secondary, or pregnancy-related characteristics of biological sex as well as identity, experience, or social status-related dimensions of social gender, and sexual and gender minority categorizations (Bauer 2023). In addition, research studies often employ undifferentiated sex/gender measures without a clear dimension specified, such as in administrative databases (Bauer 2023).

Borrowing the idea of entanglement from other fields, Springer et al. (2012) describe how sex and gender are interwoven in ways that often do not allow their effects to be disaggregated. Thus, sex and gender are not "pristine categories" with clearly divisible biologic and social effects. As a category, sex is not itself a biological mechanism (Springer et al. 2012), and sex hormones, often assumed to be the dimension of sex that plays a causal role, are also impacted by gender-based social context (Hyde et al. 2019) and structural factors. Recent scholarship has expanded on this idea of gender/sex as an entangled phenomenon (DuBois and Shattuck-Heidorn 2021; van Anders 2024).

Thus, we need to understand that research focused on sex differences or current efforts to promote sex as a biological variable (NIH 2015) will always capture entangled effects of gender. Table 7.1 lists some research-relevant dimensions of biological sex, social gender, undifferentiated sex/gender data, and gender minority cross-classifications. From the dimensions of sex included, it is apparent that most dimensions of biological sex may be influenced by gender but also share connections with other dimensions of sex. For example, a pregnancy constitutes a sexed condition requiring a uterus, with creation of a new sexed organ (placenta, a combination of maternal and fetal tissues), and producing potential chromosomal sex changes through microchimerism. The likelihood of a pregnancy's commencement and continuation depend on a host of social factors that shape partnerships, contraception, abortion, nutrition, infection, and immune or inflammatory processes. How someone is treated while pregnant is also highly gendered and may create friction with or contradict one's own gender identity. Thus, the "biological" sex effects of a

Table 7.1 Sex and gender multidimensionality at the individual level: A conceptual tool for health researchers, after Bauer (2023).

Dimensions of Biological Sex		
Chromosomal sex	Reproductive sex	Intersex status
Sex assigned at birth	Organ-specific status	Pregnancy
Hormonal milieu	Sexed physiology	
Dimensions of Social Gender		
Gender identity	Gender role	Internalized gender stigma
Intersex identity	Metaperceived gender	Gender ideology
Lived gender	Masculinity and/or femininity	Enacted gender stigma/ discrimination
Undifferentiated Sex/Gender		
Sex/gender in administrative databases	Computer (AI)-classified sex/gender	Researcher-perceived sex/ gender
Undifferentiated survey item sex/gender		
Gender Minority Cross-Classifications		
Gender identity ≠ sex assigned at birth	Lived gender ≠ sex assigned at birth	

pregnancy can be impacted by and impact multiple dimensions of sex. Moreover, given the causal interactions with multiple dimensions of social gender and gender minority status, they can never be understood as purely biological, but will always be biopsychosocial.

4 Intersectional Entanglement

Entanglement is not unique to sex and gender, and it is not limited to sex and gender. In studying racial inequity, it may be impossible to disentangle racism from colonialism, islamophobia, antisemitism, colorism, xenophobia, ethnic biases, and discrimination related to language, accent, or dialect. Socioeconomic status is the entanglement of education, income, wealth, and other resources.

Sex and gender are entangled with all of these. We know that experiences of racism are gendered (or experiences of sexism are racialized) in ways that produce characteristically different types of experiences at different race/gender intersectional locations (Brown et al. 2017; Keum et al. 2018; Lewis and Neville 2015; Liu et al. 2018). We know that adjusting for bias in how we define "sex" has affected findings in sexual and reproductive cancer research in ways that also substantially affect magnitude of racial disparities and age trends (Bauer 2023; Beavis et al. 2017; Hammer et al. 2017). We know that established "sex differences," such as mathematics performance, vary across age, ethnicity, and country (Hyde et al. 1990). In fact, we will find that in practice we cannot disentangle sex and gender from race, ethnicity, social class, sexual orientation, and disability, amongst other social groupings.

To extend the pregnancy example just mentioned, many of the social factors that impact pregnancy can be understood not only as gender-related factors but as being related to other social categories of difference and corresponding social processes (DuBois et al. 2024; Thayer and Kuzawa 2015). The processes that lead to conception and continuation of a pregnancy are dependent on culture, religion, sexual orientation, illness, violence, and access to relevant health services, which are not solely functions of sex and gender. Thus, any sex and gender effects that are estimated in data analysis will be patterned across other social factors that impact equity of health services, sexual partnership patterns, and meanings attached to conception, pregnancy, and family status.

This intersectional entanglement cannot be divorced from concerns with social power. The social patterning of health and other resources across sex, gender, and other categories is shaped by social power. As one of its core ideas, social power is the *sine qua non* of intersectionality (Bowleg and Bauer 2016). Intersectionality explicitly concerns the ways in which social power shapes and constrains experiences across multiple axes of oppression in ways that are specific to intersectional social locations—the ways that race/ethnicity, sex/gender, social class, and sexual orientation are understood and treated in social context (Collins 1990; Davis 1983). This structural patterning of oppression is then reflected in health and social inequalities.

We may ask whether the complexity of studying sex and gender, given their entanglement (Springer et al. 2012), is already uncapturable, and whether it is truly necessary to then also consider sex and gender in the context of race, ethnicity, age, sexual orientation, class, and other categories of difference across which social power is structured. Acknowledging this fundamental structuring of power, and the ways it becomes embodied in individuals whose lives are shaped by these systems of power, produces a situation of fact. Just as there is no "pure" effect of gender or sex (Springer et al. 2012), so too do sex and gender always contain aspects of ethnicity, race, religion, age, sexual orientation, ability, and social class. If ignored, we collapse much meaningful heterogeneity into our understanding of sex and gender. Importantly, we also miss critical opportunities that could lead to a more accurate understanding of their impacts.

Intersectional entanglement cannot be escaped, but it does provide an opportunity to better understand health and to understand the processes of what Collins (2019) calls relationality, processes through which dimensions of difference and social power interact. Collins outlines three approaches, all of which are highly useful for researchers: addition, articulation, and co-formation (Collins 2019). Her concept of relational addition provides a tool for researchers already focused in a specific area, in that she asks: What happens when you add another axis? For example: What happens when you add race, ethnicity, language, and culture to sex and gender? What happens when you add social class, education, income, and resources to sex and gender? What happens when you add ability and disability, sexual orientation, family status, or stigmatized conditions/identities such as addiction, mental health diagnoses, or HIV status?

Relationality also includes co-formation—the process through which two or more aspects of a self are co-formed into a whole that can no longer be disaggregated (Collins 2019). Relationality is key to understanding intersectional entanglement and how it becomes a potentially indivisible part of oneself.

5 Embodiment of Intersectional Entanglement

Springer et al. (2012) hold that sex is not a biological process. Neither is race nor social class. Most dimensions of sex, however, can play specific roles in biological processes, and differences that are not biological at birth may be produced biologically over the life course through gendered, raced, and/or classed processes. As an example, let us consider how early childhood experiences of violence or involuntary bodily alterations become embodied. These processes may directly alter the body, for example in female genital cutting (FGC), but can also have other physiologic effects. Einstein (2008) considers the effects of FGC—which includes female circumcision practices, surgeries on intersex infants, and other vulvar surgeries that may similarly cut nerves and muscle—on not only the genitals but also the central nervous system. FGC affects the brain in ways that are related to genital trauma.

Heim et al. (2013) similarly found central nervous system differences in those who experienced violence as children, including differences between those who experienced physical violence and sexual violence. While childhood physical and sexual assault are not unique to any gender, they are patterned across society in ways that reflect power and vulnerability, for example across disability, family status, and gender. Collins (2019) theorizes that violence itself may be considered a saturated site for intersectionality, one in which interwoven social power relations are so entwined that studying this area may allow us to reflect a picture back onto a society that exposes new information about power relations in that society generally. She writes that patterns of violence encode rich information about power structures, as "people don't go willingly to their assigned places" (Collins 2019). Thus, even if abuse produced the same results across sexes/genders, we would still see population-level patterning, wherein those more likely to experience these forms of violence would as a group have different patterns of brain function. This same type of effect can be expected for less extreme experiences than violence or FGC, provided that they impact the developing central nervous system and are patterned across social groups.

These examples and the earlier example of pregnancy demonstrate some of the ways in which gendered, racialized, classed, and other types of experiences may become embodied. How does social power translate into biological differences in health and the experiences related to them? There are at least five different causal pathways between social power, oppression, and privilege and biological differences in health:

1. Direct bodily changes (e.g., surgeries, violence, pregnancies, wear and tear from repeated motion)
2. Neurological changes (e.g., CNS differences)
3. Physiological stress responses and their effects
4. Epigenetics and other pathways affecting gene expression
5. Changes in behaviors, thinking, apprehension, or imagination related to navigating social marginalization

What then does an embodied intersectional entanglement mean for researchers? The first implication is that we must accept that bodies, experiences, and health are formed through sexed biology and gendered experiences that are also racialized, classed, and otherwise affected by social power structures that both delimit and constrain power and action to shape meaning. We must accept intersectional entanglement as a fundamental underlying reality. Intersectional entanglement's embodiment through a range of different processes means that understanding bodies as sexed, and people as gendered, cannot be understood independent of other dimensions of difference. As Collins (2019) noted, this type of paradigm shift can result when we add intersectionality into an established field such as sex and gender research.

Intersectional entanglement also has implications for studies that are not explicitly intersectional because gender or sex effects may vary across other groups. All quantitative gender effects or sex effects can thus be understood as average effects within a study sample or the population to which it is weighted, and thus all qualitative findings as specific to the intersectional positions of the group of persons under study. This specificity of quantitative results highlights the need to characterize study samples not only by single sociodemographics but with regard to intersectional social locations and to interpret findings accordingly.

6 Implications for Study Design

In discussing intersectionality's trajectories and emergence as an academic field, Cho et al. (2013) argue that intersectionality may be best understood as an "analytic sensibility." As such, they state that what makes an analysis intersectional is not the terminology used, but the way one thinks about sameness and difference in relation to social power. They argue for a dynamic understanding of relational processes and a framing of social categories that is not static but instead wherein categories are "always permeated by other categories, fluid and changing, always in the process of creating and being created by dynamics of power" (Cho et al. 2013:795). This framing of intersectionality as an analytic sensibility makes clear that intersectional entanglement may represent underlying truth in nature, but can also serve as a rich field for discovery regarding how these processes interact.

Intersectionality is inherent in the approach and framing of research, rather than in the specific study design or research methods used. McCall (2005) provides three approaches to incorporating intersectional complexity in research: The anticategorical approach pushes at the boundaries of established categories through questioning and deconstruction. The intracategorical approach focuses on specificity of experience within social intersections. The intercategorical approach focuses on comparison across intersections.

These are most often considered under what Choo and Ferree (2010) call a group-centered approach: one that starts with social position or status groups as an organization for the study of social power, privilege, and marginalization. While this approach may be common, Choo and Ferree (2010) distinguish it from two other approaches that may serve to better keep a focus on social power: a process-focused approach that centers processes of oppression and an institutional approach that focuses on structural factors.

While McCall's approaches may also be used for processes of oppression and privilege, they do not often include an institutional or structural approach; structural approaches may be added as a fourth approach to intersectional complexity. Table 7.2 provides a typography of approaches in intersectional complexity, along with a set of questions for researchers to ask their teams.

The framing of *anticategorical* processes—"conceiving of categories not as distinct but as always permeated by other categories, fluid and changing, always in the process of creating and being created by dynamics of power"—emphasizes what intersectionality does rather than what intersectionality is (Cho et al. 2013:795). This approach might call into question the way that categories of sex (and its dimensions) and gender (and its dimensions) are conceptualized, highlighting the ways that their boundaries are defined by racialized (and racist) processes, class dynamics, and gendered conceptualizations of sexual orientation. Anticategorical studies are most often qualitative, though quantitative studies can strategically use categorization to study that categorization (e.g., Bauer and Jairam 2008; Bauer and Brennan 2013; Bauer et al. 2017, 2020).

Intersection-specific experiences are those that may be unique, or have unique character, for those whose lives are lived out at a particular social intersection. This relates to McCall's conceptualization of *intracategorical* complexity as an approach (McCall 2005) and may involve qualitative studies that explore the depth of

Table 7.2 Questions for sex and gender researchers to ask in incorporating intersectionality. For further discussion of anti-, intra-, and intercategorical approaches, see McCall (2005).

Approach to Intersectional Complexity	Focus	Questions for Researchers
Anticategorical	Deconstruction of categories	In what ways are the dimensions of sex/gender we are studying shaped by race/ethnicity, religion, social class, sexual orientation, or ability? Vice versa? What influences the borders of these classifications?
Intracategorical	Specificity of experience within a particular intersectional position	What happens if we add in race to our consideration of sex/gender? Social class? Sexual orientation? Are there intersection-specific constructs that need to be measured?
Intercategorical	Comparison across intersectional groups	Given intersectional entanglement, which aspects of sex/gender do we want to make visible versus leaving collapsed within a group? Do we have adequate sampling to produce meaningfully precise measures for those at each intersection?
Structural	Systems of oppression, including within and between group differences	How do structural contexts shape the opportunities and experiences for those who live within them? Can structural approaches to power and intersectionality be incorporated? Can this be combined in a multilevel approach with individual data? Can we measure how some groups may be disadvantaged but others potentially advantaged by structural power?

experience for those at an intersection (e.g., young Black gay men), or quantitative subgroup analyses, or community studies.

Research shows that conceptualizations of similar concepts may vary dramatically across intersections. For example, the mixed-methods study that developed and validated the Gender Racial Microaggressions Scale for Asian American Women (GRMSAAW) included four subscales: (a) ascription of submissiveness, (b) assumption of universal appearance, (c) Asian fetishism, and (d) media invalidation (Keum et al. 2018). Together, they capture the range of gendered racial microaggressions experienced by Asian American women. The specificity of this measure can be clearly seen in sharp contrast if we consider another earlier intersection-specific measure: the Gendered Racial Microaggressions Scale for Black women (GRMS), also developed using a mixed-methods approach. Like the GRMSAAW, GRMS sought to measure intersection-specific racial microaggressions in a group of American women and produced four subscales: (a) assumptions of beauty and sexual objectification, (b) silenced and marginalized, (c) strong Black woman stereotype, and (d) angry Black woman stereotype (Lewis and Neville 2015). Note, however, that there is no overlap in subscales between the GRMS and GRMSAAW. This intracategorical specificity highlights the potential need for intersection-specific measures to capture processes that are not relevant or measurable across other intersectional groups.

Mixed-methods research methods, which sequence or triangulate between multiple qualitative and quantitative data sources (Creswell and Clark 2017), have high potential for advancing knowledge on intersectional entanglement. Intracategorical approaches can pair extremely well with mixed-methods research. Mixed methods are often used in developing intracategorical measures, as previously described, wherein an initial qualitative phase is used to generate description of experience, identify themes, and generate survey items for a potential new measure which is then validated quantitatively. However, mixed-methods approaches that take a sequential quantitative then qualitative explanatory approach also have good potential for intracategorical study, wherein findings from an intersection-specific quantitative analysis can be explained and illustrated through follow-up qualitative interviews or focus groups.

Agénor (2020) notes that intersectionality works well with community-based participatory research (CBPR) approaches, and this may be particularly true for intracategorical approaches to specificity of experience. In particular, where a social intersection of interest matches up with individual and community identity, CBPR provides an opportunity to work within a community to make their experiences visible in research-informed settings.

Intercategorical descriptive approaches focus across intersectional categories, measuring or describing difference (McCall 2005). The focus here is on comparison. This approach can be seen as advancing health disparities research, and there is a range of statistical methods that can accommodate comparisons across larger numbers of groups (Mahendran et al. 2022a, b). While less often seen in quantitative intersectional research to date (Bauer et al. 2021), intercategorical analytic studies also provide opportunities to understand intersection-specific causal pathways (Bauer and Scheim 2019; Bright et al. 2016).

In practice, it is not always possible to distinguish clearly between anticategorical, intracategorical, and intercategorical approaches (Bauer et al. 2021; Guan et al. 2021). In contrast, structural approaches are more readily identifiable as they draw on group-level data such as policies, institutional practices, and population economic measures. As research and methods on structural racism (Adkins-Jackson et al. 2021; Bailey et al. 2021) and structural sexism (Homan 2019) advance, work explicitly addressing structural intersectionality is beginning to take place (Homan et al. 2021). Structural intersectionality approaches highlight social inequalities within large-scale institutions and facilitate the understanding of complex systems of social marginalization (Homan et al. 2021).

7 An Intersectional Future for Sex and Gender Research?

Intersectionality scholars have maintained a focus on sex and gender, though not necessarily coming from the disciplinary traditions of sex difference research, feminist science, or women's health. By adding intersectionality into the established fields of sex and gender research, we can gain an understanding of intersectional entanglement that expands current ideas of sex and gender entanglement to better represent the biopsychosocial nature of the ways that social power becomes embodied over the life course. This improves interpretation of average effects of sex and/or gender and their dimensions as specific to the characteristics and social power structures of populations under study. It also provides opportunities for explicit intersectional study design to generate new—potentially paradigm-shifting—perspectives on intersectional entanglement, human experience, and health equity.

References

Adkins-Jackson PB, Chantarat T, Bailey ZD, and Ponce NA (2021) Measuring Structural Racism: A Guide for Epidemiologists and Other Health Researchers. Am J Epidemiol 191 (4):539–547. https://doi.org/10.1093/aje/kwab239

Agénor M (2020) Future Directions for Incorporating Intersectionality into Quantitative Population Health Research. Am J Public Health 110 (6):803–806. https://doi.org/10.2105/AJPH.2020.305610

Bailey ZD, Feldman JM, and Bassett MT (2021) How Structural Racism Works: Racist Policies as a Root Cause of U.S. Racial Health Inequities. N Engl J Med 384 (8):768–773. https://doi.org/10.1056/NEJMms2025396

Bauer G, and Jairam J (2008) Are Lesbians Really Women Who Have Sex with Women (WSW)? Methodological Concerns in Measuring Sexual Orientation in Health Research. Women Health 48 (4):383–408. https://doi.org/10.1080/03630240802575120

Bauer GR (2014) Incorporating Intersectionality Theory into Population Health Research Methodology: Challenges and the Potential to Advance Health Equity. Soc Sci Med 110:10–17. https://doi.org/10.1016/j.socscimed.2014.03.022

——— (2023) Sex and Gender Multidimensionality in Epidemiologic Research. Am J Epidemiol 192 (1):122–132. https://doi.org/10.1093/aje/kwac173

Bauer GR, Braimoh J, Scheim AI, and Dharma C (2017) Transgender-Inclusive Measures of Sex/Gender for Population Surveys: Mixed-Methods Evaluation and Recommendations. PLoS One 12 (5):e0178043. https://doi.org/10.1371/journal.pone.0178043

Bauer GR, and Brennan DJ (2013) The Problem with "Behavioral Bisexuality": Assessing Sexual Orientation in Survey Research. J Bisex 13 (2):148–165. https://doi.org/10.1080/15299716.2013.782260

Bauer GR, Churchill SM, Mahendran M, et al. (2021) Intersectionality in Quantitative Research: A Systematic Review of Its Emergence and Applications of Theory and Methods. SSM Popul Health 14:100798. https://doi.org/10.1016/j.ssmph.2021.100798

Bauer GR, Mahendran M, Braimoh J, Alam S, and Churchill SM (2020) Identifying Visible Minorities or Racialized Persons on Surveys: Can We Just Ask? Can J Public Health 111 (3):371–382. https://doi.org/10.17269/s41997-020-00325-2

Bauer GR, and Scheim AI (2019) Methods for Analytic Intercategorical Intersectionality in Quantitative Research: Discrimination as a Mediator of Health Inequalities. Soc Sci Med 226:236–245. https://doi.org/10.1016/j.socscimed.2018.12.015

Beavis AL, Gravitt PE, and Rositch AF (2017) Hysterectomy-Corrected Cervical Cancer Mortality Rates Reveal a Larger Racial Disparity in the United States: Corrected Cervix Cancer Mortality Rates. Cancer 123 (6):1044–1050. https://doi.org/10.1002/cncr.30507

Bowleg L (2008) When Black + Lesbian + Woman ≠ Black Lesbian Woman: The Methodological Challenges of Qualitative and Quantitative Intersectionality Research. Sex Roles 59 (5):312–325. https://doi.org/10.1007/s11199-008-9400-z

——— (2012) The Problem with the Phrase "Women and Minorities": Intersectionality, an Important Theoretical Framework for Public Health. Am J Public Health 102 (7):1267–1273. https://doi.org/10.2105/AJPH.2012.300750

Bowleg L, and Bauer GR (2016) Invited Reflection: Quantifying Intersectionality. Psychol Women Q 40 (3):337–341. https://doi.org/10.1177/0361684316654282

Bright LK, Malinsky D, and Thompson M (2016) Causally Interpreting Intersectionality Theory. Philos Sci 83 (1):60–81. https://doi.org/10.1086/684173

Brown DL, Blackmon S, Rosnick CB, Griffin-Fennell FD, and White-Johnson RL (2017) Initial Development of a Gendered-Racial Socialization Scale for African American College Women. Sex Roles 77 (3):178–193. https://doi.org/10.1007/s11199-016-0707-x

Cho S, Crenshaw KW, and McCall L (2013) Toward a Field of Intersectionality Studies: Theory, Applications, and Praxis. Signs 38 (4):785–810. https://doi.org/10.1086/669608

Choo HY, and Ferree MM (2010) Practicing Intersectionality in Sociological Research: A Critical Analysis of Inclusions, Interactions, and Institutions in the Study of Inequalities. Sociol Theory 28 (2):129–149. https://doi.org/10.1111/j.1467-9558.2010.01370.x

CIHR (2018) Sex, Gender and Health Research. https://cihr-irsc.gc.ca/e/50833.html

Collins PH (1990) Black Feminist Thought: Knowledge Consciousness, and the Politics of Empowerment. Routledge, New York

———— (2019) Relationality within Intersectionality. In: Intersectionality as Critical Social Theory. Duke Univ Press, Durham, pp 225–252

Collins PH, and Bilge S (2020) Getting the History of Intersectionality Straight? In: Intersectionality. Key Concepts, 2nd edn. Polity Press, Cambridge, pp 72–100

Combahee River Collective (1977) The Combahee River Collective Statement. In: Smith B (ed) Home Girls: A Black Feminist Anthology. Kitchen Table: Women of Color Press, New York, pp 272–282

Crenshaw KW (1989) Demarginalizing the Intersection of Race and Sex: A Black Feminist Critique of Antidiscrimination Doctrine, Feminist Theory and Antiracist Politics. Univ Chic Leg Forum 1989 (1):8. https://chicagounbound.uchicago.edu/uclf/vol1989/iss1/8

———— (1991) Mapping the Margins: Intersectionality, Identity Politics, and Violence against Women of Color. Stanford Law Rev 43:61. https://www.jstor.org/stable/1229039

Creswell JW, and Clark VLP (2017) Designing and Conducting Mixed Methods Research. Sage, Thousand Oaks

Davis AY (1983) Women, Race & Class. First Edition edn. Vintage, New York

DuBois LZ, Puckett JA, Jolly D, et al. (2024) Gender Minority Stress and Diurnal Cortisol Profiles among Transgender and Gender Diverse People in the United States. Horm Behav 159:105473. https://doi.org/10.1016/j.yhbeh.2023.105473

DuBois LZ, and Shattuck-Heidorn H (2021) Challenging the Binary: Gender/Sex and the Bio-Logics of Normalcy. Am J Hum Biol 33 (5):e23623. https://doi.org/10.1002/ajhb.23623

Einstein G (2008) From Body to Brain: Considering the Neurobiological Effects of Female Genital Cutting. Perspect Biol Med 51 (1):84–97. https://doi.org/10.1353/pbm.2008.0012

Guan A, Thomas M, Vittinghoff E, et al. (2021) An Investigation of Quantitative Methods for Assessing Intersectionality in Health Research: A Systematic Review. SSM Popul Health 16:100977. https://doi.org/10.1016/j.ssmph.2021.100977

Hammer A, Kahlert J, Rositch AF, et al. (2017) The Temporal and Age-Dependent Patterns of Hysterectomy-Corrected Cervical Cancer Incidence Rates in Denmark: A Population-Based Cohort Study. Acta Obstet Gynecol Scand 96 (2):150–157. https://doi.org/10.1111/aogs.13057

Heim CM, Mayberg HS, Mletzko T, Nemeroff CB, and Pruessner JC (2013) Decreased Cortical Representation of Genital Somatosensory Field after Childhood Sexual Abuse. Am J Psychiatry 170 (6):616–623. https://doi.org/10.1176/appi.ajp.2013.12070950

Homan P (2019) Structural Sexism and Health in the United States: A New Perspective on Health Inequality and the Gender System. Am Sociol Rev 84 (3):486–516. https://doi.org/10.1177/0003122419848723

Homan P, Brown TH, and King B (2021) Structural Intersectionality as a New Direction for Health Disparities Research. J Health Soc Behav 62 (3):350–370. https://doi.org/10.1177/00221465211032947

Hyde JS, Bigler RS, Joel D, Tate CC, and van Anders SM (2019) The Future of Sex and Gender in Psychology: Five Challenges to the Gender Binary. Am Psychol 74 (2):171. https://www.researchgate.net/publication/326500725_The_Future_of_Sex_and_Gender_in_Psychology_Five_Challenges_to_the_Gender_Binary

Hyde JS, Fennema E, and Lamon SJ (1990) Gender Differences in Mathematics Performance: A Meta-Analysis. Psychol Bull 107 (2):139–155. https://www.researchgate.net/publication/21016828_Gender_Differences_in_Mathematics_Performance_A_Meta-Analysis

Keum BT, Brady JL, Sharma R, et al. (2018) Gendered Racial Microaggressions Scale for Asian American Women: Development and Initial Validation. J Couns Psychol 65 (5):571–585. https://doi.org/10.1037/cou0000305

King DK (1988) Multiple Jeopardy, Multiple Consciousness: The Context of a Black Feminist Ideology. Signs 14 (1):42–72

Krieger N (2003) Genders, Sexes, and Health: What Are the Connections—and Why Does It Matter? Int J Epidemiol 32 (4):652–657. https://doi.org/10.1093/ije/dyg156

Krieger N, and Zierler S (1997) The Need for Epidemiologic Theory. Epidemiology 8 (2):212–214. https://pubmed.ncbi.nlm.nih.gov/9229218/

Lewis JA, and Neville HA (2015) Construction and Initial Validation of the Gendered Racial Microaggressions Scale for Black Women. J Couns Psychol 62 (2):289–302. https://doi.org/10.1037/cou0000062

Liu T, Wong YJ, Maffini CS, Mitts NG, and Iwamoto DK (2018) Gendered Racism Scales for Asian American Men: Scale Development and Psychometric Properties. J Couns Psychol 65 (5):556–570. https://doi.org/10.1037/cou0000298

Mahendran M, Lizotte D, and Bauer GR (2022a) Describing Intersectional Health Outcomes: An Evaluation of Data Analysis Methods. Epidemiology 33 (3):395–405. https://doi.org/10.1097/EDE.0000000000001466

——— (2022b) Quantitative Methods for Descriptive Intersectional Analysis with Binary Health Outcomes. SSM Popul Health 17:101032. https://doi.org/10.1016/j.ssmph.2022.101032

McCall L (2005) The Complexity of Intersectionality. Signs 30 (3):1771–1800. https://doi.org/10.1086/426800

NIH (2015) NOT-OD-15-102: Consideration of Sex as a Biological Variable in NIH-Funded Research. https://grants.nih.gov/grants/guide/notice-files/not-od-15-102.html

Springer KW, Stellman JM, and Jordan-Young RM (2012) Beyond a Catalogue of Differences: A Theoretical Frame and Good Practice Guidelines for Researching Sex/Gender in Human Health. Soc Sci Med 74 (11):1817–1824. https://doi.org/10.1016/j.socscimed.2011.05.033

Thayer ZM, and Kuzawa CW (2015) Ethnic Discrimination Predicts Poor Self-Rated Health and Cortisol in Pregnancy: Insights from New Zealand. Soc Sci Med 128:36–42. https://doi.org/10.1016/j.socscimed.2015.01.003

van Anders SM (2024) Gender/sex/ual Diversity and Biobehavioral Research. Psychol Sex Orientat Gend Divers 11 (3):471–487. https://doi.org/10.1037/sgd0000609

8

Gender/Sex Dynamics in Human Biomedical and Clinical Research

Robert-Paul Juster, Lisa Bowleg, Lu Ciccia, Joshua B. Rubin,
Carla Sanchis-Segura, Susann Schweiger, Eric Vilain, and Tonia Poteat

Abstract This chapter explores the intertwined associations between sex, gender, and intersectionality, and their roles in biomedical and clinical research. Stressing the importance of clear definitions, interdisciplinary collaboration, practical applications, and inclusivity, this chapter discusses the advantages of individual phenotyping and personalized medicine while recognizing the complexities these concepts present in clinical settings. The challenges of defining and measuring gender and sex are addressed as is the need to be inclusive of a diversity of identities and disciplinary perspectives, the limitations of rigid categories, as well as the importance of nuanced frameworks, clear communication, and fully informed patient choices. Scientists, policy makers, and stakeholders share responsibility for incorporating sex and gender considerations into research and health care. This calls for applying an intersectional approach and recognizing that various perspectives on sex and gender need not be in opposition but rather complementary. Just as the coexistence of different viewpoints enabled this chapter, the discussions presented allow for a nuanced understanding of the topic well beyond disciplinary silos.

Keywords Gender/sex entanglement; intersectionality; biomedicine, precision medicine, social policy

Robert-Paul Juster (✉)
Dept. of Psychiatry and Addiction, University of Montreal, Montreal, QC H3T 1J4, Canada
Email: robert-paul.juster@umontreal.ca

Affiliations for the coauthors are available in the List of Contributors

© Frankfurt Institute for Advanced Studies (FIAS) 2025
L. Z. DuBois et al. (eds.), *Sex and Gender*, Strüngmann Forum Reports,
https://doi.org/10.1007/978-3-031-91371-6_8

Group photos (top left to bottom right) Robert-Paul Juster, Lisa Bowleg, Lu Ciccia, Josh Rubin, Susann Schweiger, Tonia Poteat, Carla Sanchis-Segura, Eric Vilain, Josh Rubin, Lu Ciccia, Robert-Paul Juster, Lisa Bowleg, Tonia Poteat, Carla Sanchis-Segura, Eric Vilain, Susann Schweiger, Lu Ciccia, Robert-Paul Juster. Photos by Norbert Miguletz.

1 Introduction

This chapter summarizes a series of discussions that took place during the Ernst Strüngmann Forum on Sex and Gender Entanglement. As our working group explored the complexities of sex, gender, and intersectionality in biomedical and clinical research, discussion centered on the difficulties of clear definitions for sex and gender across disciplines and the challenges of interdisciplinary conceptualizations of gender/sex entanglement. We also explored practical applications of research focused specifically on human participants and considered fundamental research using cell or animal models, where gender is often quite difficult to conceptualize in translational terms. Notwithstanding our multidisciplinary backgrounds, we shared some core values embodied within our diverse research areas and a collective commitment to engage in respectful and constructive discussions, despite differences of opinion.

Our diverse research backgrounds extended from biomedical science, feminist studies, HIV research, neuroscience, and oncology to public health. Several scholars are champions that have applied sex and gender concepts to advance research and practice in areas such as cancer treatment, neurological studies, and precision medicine. Respectful of different perspectives emerging from the Global North and Global South (e.g., respectively so-called "developed" vs. "developing" countries; Mareï and Savy 2021), there is a clear need for localization in international research. The exclusion of cross-cultural considerations ultimately limits the representation of diverse populations, many of whom are marginalized and can remain invisible. Indeed, not all communities are equally represented in biomedical research, and forums such as this one provide invaluable opportunities to provide an international exchange that fosters sustained collaboration. Otherwise, such barriers can lead to skewed knowledge production, which will limit the very definitions of key concepts.

Our initial introductions and ensuing conversations revolved around addressing clear research objectives, as defined by the questions listed below, funding priorities, and the application of intersectionality as a cross-cutting principle integral to sex and gender science. This led to a more nuanced discussion on how research and practice consider sex, gender, and intersectional power dynamics in biomedicine. This topic is important because progress is needed to address the limitations of sex-centered policies and contribute to a more comprehensive understanding of diverse lived experiences. There is a danger of only considering sex-centered policies in biomedical research, since this does not adequately address the complexities of gender. Other factors (e.g., weight and height) vary but are not defined by sex, yet they impact the functioning of our bodies. Finally, the lived experience of marginalized groups—race and ethnic minority groups as well as the lesbian, gay, bisexual, transgender, and queer (LGBTQ+) communities—has an impact on health and well-being, yet this is often not considered in sex-focused policies (IOM 2011).

In setting our discussion agenda to address the interplay of sex, gender, gender/sex entanglement, and additional biological factors in clinical and biomedical research, we formulated the following guiding questions:

- Discuss the meaning of sex, gender, and intersectionality in the context of clinical and biological sciences. What are our own personal biases?

- What do the categories of binary sex (male, female) and gender (man, woman, trans, nonbinary) enable us to achieve? What harms and biases can result?
- How can gender and sex be better integrated with biomedical research (e.g., neuropsychiatry, epigenetics, clinical research)?
- How can sex and gender be integrated into social sciences (e.g., public health, sociology, psychology)?
- How can gender experience and intersectionality, involving the dynamism within individuals and across historical and cultural contexts, be better integrated into health research?

Our ensuing discussions were intense and sometimes heated. We persevered, however, in good faith, spurred on by the importance of these issues. We offer this summary to communicate insights that emerged from our discussions and to encourage others to take further steps.

2 The Meaning of Sex, Gender, and Intersectionality in Clinical and Biological Sciences

In the context of clinical biomedical science, our discussion of sex, gender, and intersectionality proved to be a complex conversation. Using examples from our own work and perspectives, we explored how gender/sex entanglement impacts clinical research and patient care, genetics, and intellectual disabilities, and discussed implications for research and medical practice. Please note that a substantial portion of this work represented personal views that are not always linked to cited literature.

2.1 Definitions of Sex and Gender

Given the diversity in our backgrounds, we explored whether we shared clear definitions for terms like sex, gender, and intersectionality. The exercise of reaching consensus was immediately seen as challenging, with some arguing that it was exceedingly time-consuming for such a forum. There was some disagreement on whether it is productive or counterproductive to have clear definitions of sex, gender, and intersectionality across our respective disciplines as a point of reference, yet there is a risk in getting caught in disciplinary silos. We arrived at general definitions of sex and gender that are overall consistent with those from our colleagues throughout this book. Nevertheless, this conversation regarding consensus of definitions raised concerns about the oversimplification of sex and gender in clinical research. Indeed, how are social aspects of "gender" to be considered alongside biological "sex" factors?

Despite some initial resistance to the idea of disciplinary definitions in our initial introductions, each participant provided, in turn, their own brief definitions of sex and gender to situate perspectives. Overall, there was consensus that sex is generally considered a biological variable, whereas gender is generally perceived as a sociocultural variable (Greaves 2011; Johnson et al. 2007). There was also common agreement that this distinction may have been historically useful to advance sex and gender science as well as the strategic planning of granting agencies that fund health research (Clayton and Tannenbaum 2016; Duchesne et al. 2017; Tannenbaum et al. 2016). Still, it is possible that distinguishing sex and gender may no longer be as useful as it once was. Indeed, this categorization is often unrealistic when one considers how inexorably intertwined sex and gender are conceptually from the level of cells to communities (Junker et al. 2022; Juster et al. 2016). Moreover, the idea of limiting binary views of two sexes in biomedical sciences prevents more inclusive thinking regarding different biological observations (e.g., variations in sexual characteristics and intersex people). This discussion raises questions about the relationship between biological sex and gender identity, suggesting that they are not necessarily interconnected in a straightforward manner. As a group, we agreed that scholars should endeavor to evolve definitions and complexities in these terms, consistent with the notion of gender/sex entanglement where both constructs are intertwined, as detailed below.

While the usage of certain terms like "gender" and "sex" was questioned, these terms might have evolved over time and could carry specific cultural and histori-cal connotations. For example, the German language classifies nouns according to masculine, feminine, and neutral gender forms, whereas French utilizes masculine and feminine forms as well as a very specific use of the word "genre" that is too often conflated with sex. We delved further into the need for not just critiquing mas-culinities but also the consideration of femininities in gender transformative work. Terminology in the field of gender and sex can evolve over time and vary across dis-ciplines. Using outdated terms might not be representative of current understanding, although here too disciplinary differences abound. It is thus important to recognize that any system of measurement, including gender and sex, is underpinned by theo-retical assumptions and decisions made by researchers.

2.2 Intersectionality

Originating with the Black feminist writings of Crenshaw (1989), the concept of intersectionality involves considering additional variables (e.g., race, ethnicity, age, socioeconomic status, gender identity, sexuality, and other factors) to understand how power operates within different dimensions of sex and gender. According to Bowleg (see also Chapter 9), understanding intersectionality requires us to con-sider how various social systems (e.g., race, gender, socioeconomic status, and other identities) interact and influence a person's experiences and opportunities. This concept is central to how clinical research and health care can be enhanced. Social power is influenced by a complex interplay of social factors that often op-press minoritized communities. Thus, clinical research needs to consider and ac-count for intersectionality and the diverse experiences of communities, nationally and internationally. Researchers must look beyond a binary understanding of sex and gender and consider the various ways they intersect to affect health and well-being, together and synergistically (for further discussion, see Chapter 7).

As a theoretical and analytical framework, the concept of intersectionality is a critical framework that highlights how interlocking systems of structural oppression (e.g., racism, cisgenderism, sexism, class exploitation, heterosexism, *and* ableism) shape social, economic and health outcomes for groups historically marginalized at multiple sociodemographic intersections (e.g., ethnicity, sex/gender, socioeconomic status, disability status). For seminal literature on intersectionality, we refer the reader to the following collection of sources: Bowleg (2012), Combahee River Collective (1977), Collins (1990), Collins and Bilge (2020), and Crenshaw (1989, 1991). Power is foundational to intersectionality (Bowleg and Bauer 2016; Cho et al. 2013; Collins and Bilge 2020). Intersectional perspectives explore the significance of power and cultural context in understanding research outcomes. Given this recognition of power, there was shared interest in thinking of ways to make research more inclusive and equitable during our discussions. The intersectional perspective thus helps us better understand the cross-cutting intersections of factors such as sex, gender, genetic ancestry, racial or ethnic status, sexual and gender minority status, age, and disability, to name but a few.

One major challenge that we identified concerns the complexity of integrating intersectionality in fundamental scientific research. There are important examples in biomedicine where intersectionality and power dynamics impact the who, what, when, where, why, and how of scientific methods. The case of Henrietta Lacks and HeLa cells were introduced as an example of how power imbalances impact biomedical research (Lyapun et al. 2019). Henrietta Lacks was a Black woman in the United States who died in 1951 of cervical cancer. A tissue biopsy of her cancer cells was taken for research without her or her family's consent. This theft yielded the first immortal cell line, called HeLa cells (Lucey et al. 2009). Since then, the HeLa cell line has been an essential tool for generations of scientists and has been key to numerous biomedical breakthroughs for decades. Despite the accolades achieved and wealth accrued by scientists through use of HeLa cells, neither Ms. Lacks nor her family received any material benefit from the use of the cells obtained without consent. In addition to highlighting the intersectional, ethical, and power-related issues in biomedical research, this case also draws attention to the specific epigenetic effects that may travel through HeLa cell lines (Müller 2020).

2.3 Sex/Gender Entanglement

Our discussions delved into the complexities of understanding the entanglement of sex and gender, also referred to as sex/gender or gender/sex entanglement (see Chapter 1). In discussing the complexities of sex/gender entanglement, the challenges of applying intersectionality to biomedical research and clinical approaches were immediately apparent, particularly as it pertains to individual patient phenotypes. The distinction between sex and gender was debated and personal perspectives were provided that are not necessarily related to any specific literature. Some argued that focusing on gender and gender identity is more relevant in many biomedical, clinical, and sociopolitical contexts than focusing on categories strictly based on sex as a binary. For instance, broader macro-level forms of institutionalized gender are intertwined with power inequalities and intersectionality that go beyond sex.

One participant proposed that a highly problematic aspect in science is the misunderstanding of sex and gender distinctions, as if there was a meaningful divide when in fact one cannot easily distinguish sex and gender nor really separate them from one another. In fact, this divide has arguably been constructed upon the false distinction of "biological" versus "social" causality. Hence, this is simply a new application of the outdated arguments of the "nature versus nurture" debate (Keller 2010). To provide a way out of this conundrum, we propose that research should better articulate the operational definitions of sex and gender distinctions and entanglements in clear terms. If this is to be based on the distinction between organisms and persons, then sex should be related to organisms whereas gender should be related to persons. Importantly, we are not suggesting that gender/sex entanglement be necessarily broken down into constituent parts but rather that researchers be precise and descriptive about what aspects and measures they are referring to when describing dimensions of sex and gender (see Chapter 5).

Regarding sex, we also discussed whether a distinction should be made about what sex refers to. Is it chromosomal sex, anatomy, hormones, or variations therein? What is often referred to as "sex" in science is used to refer to the biological function of producing, for example, big or small gametes. In humans, this occurs in an almost binary and mutually exclusive way. On the other hand, what sex *implies* or is *assumed* to mean for most people is what constitutes as an organism. For example, male/female organisms are represented as a developmental bias that idiosyncratically affects the construction and activity of some organs and systems in a highly contingent way. Although this originates developmentally from the biological needs of sexual reproduction for our species, these complex biological processes operate via the same molecular agents involved in the specification of gamete production. In this way, the results are rarely binary or mutually exclusive. Therefore, sex-biased development does not result in "two kinds of organisms" but rather a series of differentially sex-biased features in each organism. Some of these features (i.e., those directly related to reproduction) show an accurate mapping on the sex of the organism, while others show a much more blurred relationship with the sex categories derived from gamete production. They are also much more affected by other factors like environment conditions that influence individual variability.

Gender, by contrast, relates to people, not to organisms. Persons are emergent products of human organisms comprised of cells that are shaped by the social communities and environmental conditions surrounding them. Emergent entities, patterns, or regularities are those that arise out of the interaction of more fundamental signals (e.g., SRY gene, *in vitro* testosterone effects) which supposably predate socialization. Still, emergent entities are different from, for example, single cells because they exhibit new properties. As such, emergent entities are not reducible to constituent parts made of cells and tissues (Noble 2012; O'Connor 2020). In other words, emergent entities and properties represent "wholes" that are greater than the sum of their parts. Accordingly, there is no person without organism nor gender without sex. The former, however, cannot be reduced to the latter. Therefore, this conceptualization assumes an inherent and irreducible entanglement of sex and gender while also acknowledging their possible differentiation and emphasizing their reciprocal irreducibility.

From this perspective, gender is operatively defined as a multilayered series of intertwined but not necessarily aligned perceptions or reflections about sex. For example, for one of these perceptions (gender identity), the same individual is the subject and the object of the perception; in others, the individual provides the object that is perceived by other individuals (i.e., gender expression). A third major class of perceptions (e.g., gender imagery, gender relations) are those harbored by higher-order interindividual aggregations (e.g., societies and cultures). For example, gender imagery refers to representations of gender in symbolic language and artistic productions (Fausto-Sterling 2012:7). Here, the object of perception is not the person as an individual but as a member of a socially and culturally defined group. Finally, we propose that any possible scientific knowledge about sex is also a gendered perception. That is, while sex exists independently of gender and of the human species, any conceptualization and understanding of sex is a human construction (Haraway 1988). As such, it is necessarily a gendered (i.e., a culturally or socially mediated) perception of what sex really is.

2.4 Applications of Sex and Gender Considerations in Clinical Research

In debating the utility of categorizing individuals based on sex and/or gender in clinical research, some argued for deeper phenotyping (e.g., N-of-1 trials, described below) of patients to better tailor treatments, whereas others expressed concerns about the practicality and effectiveness of such an approach, particularly in clinical trials. Examples were provided where traditional binary sex categorization in clinical algorithms fail to consider variations in anatomy, hormone levels, and other factors. This highlights the potential limitations of binary categorization in health care and points to the following open areas for future consideration:

- *Individual Phenotyping and N-of-1 Trials*: In the study design of $N = 1$ trials, every patient receives a unique combination of treatments based on deep phenotyping with repeated measurements. This approach is of great interest in precision medicine (Lillie et al. 2011) and necessarily implicates the generation of a complex picture for each patient, comprising individual variables and perhaps aspects or dimensions representative of intersectionality. In our discussion of the prevalence of male-only animal models in research, we emphasize the need to consider both female and male subjects together to afford a better understanding of sex- and gender-related factors in disease and treatments.
- *Relevance of Sex and Gender in Clinical Outcomes*: Since factors such as metabolism, which can vary between male and female sexes, can play important roles in treatment response and overall clinical outcomes, they are important to consider in biomedical decision making. Knowledge of the assigned sex at birth (based on appearance of external genitals) of patients alongside gender identity can influence medical decisions. In addition, it is especially critical to consider, for example, chromosomal sex in X chromosome-linked diseases. Since phenotypes of X-linked disorders are

often substantially milder in XX than in XY individuals, prenatal diagnostics are only offered in some cases for male fetuses, not for female fetuses. However, such clinical decisions need to be based on sound evidence and bears the risk of misguided decisions.

- *Entanglement of Sex and Gender in Psychosocial Stress Research*: The entanglement of sex and gender influences clinical outcomes as well as patterns of diseases across different populations. For example, heart disease now kills more women in North America than men, a shift that did not exist 50 years ago. The impact of psychosocial stress on health and its role in understanding sex- and gender-related health disparities was cited as a concrete example (Juster et al. 2019; Kajantie and Phillips 2006). Stress is considered a key variable in clinical research, given that it is a key mechanism involved in numerous health outcomes. As an example of sex and gender considerations in stress research, women are more likely to self-report distress on psychosocial questionnaires when exposed to laboratory-based stressors, although men produce more stress hormones in these stressful situations (Kajantie and Phillips 2006). Is this difference based on sex or gender, or gender/sex entanglement? To answer such questions, science must improve the measurement of sex and gender in biomedical research and health care. Understanding the stress mechanisms shaped by sex-linked factors and differences, gender experience, and gender diversity is essential for providing better solutions. To do so, there should also be considerations of sex differences as well as sex-specific effects and interactions. With regard to sex-specific effects, the disaggregation of analysis by sex is one way to go beyond simple sex differences based on comparisons of binary groups and is encouraged to promote rigor and reproducibility in health research (Clayton and Tannenbaum 2016).

- *Bias and Variability*: Different fields use bias as a challenge or as an opportunity. Social psychology, for instance, is dedicated to experimentally elucidating the influence of social stereotypes and bias in relation to human behavior (Hehman et al. 2014). By contrast, research into the influence of gender and sex in biomedical systems is severely underrepresented. Thus, there is a crucial need for increased attention and research funding in this area, as this constitutes a central barrier to advancement in biomedicine. Bias, both in clinical research as well as in the interpretation of patient data, is a significant concern. In particular, gender biases are significant in who conducts biomedical and clinical research as well as science, technology, engineering, and mathematics more broadly (Sebastian-Tirado et al. 2023). We underscore the importance of addressing these biases to improve research outcomes. Gender bias can influence the course and severity of diseases (Johnson et al. 2007), and future clinical research is needed to address this. In addition, the environment and knowledge base that individuals have expertise in will play a crucial role in determining the relationship between specific diseases. In parallel to advancing clinical practice approaches, a more comprehensive understanding of how sex and gender impact health is needed more broadly in science and technology.

2.5 Summary

There are undoubtedly opportunities and challenges that must be faced if an intersectionality framework is to be incorporated into clinical and biomedical science. Potential benefits include individual phenotyping and personalized medicine, although both will be complex to realize in real-world clinical settings. Nonetheless, as a guiding principle of best practice, we stress the importance of taking an intersectional approach when conducting clinical research and providing medical care. We identified critical challenges for direct applications to cellular research that are difficult to make and require further delineation. Indeed, we call for improved measurement and consideration of sex- and gender-related variables in research and specifically health-care decision making. In particular, gender/sex entanglement is challenging to clinical care. The complexities involved in gender/sex entanglement also require far more research and funding. Discussion and concern regarding genetic data, in particular, underscores the role of societal factors in health conditions with strong ethical and moral implications.

3 Categories of Binary Sex and Gender

Our discussion centered on various aspects related to gender and sex measurement, the need for categories in many statistical approaches, and the complexities inherent in these concepts. What do categories enable us to achieve and, importantly, what harms and biases can result? While this question centers the discussion on binaries of male/female, girl/boy, women/men, cisgender/transgender, we explored these constructs according to categories as well as along continuums. Below we summarize the key points and themes that emerged.

3.1 Diversity of Gender and Sex Measurements

There are diverse methods of measuring gender and sex, which range from binary categories to more continuous (e.g., Bauer et al. 2017) and nuanced approaches. While outdated and controversial, the Bem Sex Role Inventory is a measure of gender roles, represented as stereotypically masculine and feminine personality traits (Bem 1974, 1977, 1981). More recently, with the aim of better representing gender identity, the inclusion of transgender and nonbinary items has been emphasized through a popular measure by Bauer and colleagues (Bauer et al. 2017; Kozee et al. 2012), and has gone through various iterations to be maximally representative. Note that this instrument does not have an item for intersex people, which we encourage researchers to include. Moving into even broader gender domains (e.g., gender relations), a newer instrument developed by gender scientists at Stanford has emerged and has been cross-culturally adapted to Canada and Spanish-speaking countries (Abdel-Sayyed et al. 2024; Díaz-Morales et al. 2023; Nielsen et al. 2021).

A key point that emerged from our discussion was the use of continuum scales to capture gender and sex, as they allow for more representative data to consider the nuances of gender (for further discussion, see Chapter 10). Even sex can be considered continuously (Williams et al. 2023). For example, the measure of hormones such as estrogen, testosterone, and progesterone (Juster et al. 2016) are known to influence as well as interact with stress and health (Viau 2002). Whenever possible, we encourage researchers to use continuous measures and avoid dichotomizing results. This can be accomplished, for example, by using Bauer's instrument of gender identity, which poses questions that allow participants to express their feelings about their gender dynamically as well as by using measures of lived experience and gender embodiment (DuBois et al. 2021b). This continuous approach assesses dimensions over time and has also been done with measures of sexual orientation evaluated currently, in the past, and under an ideal setting (Klein et al. 1990). The intersection of gender diversity and other factors (e.g., race and sexuality) is complex and often omitted when considering sex, gender, and sexual orientation collectively. We therefore recommend the development and use of continuous measures of sex, gender, and intersectionality to represent continuums.

3.2 Agency of Individuals

Our discussion highlighted the limitations of rigid categories, which can generate dissonance in people and result, even inadvertently, in the negative consequences of pathologization or stigmatization. We touched on the statistical power of different approaches, focusing on the trade-offs between continuum scales and categorical measures and noted the importance of considering the distribution of data as an essential factor. This led to broader questions about the critique of gender categories and the impact these categories may have in shaping research methodologies and assumptions.

It is also imperative to consider *how* individuals respond to sex and gender categorization and specifically how transgender and nonbinary people want to be asked about any categorizations (Puckett et al. 2020). We stress the importance of recognizing gender diversity and the need to avoid pathologizing individuals. We also emphasize the necessity to move toward nonbinary and gender diverse scales (described further below) that capture unique experiences not common to majority groups. Relatedly, it is important to consider gender identity, for example, when returning information on hormone concentrations to participants as these can inadvertently have negative effects if this information is then interpreted and/or used to define what is male-typical or female-typical. Furthermore, where do transgender and nonbinary individuals fit into these binary systems for clinical reference ranges, which are based on presumably cisgender people (DuBois et al. 2021a)? Despite a movement toward returning biological results to those who participated in a study as a right and responsibility (NASEM 2018), individuals must have agency and be encouraged to decide whether they want this information. This is especially challenging in the context of genetic consultation. Here, future research is needed to address the requisite qualitative dimensions that will capture participants' responses and feelings toward gender and sex categorizations.

3.3 Evolutionary Contexts

It is vital to position our understanding of sex and gender in a broader context of evolution and reproduction, as described by natural selection (Darwin 1859) and sexual selection (Darwin 1871). Doing so forces us to reconsider a fundamental question at the heart of gender/sex entanglement: Can we make sense of biology without culture? To address this question, let us consider the phenomena of plasticity and epigenetics from two perspectives. The first concerns ontogeny. Current knowledge in molecular biology has allowed us to conclude that, since we are gestational processes, we exist in a biomaterial dialogue with the environment—one that reaches even the processes of genital differentiation (Ciccia 2023). The second refers to the transgenerational effects of epigenetic regulations that give rise to changes in certain biological configurations—effects that occur when changes occur in sex cells. Transgenerational effects imply that experiences are imprinted into the epigenetic signature of individuals and can be passed on to subsequent generations (Tollefsbol 2019). In contrast to biological determinism, which is inherent to genetics, this epigenetic mechanism shapes the expression of genes by social experiences as key environmental determinants. Every human being has incredible plasticity through epigenetic processes which dance synchronously with our social lives as well as contingent and situated experiences (Ciccia 2023). This adds another level of complexity, as social practices can affect not only the individual but also their descendants.

3.4 Complexity of Biological Variation

In biological anthropology, the complexity of biological variations is of great interest, particularly concerning sex-linked characteristics, such as hormones, and how they intersect with gender and gender identity. As discussed above, categorizing individuals into binary sex groups oversimplifies intricate biological and social realities and gender/sex entanglement (DuBois and Shattuck-Heidorn 2021). In clinical research, this poses challenges when individuals with variations of sexual characteristics, otherwise called intersex people, are involved (Sandberg and Vilain 2022; Timmermans et al. 2019). Yet categorizing people based on the traditional sex binary framework and imposing associated norms is misleading and may lead to oversimplified and inaccurate findings. In turn, this data may not be applicable to diverse individuals based on gender identity as well as biological sex-based variations, according to chromosomes, genitals, hormonal milieu, and other characteristics.

In terms of differences between causal relationships and correlations as well as the effects of categorical thinking in research and assumptions about causality, it is problematic to assume a direct and simple relationship between a hormone (e.g., testosterone) and specific behaviors (e.g., aggression) or traits (e.g., relationship status). Research shows, for example, how social contexts interact with biological processes in bidirectional, intercorrelated ways among men, women, and sexual minorities (van Anders et al. 2015; van Anders and Watson 2006). The need to consider those correlations between certain biological features (e.g., hormones) and behaviors may not necessarily imply direct causality. A key example is how a causal correlation between testosterone levels and aggressive behavior is often assumed. In reality, what is often observed is that people who tend to have higher endogenous testosterone levels (i.e., cisgender males) are more likely to exhibit violent behavior. Conceptualizing this association as a correlation allows us to address the inherent complexities and to investigate how biology and society are interconnected. The power of scientific discourse and societal expectations also shapes how individuals perceive themselves and their unique past experiences. In other words, scientific findings have cultural and social implications well beyond the research questions they were meant to address. Thus, it is vital to communicate research findings carefully to avoid misrepresentation that could perpetuate the status quo. Consistent with an intersectional approach, misrepresentation could exacerbate existing inequities and contribute to the invisibility of nonnormative bodies, experiences, and identities in clinical understanding.

3.5 Reification, Gender/Sex Norms, and Identity Considerations

The concept of reification, where an inflexible binary understanding of gender/sex norms is reinforced, can lead to the exclusion and pathologization of groups of people who, for example, fall outside of dominant gender/sex norms. This concept suggests that oversimplified frameworks like "male versus female" affect how people perceive themselves and their capabilities. Indeed, the tension between biological considerations and individual identity is better thought of as spectrums rather than bipolar ends of categories, a point that even Bem made when she created the infamous gender role instrument (Bem 1974). As such, when assessing health and well-being along numerous dimensions, we need to look beyond fixed categories and consider instead the baseline levels and individual needs.

The value of having interdisciplinary perspectives when addressing complex issues related to sex, gender, and health is of paramount importance, since individual considerations traverse multiple disciplinary lines. This encourages us to think beyond traditional binary frameworks. Notwithstanding, we acknowledge the importance of individual variability. For example, individual differences in response to medical interventions (e.g., psychopharmacological medications) can be shaped by gender (e.g., prescription patterns) and impact biological functioning, for instance, through inhibition of the hypothalamic-pituitary-gonadal/adrenal axes via anti-inflammatory medications (Crofford et al. 1999). Reification and consideration of gender norms underscores the reality that one size does not fit all. Indeed, some individuals will or will not respond to the same treatments based on a myriad of constitutional (e.g., body size differences) and experiential (e.g., stigma) factors.

3.6 Summary

Measuring gender and sex is anything but straightforward, as it needs to account for and be inclusive of diverse identities and experiences. It must also avoid rigid categorization and be sensitive to the ongoing evolution of terminology if it is to be valid for statistical considerations. It is important to understand the theoretical assumptions behind measurement systems and to recognize the diversity of human experiences related to gender and sex. Overall, we stress the need for a more nuanced and complex understanding of sex and gender as well as the importance of avoiding oversimplified frameworks in both clinical research and medical practice. This highlights the powerful influence that societal discourse exerts on the infinite ways that individuals experience their own gender and bodies.

4 Integrating Gender and Sex into Biomedical Research

Centering on biologically oriented perspectives of sex and gender, we explored how these concepts can be integrated into neuropsychiatry, epigenetics, and clinical research to improve medical interventions. In our discussions, we debated whether focusing on biological mechanisms or categorizing based on sex and gender differences is effective for clinical practice. As many fields move toward precision medicine, researchers must work with methodologists and statisticians to form a deeper understanding of mechanisms and precise research questions that will minimize bias. Several key themes emerged from our discussions.

4.1 Sex-Based Differences in Biomedical Research

Even within the same biological pathways, certain biological functions differ in males compared to females, contributing to sex-linked differences in how diseases develop and how people respond to treatment. For example, the frequency of physical activity, often gendered, can increase hepatic metabolism and affect the rate at which drugs are metabolized. This means that a cisgender female athlete may have a higher metabolic rate than a cisgender sedentary male. However, this fact, fundamental to the study of pharmacokinetics (i.e., the administration, distribution, metabolization, and elimination of a drug), is not captured in the idea of sex.

Differences in the level of analysis for dimensions of sex as a biological variable also have implications for cancer research. When considering gonadal or anatomical sex, individuals can have different susceptibilities to male-biased cancer risk (e.g., testicular, prostate) or female-biased cancer risk (e.g., ovarian, cervical, breast). In another example, the impact of sex on cancer risk highlights that individuals with different sex chromosomes (e.g., XX and XY) can confer different risks. In cellular research, scientists are actively studying cellular and genetic differences between individuals with different sex chromosomes (Dorak and Karpuzoglu 2012). Their goal is to uncover the cellular mechanisms that contribute to sex differences in cancer risk and response to treatment. Beyond cancer, sex differences in health outcomes that focus on sex as a binary biological variable exist for many conditions in biomedicine, yet gender as a dimensional sociological variable is often omitted despite evidence of its impact. Indeed, gender is a social and structural variable comprised of multiple domains (e.g., gender identity, gender expression, gender roles, institutionalized gender) that influence health (Barr et al. 2024).

4.2 Mechanism versus Categorization

Identifying differences between males and females can be a useful starting point for understanding the underlying biological mechanisms that may drive differences in clinical outcomes and have potential implications for medical interventions. By focusing on the mechanisms behind entangled sex and gender differences, as they relate to categorizing interventions and treatments (Lee et al. 2023; Pape et al. 2024), we explored whether categorization alone is sufficient to ensure safety and effectiveness in clinical applications of observational data. Here, we used the clinical experience of drug-eluting cardiac stents (Coughlan et al. 2023): females carry a greater risk for cardiac death in the immediate period following stent placement. Is this categorical observation sufficient to alter clinical practice? What changes in practice are required to allow for additional research that can deepen our knowledge? Answering these questions might permit a more nuanced approach to patient stratification for treatment and provide additional insights into the mechanisms underlying the observation, which in turn could enhance safety and effectiveness. Future studies need to address whether a biological mechanism-oriented approach is the next logical step in research or if categorization is sufficient to ensure safety and effectiveness.

4.3 Incentives, Statistical Considerations, and Equivalence

The incentives currently available to pharmaceutical companies and researchers in science and technology aim at keeping people healthy, yet they often rely on sex/gender categorization to win treatment approval. Indeed, there is often rather limited motivation to delve into the underlying biological mechanisms that may drive clinical outcomes, especially as these are exponentially more complex when considering sex and gender. However, understanding biological mechanisms is crucial for precision medicine. To minimize bias in their analysis, researchers should ideally start with precise research questions and develop a clear-cut framework for any group-based comparisons. Alternatively, examining similarities in biological mechanisms and processes across subgroupings of people is also essential. Consider, for example, the gender similarities hypothesis, which argues that males and females are similar on most, but not all, psychological variables (Hyde 2005). To test equivalence in drug development and clinical trials, it is thus important to consider sex and gender similarities.

The concept of intersectionality further complicates analysis, given its limited application to methodological approaches in biomedical research. There are challenges in statistical analysis, including the need to develop new statistical techniques to analyze complex, uncontrolled data in biomedical research. Sanchis-Segura and Wilcox (2024) discuss many of these issues, including the need to develop and/or use new methods, the limitations of statistical significance testing, and the need to consider meaningful effect size indexes. The terms "similarity testing" and "equivalence testing" are sometimes used interchangeably. Equivalence testing ultimately involves determining the range of equivalence that is acceptable, such that a difference between groups is either biologically meaningless or clinically significant. The absence of rigorous equivalence testing in many scientific fields, including psychology and clinical trials, has led to a focus on statistically significant differences without considering the magnitude of those differences. There is a need to understand the variability within groups, the size of differences, and the importance of framing statistics correctly. In addition, it is important to collaborate closely with statisticians so that appropriate statistical methods can be used appropriately for analyzing sex and gender differences. Effective communication between researchers and statisticians is therefore crucial for data analysis and interpretation.

4.4 Summary

The complexities and challenges of integrating sex and gender into biomedical research are immense. We need to improve our understanding of biological mechanisms, develop better statistical tools, and engage in effective interdisciplinary collaboration. Integrating sex and gender equivalence in scientific research and health care requires nuanced approaches and clear communication. Researchers should strive to provide fully informed choices to patients and individuals, acknowledging the limitations and uncertainties in scientific findings. It is the combined responsibility of individual scientists, policy makers, stakeholders, and the scientific community to ensure that sex and gender considerations are integrated into research and health care. For this to happen, nuanced communication, collaboration with stakeholders, and ongoing critical thinking are necessary. Different stakeholders approach the same information from varying perspectives, depending on their role and objectives. This reality makes it essential to tailor communication and interpretation accordingly to avoid misrepresentation.

5 Integrating Sex and Gender into the Social Sciences

Connected to the previous section on biomedical and clinical research, the challenges involved in integrating gender and sex into biomedical research may necessitate stronger interactions with the social sciences. Here, the concept of entanglement emerged as a central theme, emphasizing the interconnectedness and complexity of sex and gender and their entanglement across different aspects of human life and societal/structural contexts. This multifaceted discussion raises various themes.

5.1 New Materialism and Complexity

New materialism was introduced as a framework to examine the entanglement of social, cultural, and biological elements (Pearson 2011). Within this framework, biology is placed in a central position. It provides a reconceptualization of our biology, made possible by the phenomena of plasticity and epigenetics. The flexibility and possibility of change that characterize plasticity and epigenetics result in epistemological tools to further investigate how power structures can be expressed biologically.

Complexity is a recurring theme; both biology and social sciences are complex and not reducible to single factors. The paradigmatic example of this complex interweaving is the brain. The crossroads between neurosciences, gender studies, and new feminist materialisms have been fundamental in exposing differences between cisgender men and women in brain architecture, as well as in certain patterns of neuronal activation. These differences, however, do not reflect simple binary sex differences but rather differences that result from sex/gender entanglements. In other words, we do not know whether observed differences in the brain result from social practices, sex-linked biological factors, or both (Kaiser 2016). Moreover, all of our biology is subject to social conditioning. A clear example of this is reflected in embodied stress and health equity research. For example, higher cortisol levels are associated with gender minority stress experienced by transgender and nonbinary people because of cisheteronormativity and gender-based norms and stigma (DuBois et al. 2024). The consequences of this stigma involve physical abuse and violence for this population, which modulates stress hormone profiles that can contribute to health disparities. These cases demonstrate the importance of interdisciplinary collaboration between biologists and social scientists—an approach that we consider necessary to understand the intricate relationship between biology and society.

5.2 Structural Stigma and Sexism

Research on structural stigma related to sexual minorities and gender issues is an area of growing interest. The work of Mark Hatzenbuehler, in particular, examines how laws and policies impact the mental and physical health of marginalized groups, particularly within the LGBTQ+ community (Hatzenbuehler 2009, 2014, 2018; Hatzenbuehler et al. 2009a, b, 2013). State-level laws can be used to index structural stigma. More recently, the use of biomarkers (physical indicators) to study the effects of social policies on health implies a connection between societal factors, such as laws and policies, and their impact on individuals' physiological well-being (DuBois and Juster 2022; DuBois et al. 2024; Hatzenbuehler and McLaughlin 2013; Juster et al. 2024). Similarly, structural sexism has been indexed by Patricia Homan to show how gender inequalities relate to measurable geospatial disparities in women's health (Homan 2019; Homan et al. 2021). The importance of considering the sociocultural dimensions of sex and gender in research and policy is paramount to the advancement of research on the structural determinants of health.

These dimensions play a crucial role in understanding and addressing complex issues central to gender/sex entanglement and intersectionality.

5.3 Interdisciplinary Collaboration in Research

To advance research on sex, gender, and intersectionality, effort must be given to increase interdisciplinary research and collaboration, quite simply because the issues embedded within gender and sex are multifaceted and thus require insights from various fields. There is a recognition that traditional academic departments and disciplines, such as psychology, often stick to conventional research methods, topics (e.g., anxiety), and conceptualizations of individuals as the primary unit of analysis. These fields tend to rely on simple analyses of data without critically considering aspects like gender or sex at a more community level. The examples provided above of structural stigma and sexism move beyond the individual level.

Multiple challenges must be faced when concepts of sex and gender are integrated across different disciplines and along various dimensions: from an individual level all the way to the population level. These include, for example, variations in language, methodology, and engagement with the issues at hand. The prevailing academic approach is limited by disciplinary boundaries. We believe that academics should be encouraged to think beyond traditional boundaries and consider how their research intersects with other disciplines that may well be outside their comfort zones of expertise. Indeed, the spirit of this Ernst Strüngmann Forum enabled this synergy, reflecting the importance of speaking common languages of sex and gender despite distinct disciplinary dialects. Furthermore, developing transdisciplinary measures and emergent properties is central to this objective. For instance, self-rated health ("how healthy are you on a scale of 1 to 5") combines biological, social, psychological, behavioral, and even spiritual dimensions of a person's life (Picard et al. 2013). In a transdisciplinary spirit, such measures can capture complex phenomena that cannot be adequately explained within a single discipline.

5.4 Transcending Material and Symbolic Inequalities

In the context of public health and gender studies, material inequalities relate to structural, social, and economic factors. The term symbolic inequalities is also used to characterize issues of power and normativity (Christensen 2023). Addressing these issues is essential. There is a clear need to consider the global relevance of research while also being mindful of local contexts. For instance, it is essential to create variables and factors that can be imported and modified to fit local contexts, rather than to impose global frameworks. Nevertheless, there is ongoing debate as to whether global frameworks can be imported and modified to suit local contexts, or whether entirely local frameworks need to be developed. Despite our discussion of the challenges and advantages of each approach, we are unable to offer a solution. Concern was raised about underrepresentation of certain groups in academic spaces and the need to include voices from regions or communities not typically involved in global conversations. For example, scholars from Latin America, Africa, and Asia are sometimes poorly represented in academic spaces. The significance of concepts in academic discourse are needed to make global health issues visible. Indeed, this underrepresentation results from structures of violence. In this sense, the notion of intersectionality can be used to investigate which variables are relevant in certain contexts, and which may not be relevant in others.

5.5 Complexity and Biological Essentialism

There is a balance between addressing the complexity of sex and gender while avoiding essentialism, which can lead to harm, stigma, and misrepresentation of individuals and groups. Concern was raised about the risk of essentializing complex concepts like sex and gender, and we stress the importance of considering the sociocultural dimensions when addressing sex and gender. Equally, when addressing the complexities of these issues, one should avoid oversimplification. There are inherent limitations in using certain terms and vocabularies, as they can carry different meanings and create miscommunication. Thus, the role of language in mediating discussion is vital, as is the importance of finding common ground in terminology.

5.6 Summary

Integrating the concepts of sex, gender, and entanglement across disciplines, particularly between the social sciences and biological sciences, must be viewed as an ongoing process. Unfortunately, there is no simple remedy that might improve this process. Indeed, our conversations on this topic often broke down when we discussed what was meant by sex and gender measurement from different levels of analyses. This highlights the difficulty of using specific disciplinary terminologies in interdisciplinary dialogues. As a community, we need to establish common ground in terminology to bridge disciplinary gaps, to avoid oversimplification and essentialism, as well as to facilitate productive communication. Interdisciplinary, multidisciplinary, and transdisciplinary research involves multiple layers of complexity and different levels of analysis. To move forward, we must meet the inherent challenges and develop an inclusive, globally aware approach that accounts for both local and universal concerns.

6 Integrating Gender Experience and Intersectionality into Health Research

How can gender experience and intersectionality—which involves dynamic interactions across individuals as well as across historical and cultural contexts—be better integrated into health research? The first step in addressing this issue involves gathering data with an intersectional lens, sharing it in a way that continually supports individual and collective learning, and the significance of population statistics when making health-care decisions. In this section, we highlight further themes that emerged from our discussions.

6.1 Intersectionality

Cross-cutting perspectives are inherent to the notion of intersectionality, discussed in detail by Lisa Bowleg (see Chapter 9). Intersectionality concerns the impact of health inequities on marginalized groups that have historically been oppressed. The challenge of trying to isolate variables (e.g., sex and gender) from intersecting factors (e.g., racial/ethnic minority status, genetic ancestry, sexual and gender minority status, and ability) in research is multifaceted. For instance, single-axis perspectives, where a single factor (e.g., racial/ethnic group) is studied absent its intersection with the other factors, create limitations and often overlook how multiple, interlocking systems of oppression (e.g., racism, sexism, heterosexism, cisgenderism, and ableism) combine to shape social, economic, and health experiences. From an intersectionality perspective, sociodemographic variables are interconnected and mutually co-constituted, and thus cannot be separated or added.

6.2 Discrimination and Resilience

In our discussions, we adopted the definition of resilience as a dynamic process that promotes positive adaptation among individuals exposed to severe adversity, trauma, and stress (Cicchetti and Garmezy 1993; Luthar et al. 2000; Masten et al. 1990; Rutter 2012). Resilience is often used to describe individuals' responses to adversity but may not account for the experiences of those who have adapted to systemic discrimination. In accordance, the resilience literature has been subject to criticism (Suslovic and Lett 2024) as resilience theories often fail to acknowledge the structural barriers that prevent people from forming protective profiles (Shaw et al. 2016).

There are excellent examples of resilience from an intersectional perspective in the stress physiology literature. A fascinating research program developed by Brody and colleagues has investigated John Henryism theory among Black youth into emerging adulthood (Brody et al. 2013). Legend has it that John Henry, an industrious Black railroad worker in the late 1800s, challenged and defeated a steam-powered drill in a steel-driving contest but died from exhaustion afterward. The John Henryism hypothesis (James 1994) refers to behaviors involving intense coping and hyperarousal, which fuel focused concentration and the physical energy needed to succeed against overwhelming odds (Brody et al. 2013). Resilience may thus be present in key life domains (e.g., externalizing behaviors, academic performance) but exact a physiological price (e.g., allostatic load representing the "wear and tear" of neuroendocrine, immune, metabolic, and cardiovascular systems; McEwen and Stellar 1993). This proposition was tested among rural Black

youths with low socioeconomic status and teacher-reported self-control and competence, ages 11–13 years. At age 19, depressive symptoms, externalizing behaviors,
and allostatic load levels were measured. Results showed that low socioeconomic
preadolescents with high competence had fewer adjustment problems but higher
allostatic load at age 19, thus supporting the John Henryism theory and suggesting
resilience may be "skin deep" (Brody et al. 2013).

Adaptation in domain-specific areas may clearly come at a physiological cost.
This emphasizes that resilience is not a fixed status or a universal state of "resil-
iency" but a range of adaptations to life's challenges, encompassing both positive
and negative aspects. Since it is difficult to measure discrimination objectively, we
stress the importance of considering both subjective experiences and objective data,
such as biomarkers. However, although biomarkers can offer insights into physiological responses to social stressors, they present ethical and methodological concerns (DuBois et al. 2021a).

When considering intersectionality, it is crucial to recognize how multiple, overlapping social identities (e.g., race, gender, and socioeconomic status) interact to
shape an individual's experience of resilience and discrimination. These intersecting identities can influence how people adapt to challenges and how discrimination
impacts their mental and physical health. It is therefore vital to incorporate an intersectional perspective to fully understand the complex, nuanced nature of resilience
and its measurement in diverse populations. A clear example that illustrates the need
for an intersectional perspective in the field of health can be found in the studies
conducted by Nancy Krieger, who analyzed the impact of Jim Crow laws (when
racism was legal between 1876–1965 in certain states of the United States) on the
prevalence of breast cancer in black cisgender women. Comparing the incidence
of a specific type of cancer (estrogen receptor negative) between Black and White
cisgender women born under the laws as well as after they were repealed, Krieger
found a relationship between the place of birth (where such laws operated) and an
increased incidence rate of this cancer in Black, but not White, cisgender women;
the correlation was stronger for Black cisgender women born before 1965 (Krieger
et al. 2017, 2018).

The sex category usually refers to the idea of pre-social variables that are detached from the environment. At the same time, they are interpreted as fundamental for a better understanding of disease prevalence, development, and treatment.
Ciccia (2024) argues that such a characterization implies a series of biases that
derive from a biological mechanistic reading of the processes of sexual differentiation and disease as well as the sex-prevalence relationship (Ciccia 2024). To this
end, the majority of variables considered to be of clinical relevance are defined by
the attributes we associate with the category sex. Different disciplines and feminist
epistemologies have exposed the deterministic biases that permeate the predominant scientific discourse. However, Ciccia believes that none of the criticisms considers the cause-effect logic, on which physical and symbolic gender relations are
structured, as a problem.

In this sense, a temporal linearity is naturalized that places the biological before
typically gendered behaviors, such as aggression and competition. Such naturalization results from not sufficiently problematizing the mind-body relationship from
the perspective of gender studies. Here, Ciccia proposes a crossover between the

concept of event and anomalous monism (Davidson 1980), with certain contributions from the new feminist materialisms. This position argues that there is a temporal synchronization between our biological states and our psychological states. At the same time, the irreducibility of the mental must be emphasized. Interpreting gendered behaviors in this way enables an ontology of the body that dilutes the cause-effect logic, an androcentric logic inherited from modern science founded on inherently cisheteronormative biological perspectives (Ciccia 2022).

6.3 Summary

Intersectionality is an evolving and complex construct that involves multiple dimensions, experiences, and sources of inequality. Intersectional frameworks raise ethical considerations regarding the use of data, biomarkers, and genetic information. This points to the importance of avoiding deterministic or discriminatory practices. Thus, it is necessary to recognize that correlations do not imply causation just as the ability to predict correlated findings correctly does not indicate a mechanistic understanding. Indeed, two highly correlated features can be independent consequences of a common initiating event. Overall, the importance of recognizing the complexity of intersectionality, conducting empirical research, and effectively communicating research findings in both clinical and research settings underscores the need to move beyond generalizations and address the lived experiences of individuals within their unique intersectional contexts.

7 Final Thoughts

Three main perspectives emerged from our discussion on gender/sex entanglement:

- Sex and gender are distinct concepts, yet they are also entangled.
- It is not possible to consider sex and gender separately; they are interconnected.
- It is not possible to discuss sex without considering cultural and social influences.

Still, we did not reach consensus on the best approach that health research should take to understand sex and gender constructs. Conceptually, we explored the use of "sex/gender," which focuses on (biological) sex as a variable, and the use of "gender/sex," which focuses on gender as a sociocultural variable. These distinctions and definitions, however, struck us as somewhat simplistic. Moreover, this does not solve the problem of entanglement and disciplinary usages. Although our discussions at times appeared to favor distinguishing sex as related to organisms and gender as related to persons, we were not satisfied with this division either. Indeed, we remain hopeful that gender/sex entanglement can be better conceptualized in fundamental research, where it seems simpler to focus on sex as a biological variable. When intersectionality is added, the complexity of resulting interactions reinforces the idea that the "whole is greater than the sum of its parts" in gender/sex entanglement is an emergent entity. More translational thinking and research regarding gender is needed to force us to consider cells within gendered contexts.

Interestingly, we found that our shared perspectives were not necessarily in opposition but rather represented different ways of describing concepts related to sex and gender. While these different perspectives might seem at odds or even in competition, they may be able to coexist as complementary viewpoints. It is important to recognize that these represent different perspectives rather than disagreements. The process of exploring these perspectives led to a more nuanced understanding of the topic for all of us.

In summary, we highlighted the difficulties of measuring gender and sex, the need for inclusivity in research to represent marginalized identities, the challenges of rigid categories, statistical considerations, evolving terminology, and the importance of considering intersectionality at theoretical, empirical, and clinical levels. In clinical research and medical practice, a nuanced and complex understanding is imperative, and we must avoid oversimplified frameworks. We advocate for nuanced approaches and clear communication, providing fully informed choices to patients and individuals. The responsibility to incorporate sex and gender considerations lies with individual scientists, policy makers, stakeholders, and the scientific community as a whole. Effective communication, collaboration with stakeholders, and ongoing critical thinking are crucial for better integrating gender/sex and intersectional considerations into research and health care.

We stress the importance of gathering data with a gender/sex and intersectional lens. This is further promoted by sharing perspectives that support individual and collective learning across generations, from trainees to teachers. Ethical considerations surrounding data, biomarkers, and genetic information emerged as a recurrent theme and separated us positionally the most in teasing apart sex from gender. In particular, we recommend the need to avoid deterministic and/or discriminatory practices that include interpretive errors.

Finally, we collectively emphasize the importance of recognizing that sex, gender, gender/sex entanglement, and intersectionality represent different perspectives rather than disagreements. This recognition led us to a more nuanced understanding of the topic. The think tank process of the Ernst Strüngmann Forum was valuable in this respect, and we suggest exporting this approach to other discussions in the future. Overall, our conversations highlighted the effectiveness of the Forum's approach in promoting a constructive exchange of ideas across disciplinary divides.

Acknowledgments We thank Natalie Worden and Julia Lupp for their assistance and support. Our group is grateful to L. Zachary DuBois and Anelis Kaiser Trujillo for their invitation and leadership chairing this forum. Robert-Paul Juster is supported by the Fonds de recherche du Québec Santé and holds a Sex and Gender Science Chair from the Canadian Institutes of Health Research. We wish to acknowledge Chat GPT for our use of deductive artificial intelligence used for transcript summaries of our discussion that structured an initial draft.

References

Abdel-Sayyed A, Hoang KN, Turk T, Xu LJ, and Fujiwara E (2024) Validating the Stanford Gender-Related Variables for Health Research (SGVHR) in a Canadian Population. J Health Psychol 23:13591053241247376. https://journals.sagepub.com/doi/10.1177/13591053241247376

Barr E, Popkin R, Roodzant E, Jaworski B, and Temkin SM (2024) Gender as a Social and Structural Variable: Research Perspectives from the National Institutes of Health (NIH). Transl Behav Med 14 (1):13–22. https://doi.org/10.1093/tbm/ibad014

Bauer GR, Braimoh J, Scheim AI, and Dharma C (2017) Transgender-Inclusive Measures of Sex/Gender for Population Surveys: Mixed-Methods Evaluation and Recommendations. PLoS One 12 (5):e0178043. https://doi.org/10.1371/journal.pone.0178043

Bem SL (1974) The Measurement of Psychological Androgyny. J Consult Clin Psychol 42 (2):155–162. https://psycnet.apa.org/doi/10.1037/h0036215

——— (1977) On the Utility of Alternative Procedures for Assessing Psychological Androgyny. J Consult Clin Psychol 45 (2):196–205. https://doi.org/10.1037//0022-006x.45.2.196

——— (1981) Bem Sex Role Inventory: Professional Manual. Consulting Psychologists Press, Palo Alto

Bowleg L (2012) The Problem with the Phrase "Women and Minorities": Intersectionality, an Important Theoretical Framework for Public Health. Am J Public Health 102 (7):1267–1273. https://doi.org/10.2105/AJPH.2012.300750

Bowleg L, and Bauer GR (2016) Invited Reflection: Quantifying Intersectionality. Psychol Women Q 40 (3):337–341. https://doi.org/10.1177/0361684316654282

Brody GH, Yu T, Chen E, et al. (2013) Is Resilience Only Skin Deep? Rural African Americans' Socioeconomic Status-Related Risk and Competence in Preadolescence and Psychological Adjustment and Allostatic Load at Age 19. Psychol Sci 24 (7):1285–1293. https://doi.org/10.1177/0956797612471954

Cho S, Crenshaw KW, and McCall L (2013) Toward a Field of Intersectionality Studies: Theory, Applications, and Praxis. Signs 38 (4):785–810. https://doi.org/10.1086/669608

Christensen G (2023) Three Concepts of Power: Foucault, Bourdieu, and Habermas. Power Educ 16 (2):182–195. https://doi.org/10.1177/17577438231187129

Cicchetti D, and Garmezy N (1993) Prospects and Promises in the Study of Resilience. Dev Psychopathol 5:497–502. https://doi.org/10.1017/S0954579400006118

Ciccia L (2022) Sucesos Comportamentales: Estados Mentales, Cuerpo y Género. Debate Fem 63:3–29. https://doi.org/10.22201/cieg.2594066xe.2022.63.2311

——— (2023) (Epi)Genealogía del Cuerpo Generizado. Diánoia 68 (91):113–145. https://doi.org/10.22201/iifs.18704913e.2023.91.1993

——— (2024) ¿Por Qué Es Necesario Eliminar la Categoría Sexo del Ámbito Biomédico? Hacia la Noción de Bioprocesos en la Era Posgenómica" Dossier "Estudios Trans". Interdisciplina 12 (32):105–129. https://www.revistas.unam.mx/index.php/inter/article/view/86922

Clayton JA, and Tannenbaum C (2016) Reporting Sex, Gender, or Both in Clinical Research? JAMA 316 (18):1863–1864. https://doi.org/10.1001/jama.2016.16405

Collins PH (1990) Black Feminist Thought: Knowledge Consciousness, and the Politics of Empowerment. Routledge, New York

Collins PH, and Bilge S (2020) Intersectionality: Key Concepts. 2nd edn. Polity Press, Cambridge

Combahee River Collective (1977) The Combahee River Collective Statement. In: Smith B (ed) Home Girls: A Black Feminist Anthology. Kitchen Table: Women of Color Press, New York, pp 272–282

Coughlan JJ, Räber L, Brugaletta S, et al. (2023) Sex Differences in 10-Year Outcomes after Percutaneous Coronary Intervention with Drug-Eluting Stents: Insights from the DECADE Cooperation. Circulation 147 (7):575–585. https://doi.org/10.1161/Circulationaha.122.062049

Crenshaw KW (1989) Demarginalizing the Intersection of Race and Sex: A Black Feminist Critique of Antidiscrimination Doctrine, Feminist Theory and Antiracist Politics. Univ Chic Leg Forum 1989 (1):8. https://chicagounbound.uchicago.edu/uclf/vol1989/iss1/8

——— (1991) Mapping the Margins: Intersectionality, Identity Politics, and Violence against Women of Color. Stanford Law Rev 43:61. https://www.jstor.org/stable/1229039

Crofford LJ, Jacobson J, and Young E (1999) Modeling the Involvement of the Hypothalamic-Pituitary-Adrenal and Hypothalamic-Pituitary-Gonadal Axes in Autoimmune and Stress-Related Rheumatic Syndromes in Women. J Womens Health 8 (2):203–215. https://doi.org/10.1089/jwh.1999.8.203

Darwin C (1859) On the Origin of Species by Means of Natural Selection. John Murray, London

——— (1871) The Descent of Man and Selection in Relation to Sex. John Murray, London

Davidson D (1980) The Material Mind (1973). In: Essays on Actions and Events. Clarendon Press, Oxford, pp 245–260

Díaz-Morales JF, Esteban-Gonzalo S, Martín-María N, and Puig-Navarro Y (2023) Spanish Adaptation of the Gender-Related Variables for Health Research (GVHR): Factorial Structure and Relationship with Health Variables. Span J Psychol 26:e25. https://doi.org/10.1017/SJP.2023.25

Dorak MT, and Karpuzoglu E (2012) Gender Differences in Cancer Susceptibility: An Inadequately Addressed Issue. Front Genet 3:268. https://doi.org/10.3389/fgene.2012.00268

DuBois LZ, Gibb JK, Juster RP, and Powers SI (2021a) Biocultural Approaches to Transgender and Gender Diverse Experience and Health: Integrating Biomarkers and Advancing Gender/Sex Research. Am J Hum Biol 33 (1):e23555. https://doi.org/10.1002/ajhb.23555

DuBois LZ, and Juster R-P (2022) Lived Experience and Allostatic Load among Transmasculine People Living in the United States. Psychoneuroendocrinology 143:143:105849. https://doi.org/10.1016/j.psyneuen.2022.105849

DuBois LZ, Puckett JA, Jolly D, et al. (2024) Gender Minority Stress and Diurnal Cortisol Profiles among Transgender and Gender Diverse People in the United States. Horm Behav 159:105473. https://doi.org/10.1016/j.yhbeh.2023.105473

DuBois LZ, Puckett JA, and Langer SJ (2021b) Development of the Gender Embodiment Scale: Trans Masculine Spectrum. Transgend Health 7 (4):287–291. https://doi.org/10.1089/trgh.2020.0088

DuBois LZ, and Shattuck-Heidorn H (2021) Challenging the Binary: Gender/Sex and the Bio-Logics of Normalcy. Am J Hum Biol 33 (5):e23623. https://doi.org/10.1002/ajhb.23623

Duchesne A, Tannenbaum C, and Einstein G (2017) Funding Agency Mechanisms to Increase Sex and Gender Analysis. Lancet 389 (10070):699–699. https://doi.org/10.1016/S0140-6736(17)30343-4

Fausto-Sterling A (2012) Sex/Gender: Biology in a Social World. Routledge, New York/London

Greaves L (2011) Why Put Gender and Sex into Health Research? In: Oliffe JL, Greaves L (eds) Designing and Conducting Gender, Sex, and Health Research. Sage, Los Angeles, pp 3–14

Haraway DJ (1988) Situated Knowledges: The Science Question in Feminism and the Privilege of Partial Perspective. Fem Stud 14 (3):575–599. https://doi.org/10.2307/3178066

Hatzenbuehler ML (2009) How Does Sexual Minority Stigma "Get under the Skin"? A Psychological Mediation Framework. Psychol Bull 135 (5):707–730. https://doi.org/10.1037/a0016441

——— (2014) Structural Stigma and the Health of Lesbian, Gay, and Bisexual Populations. Curr Dir Psych Sci 28 (2):127–132. https://doi.org/10.1177/0963721414523775

Hatzenbuehler ML (2018) Structural Stigma and Health. In: Major B, Dovido JF, Link BG (eds) The Oxford Handbook of Stigma, Discrimination, and Health. Oxford Univ Press, New York, pp 105–121

Hatzenbuehler ML, Keyes KM, and Hasin DS (2009a) State-Level Policies and Psychiatric Morbidity in Lesbian, Gay, and Bisexual Populations. Am J Public Health 99 (12):2275–2281. https://doi.org/10.2105/ajph.2008.153510

Hatzenbuehler ML, and McLaughlin KA (2013) Structural Stigma and Hypothalamic-Pituitary-Adrenocortical Axis Reactivity in Lesbian, Gay, and Bisexual Young Adults. Ann Behav Med 47 (1):39–47. https://doi.org/10.1007/s12160-013-9556-9

Hatzenbuehler ML, Nolen-Hoeksema S, and Dovidio J (2009b) How Does Stigma "Get under the Skin"?: The Mediating Role of Emotion Regulation. Psychol Sci 20 (10):1282–1289. https://doi.org/10.1111/j.1467-9280.2009.02441.x

Hatzenbuehler ML, Phelan JC, and Link BG (2013) Stigma as a Fundamental Cause of Population Health Inequalities. Am J Public Health 103 (5):813–821. https://doi.org/10.2105/ajph.2012.301069

Hehman E, Ingbretsen ZA, and Freeman JB (2014) The Neural Basis of Stereotypic Impact on Multiple Social Categorization. Neuroimage 101:704–711. https://doi.org/10.1016/j.neuroimage.2014.07.056

Homan P (2019) Structural Sexism and Health in the United States: A New Perspective on Health Inequality and the Gender System. Am Sociol Rev 84 (3):486–516. https://doi.org/10.1177/0003122419848723

Homan P, Brown TH, and King B (2021) Structural Intersectionality as a New Direction for Health Disparities Research. J Health Soc Behav 62 (3):350–370. https://doi.org/10.1177/00221465211032947

Hyde JS (2005) The Gender Similarities Hypothesis. Am Psychol 60 (6):581–592. https://doi.org/10.1037/0003-066X.60.6.581

IOM (2011) The Health of Lesbian, Gay, Bisexual, and Transgender People: Building a Foundation for Better Understanding. National Academies Press, Washington, DC

James SA (1994) John Henryism and the Health of African-Americans. Cult Med Psychiatry 18 (2):163–182. https://doi.org/10.1007/bf01379448

Johnson JL, Greaves L, and Repta R (2007) Better Science with Sex and Gender: A Primer for Health Research. Women's Health Research Network, Vancouver

Junker A, Juster RP, and Picard M (2022) Integrating Sex and Gender in Mitochondrial Science. Curr Opin Physiol 26:100536. https://doi.org/10.1016/j.cophys.2022.100536

Juster R-P, Raymond C, Desrochers AB, et al. (2016) Sex Hormones Adjust "Sex-Specific" Reactive and Diurnal Cortisol Profiles. Psychoneuroendocrinology 63:282–290. https://doi.org/10.1016/j.psyneuen.2015.10.012

Juster RP, de Torre MB, Kerr P, et al. (2019) Sex Differences and Gender Diversity in Stress Responses and Allostatic Load among Workers and LGBT People. Curr Psychiatry Rep 21 (11):110. https://doi.org/10.1007/s11920-019-1104-2

Juster RP, Rutherford C, Keyes K, and Hatzenbuehler ML (2024) Associations between Structural Stigma and Allostatic Load among Sexual Minorities: Results from a Population-Based Study. Psychosom Med 86 (3):157–168. https://doi.org/10.1097/PSY.0000000000001289

Kaiser A (2016) Sex/Gender Matters and Sex/Gender Materialities in the Brain. In: Pitts-Taylor V (ed) Mattering: Feminism, Science, and Materialism. Biopolitics. New York Univ Press, New York, pp 122–139

Kajantie E, and Phillips DI (2006) The Effects of Sex and Hormonal Status on the Physiological Response to Acute Psychosocial Stress. Psychoneuroendocrinology 31 (2):151–178. https://doi.org/10.1016/j.psyneuen.2005.07.002

Keller EF (2010) The Mirage of a Space between Nature and Nurture. Duke Univ Press, Durham

Klein F, Sepekoff B, and Wolf TJ (1990) Sexual Orientation: A Multivariable Dynamic Process. In: Geller T (ed) Bisexuality: A Reader and Sourcebook. Times Change Press, New York, pp 136–151

Kozee HB, Tylka TL, and Bauerband LA (2012) Measuring Transgender Individuals' Comfort with Gender Identity and Appearance: Development and Validation of the Transgender Congruence Scale. Psychol Women Q 36 (2):179–196. https://psycnet.apa.org/doi/10.1177/0361684312442161

Krieger N, Jahn JL, and Waterman PD (2017) Jim Crow and Estrogen-Receptor-Negative Breast Cancer: US-Born Black and White Non-Hispanic Women, 1992-2012. Cancer Causes Control 28 (1):49–59. https://doi.org/10.1007/s10552-016-0834-2

Krieger N, Jahn JL, Waterman PD, and Chen JT (2018) Breast Cancer Estrogen Receptor Status According to Biological Generation: US Black and White Women Born 1915-1979. Am J Epidemiol 187 (5):960–970. https://doi.org/10.1093/aje/kwx312

Lee KMN, Rushovich T, Gompers A, et al. (2023) A Gender Hypothesis of Sex Disparities in Adverse Drug Events. Soc Sci Med 339:116385. https://doi.org/10.1016/j.socscimed.2023.116385

Lillie EO, Patay B, Diamant J, et al. (2011) The N-of-1 Clinical Trial: The Ultimate Strategy for Individualizing Medicine? Per Med 8 (2):161–173. https://doi.org/10.2217/pme.11.7

Lucey BP, Nelson-Rees WA, and Hutchins GM (2009) Henrietta Lacks, HeLa Cells, and Cell Culture Contamination. Arch Pathol Lab Med 133 (9):1463–1467. https://doi.org/10.5858/133.9.1463

Luthar SS, Cicchetti D, and Becker B (2000) The Construct of Resilience: A Critical Evaluation and Guidelines for Future Work. Child Dev 71 (3):543–562. https://doi.org/10.1111/1467-8624.00164

Lyapun IN, Andryukov BG, and Bynina MP (2019) HeLa Cell Culture: Immortal Heritage of Henrietta Lacks. Mol Genet Microbiol Virol 34 (4):195–200. https://doi.org/10.3103/S0891416819040050

Mareï N, and Savy M (2021) Global South Countries: The Dark Side of City Logistics. Dualisation Vs Bipolarisation. Transp Policy (Oxf) 100:150–160. https://doi.org/10.1016/j.tranpol.2020.11.001

Masten AS, Best K, and Garmezy N (1990) Resilience and Development: Contributions from the Study of Children Who Overcome Adversity. Dev Psychopathol 2 (4):425–444. https://doi.org/10.1017/S0954579400005812

McEwen BS, and Stellar E (1993) Stress and the Individual: Mechanisms Leading to Disease. Arch Intern Med 153 (18):2093–2101. https://doi.org/10.1001/archinte.1993.00410180039004

Müller R (2020) A Task That Remains before Us: Reconsidering Inheritance as a Biosocial Phenomenon. Semin Cell Dev Biol 97:189–194. https://doi.org/10.1016/j.semcdb.2019.07.008

NASEM (2018) Returning Individual Research Results to Participants: Guidance for a New Research Paradigm. National Academies Press, Washington, DC

Nielsen MW, Stefanick ML, Peragine D, et al. (2021) Gender-Related Variables for Health Research. Biol Sex Differ 12 (1):23. https://doi.org/10.1186/s13293-021-00366-3

Noble D (2012) A Theory of Biological Relativity: No Privileged Level of Causation. Interface Focus 2 (1):55–64. https://doi.org/10.1098/rsfs.2011.0067

O'Connor T (2020) Emergent Properties. Retrieved From https://plato.stanford.edu/entries/properties-emergent/

Pape M, Miyagi M, Ritz SA, et al. (2024) Sex Contextualism in Laboratory Research: Enhancing Rigor and Precision in the Study of Sex-Related Variables. Cell 187 (6):1316–1326. https://doi.org/10.1016/j.cell.2024.02.008

Pearson KA (2011) New Materialisms: Ontology, Agency, and Politics. Radic Philos 167:46–48

Picard M, Juster RP, and Sabiston CM (2013) Is the Whole Greater Than the Sum of the Parts? Self-Rated Health and Transdiscplinarity. Health 5 (12A):24–30. http://dx.doi.org/10.4236/health.2013.512A004

Puckett JA, Brown NC, Dunn T, Mustanski B, and Newcomb ME (2020) Perspectives from Transgender and Gender Diverse People on How to Ask About Gender. LGBT Health 7 (6):305–311. https://doi.org/10.1089/lgbt.2019.0295

Rutter M (2012) Resilience as a Dynamic Concept. Dev Psychopathol 24 (2):335–344. https://doi.org/10.1017/S0954579412000028

Sanchis-Segura C, and Wilcox RR (2024) From Means to Meaning in the Study of Sex/Gender Differences and Similarities. Front Neuroendocrin 73:101133. https://doi.org/10.1016/j.yfrne.2024.101133

Sandberg DE, and Vilain E (2022) Decision Making in Differences of Sex Development/Intersex Care in the USA: Bridging Advocacy and Family-Centred Care. Lancet Diabetes Endocrinol 10 (6):381–383. https://doi.org/10.1016/S2213-8587(22)00115-2

Sebastian-Tirado A, Felix-Esbri S, Forn C, and Sanchis-Segura C (2023) Are Gender-Science Stereotypes Barriers for Women in Science, Technology, Engineering, and Mathematics? Exploring When, How, and to Whom in an Experimentally-Controlled Setting. Front Psychol 14:1219012. https://doi.org/10.3389/fpsyg.2023.1219012

Shaw J, McLean KC, Taylor B, Swartout K, and Querna K (2016) Beyond Resilience: Why We Need to Look at Systems Too. Psychol Violence 6 (1):34–41. https://doi.org/10.1037/vio0000020

Suslovic B, and Lett E (2024) Resilience Is an Adverse Event: A Critical Discussion of Resilience Theory in Health Services Research and Public Health. Community Health Equity Res Policy 44 (3):339–343. https://doi.org/10.1177/2752535x231159721

Tannenbaum C, Schwarz JM, Clayton JA, de Vries GJ, and Sullivan C (2016) Evaluating Sex as a Biological Variable in Preclinical Research: The Devil in the Details. Biol Sex Differ 7:13. https://doi.org/10.1186/s13293-016-0066-x

Timmermans S, Yang A, Gardner M, et al. (2019) Gender Destinies: Assigning Gender in Disorders of Sex Development-Intersex Clinics. Sociol Health Illn 41 (8):1520–1534. https://doi.org/10.1111/1467-9566.12974

Tollefsbol TO (2019) Generational Epigenetic Inheritance, vol 13. Academic Press, Cambridge, MA

van Anders SM, Steiger J, and Goldey KL (2015) Effects of Gendered Behavior on Testosterone in Women and Men. PNAS 112 (45):13805–13810. https://doi.org/10.1073/pnas.1509591112

van Anders SM, and Watson NV (2006) Relationship Status and Testosterone in North American Heterosexual and Non-Heterosexual Men and Women: Cross-Sectional and Longitudinal Data. Psychoneuroendocrinology 31 (6):715–723. https://doi.org/10.1016/j.psyneuen.2006.01.008

Viau V (2002) Functional Cross-Talk between the Hypothalamic-Pituitary-Gonadal and -Adrenal Axes. J Neuroendocrinol 14 (6):506–513. https://doi.org/10.1046/j.1365-2826.2002.00798.x

Williams JS, Fattori MR, Honeyborne IR, and Ritz SA (2023) Considering Hormones as Sex- and Gender-Related Factors in Biomedical Research: Challenging False Dichotomies and Embracing Complexity. Horm Behav 156:105442. https://doi.org/10.1016/j.yhbeh.2023.105442

9

The Impossible Task of Disentangling Gender/Sex from Racialized and Other Marginalized and Oppressed Intersections

A Structural Intersectionality Approach to Health Inequities

Lisa Bowleg, Arianne N. Malekzadeh, and Katarina E. AuBuchon

Abstract Invited to answer the question, "How do gender/sex inform intersectional understandings of structural inequalities on health?" this chapter holds that the premise of the question is incorrect on two grounds. First, gender/sex and their entanglements do not inform intersectional understandings of structural inequalities on health; multiple and intersecting systems of structural oppression (e.g., racism, sexism, *and* cisgenderism) do. Second, it is impossible to separate gender/sex and their entanglements from other key intersectional positions such as racialized and sexual minoritized status. Using a structural intersectionality approach grounded in examples and data of Black and other racialized groups primarily in the United States, this chapter addresses the controversy of the single-axis assumption that gender/sex is, can, or should be disentangled from other key historically marginalized intersectional positions. It examines the unsolved problem of the relative absence of intersectionally specific language (other than gendered racism) to describe the health inequities that groups, such as Black women, men, and transgender, nonbinary, and gender-expansive people, experience because of multiple and interlocking oppression based on their intersectionally distinct positions as Black women, Black men, and Black transgender, nonbinary and gender-expansive people. Finally, the

Lisa Bowleg (✉)
Dept. of Psychological and Brain Sciences, George Washington University,
Washington DC 20052; The Intersectionality Training Institute, Philadelphia, PA 19119, USA
lbowleg@email.gwu.edu

Arianne N. Malekzadeh   Dept. of Psychological and Brain Sciences, George Washington University, Washington DC 20052, USA

Katarina E. AuBuchon   Lombardi Comprehensive Cancer Center, Georgetown University, Washington DC 20057, USA

© Frankfurt Institute for Advanced Studies (FIAS) 2025
L. Z. DuBois et al. (eds.), *Sex and Gender*, Strüngmann Forum Reports,
https://doi.org/10.1007/978-3-031-91371-6_9

open question about the lag of structural intersectionality research in the field of gender/sex and beyond is addressed.

Keywords: Intersectionality, structural intersectionality, health inequities, gendered racism, structural racism

1 Introduction: A Necessary Inversion

Invited by the Ernst Strüngmann Forum to provide a background paper for this Forum, we were asked to address the question: "How do gender/sex inform intersectional understandings of structural inequalities on health?" We instead offer what we deem to be an essential inversion of the question. The reason? From an intersectionality perspective, the premise of the question is incorrect. Akin to Ta Nehisi Coates's corrective in his book *Between the World and Me* that "race is the child of racism, not the father" (Coates 2015:7), gender/sex and their entanglements are the children of intersecting systems of structural oppression that shape health inequities, not the parents. As such, gender/sex and their entanglements do not inform intersectional understandings of structural inequalities on health; intersecting systems of structural oppression such as racism, sexism, *and* cisgenderism (to name just a few) do. Thus, the question we address is: How do structural inequalities shape intersectional health inequities? Another problem with the original question is its single-axis assumption that gender/sex is, can, or should be separated from other key intersectional positions, or prioritized.

Intersectionality, a critical theoretical framework historically grounded in the lived experiences of American Black women (Cooper 1892; Jacobs 1861/2015; Stewart 1831), Black lesbian feminist activism (Combahee River Collective 1977), and Black (Collins 1990; Crenshaw 1989, 1991; Davis 1983b; Hooks 1981; Hull et al. 1982; Lorde 1984) and Chicana (Anzaldúa 1987/2012, 1990; Moraga and Anzaldúa 1981) feminist scholarship is foundational to our argument that gender/sex cannot be disentangled from other intersectional positions. Intersectionality illuminates that gender/sex are entangled not just with each other, but with multiple and interlocking and marginalized social categories such as racialized minority status, socioeconomic position, and sexual (e.g., lesbian, gay, bisexual, asexual), gender minority status (e.g., transgender, nonbinary, gender nonconforming).

As for the gender/sex entanglement question that animates this Forum, we find gender/sex to be so enmeshed with other key intersections (i.e., racialized status) and interlocking systems of oppression (i.e., structural racism *and* heterosexism), that it is impossible—at least for us—to focus exclusively on the gender/sex enmeshment absent its other entanglements. In the United States, the country and context from which we write, sex, like gender, has historically been and *is* highly racialized. Consider, for example, pregnancy, a dimension of biological sex. Pregnancy is highly gendered (e.g., gender ideologies of girls and women as warm and nurturing, Judeo-Christian norms about the sanctity of motherhood, gender/sex stereotypes about "maternal genes"), *and* it is also very much racialized; a fact shrewdly alluded to in the subtitle of the book, *The Ethos of Black Motherhood in America: Only White Women Get Pregnant* (Harper 2021).

Whereas the historical myth of White womanhood positioned White women as innocent, virtuous, nurturing victims in need of protection from White men, Black women were granted no such protection: not prior to pregnancy, not during pregnancy, and not postpartum when required to return to field or domestic household labor, or bear the horror of having their children (many conceived through rape perpetrated by White male enslavers) sold into slavery (Davis 1983a; Hooks 1981). Centuries later, disproportionately high rates of maternal morbidity and mortality as well as infant mortality among Black women in the United States underscore the historical enmeshment of sex/gender with racism. In a 2019 *American Journal of Public Health* commentary, Owens and Fett (2019:1343) trace this contemporary intersectional health inequity to the commodification of enslaved Black women's bodies and their ability to bear children:

> Seventeenth century European exploration literature also depicted African women, in comparison with European women, as especially capable of both childbearing and field labor. The principle of *partus* [*sequitur ventral*; the legal principle that a child's status as enslaved followed the status of the mother] not only defined legal slavery but also carved out a racial [and sexed/gendered] distinction. Continuing up through the Civil War, White women's childbearing built free patriarchal lineages while southern laws forced enslaved Black women to bear children who would build capital for enslavers.

This historical backdrop of structural racism, and its entanglements with sex/gender and other forms of oppression based on class, sexuality, immigration status, and ability status (to name just some) inform our use of the terms gender/sex and sex/gender synonymously throughout this chapter. For us as intersectional scholars, the gender/sex entanglement is *always* enmeshed with other forms of interlocking structural oppression.

Intersectionality's utility to the topics of health and health inequity lies in its attention to power; namely, how multiple and intersecting systems of structural oppression (e.g., racism, sexism, cisgenderism, heterosexism, class exploitation, ageism, and ableism, to name just a few, determine health and foster health inequities) (Bowleg 2012). Alas, intersectionality has veered far and wide beyond its historical origins in Black feminist activism and organizing. As it has traveled into more mainstream environs, such as the academy, and government agencies such as the US National Institutes of Health (Alvidrez et al. 2020; Rausch 2018) and the Government of Canada (PHAC 2022), many intersectionality scholars have lambasted the "flattening" of intersectionality (Bowleg 2023; Collins 2019:253–285; May 2015). Flattening describes the myriad ways that tenets foundational to the framework (e.g., power, social inequality, and social justice) have been stripped from intersectionality as it becomes more mainstream. In the context of health inequities, flattening typically manifests as a focus on multiple "*stigmatized identities*" or "*intersectional identities*," descriptions that implicitly blame individual *identities* rather than social processes, such as intersectional stigma/discrimination or interlocking structural oppression, as the problem (Bowleg 2022).

The slippage occurs even among those who decry the flattening. Take for example the Public Health Agency of Canada's definition of intersectionality that addressed intersecting discrimination. Thereafter, the agency flattened the framework with its follow up description of intersectionality as [a framework] that "… refers to the reality that we all have many identities that intersect to make us who we are."

While it is true that everyone has multiple intersecting identities, intersectionality is not simply about identities. Intersectionality is fundamentally about attention to power and social inequality (Cho et al. 2013; Collins and Bilge 2020:1–36). In the context of health and health inequities, intersectionality highlights how power and multiple systems of interlocking structural oppression shape health and health inequities for racialized and other minoritized (e.g., gender, sexual) groups, as well as their more privileged (e.g., White, middle and upper class, heterosexual, able-bodied, cisgender men) counterparts. This focus on power necessarily draws attention to macro-level structures and systems that shape and constrain the ability of individuals to engage in health promoting behaviors.

Structural intersectionality describes how the convergence of multiple intersectional structural oppression (e.g., racism, heterosexism, sexism, cisgenderism, ableism, *and* class exploitation) creates qualitatively different experiences (e.g., health inequities) for groups historically oppressed at multiple intersections, compared with their more privileged counterparts (Crenshaw 1991). Although structural intersectionality applied to health inequities is in its infancy, it nonetheless "has considerable utility for understanding social inequalities in population health" (Homan et al. 2021:354), primarily because "intersecting systems of oppression are likely to shape health via an array of mechanisms, including differential access to economic and flexible resources (e.g., social capital, power, prestige, autonomy, self-esteem) and increased exposure to health risks such as social stressors, toxic living conditions, discrimination, stigma, and relative deprivation."

A structural intersectionality approach is also committed to critical praxis as a tool for social justice and health equity (Collins 2015; Collins and Bilge 2020). As critical praxis, this approach can promote advocacy to repeal inequitable laws, policies, and practices; implement and evaluate multilevel (e.g., individual, community, and structural) interventions that advance health equity at the population level; inform the monitoring and enforcement of anti-stigma/discrimination laws and policies that are intersectional; leverage grassroots and community-based activism to advance health equity; and develop health equity assessment tools to track the health equity effects of legal and policy changes.

Informed by intersectionality, in general, and structural intersectionality, in particular, we frame our focus on intersectional health inequities around the three concluding questions that this Ernst Strüngmann Forum invited us to ponder: one controversy, one unsolved problem, and one open question. But first, we get intersectionally reflexive about how our intersectional positions and perspectives inform this chapter. Thereafter, we address the question posed by some critics about the applicability of intersectionality beyond Black and other racialized people in the United States.

1.1 When, Where, and How We Enter: Reflexivity

Reflexivity, a hallmark of feminist and qualitative scholarship, describes the process by which scholars and researchers explicitly describe how their intersectional demographic positions, disciplinary training, values, political commitments, and biases shape their work (Berger 2015; Finlay 2002). The authors of this chapter share common identities as cisgender (all use she/her pronouns), bisexual, highly educated, able-bodied women who live in the United States. We also share doctoral training as applied social psychologists from The George Washington University, identify as critical psychologists, and bring to this work a history of collaboration on intersectionality (Bowleg et al. 2023) and the effects of structural racism and structural sexism on health (Bowleg et al. 2022; Post et al. 2024).

The lead author (Bowleg) identifies as Black and upper class, and is an immigrant to the United States, having been born and raised in The Bahamas. A Black feminist, critical health equity researcher, and leading scholar of the application of intersectionality to social and behavioral science health equity research, she has lived in the United States for almost 40 years. Her experiences as a Black woman in the United States, combined with her research programs that center on intersectionally diverse Black people in the United States—HIV prevention for Black men at different intersections of sexuality and socioeconomic status (e.g., Bowleg et al. 2013), and intersectional stress and resilience among Black lesbian, gay, and bisexual people (e.g., Bowleg 2013; Bowleg et al. 2003)—have honed her attention to how intersecting systems of structural oppression constrain opportunities for positive health and well-being and drive health inequities for historically oppressed groups. The second author (Malekzadeh) identifies as Persian American and upper class, and is the child of immigrants to the United States. She is committed to intersectional feminism and health equity research, particularly to advance the well-being of women. The third author (AuBuchon) identifies as White, German-American, queer, middle-class, and agnostic. Her experiences growing up in the southern US "Bible Belt" have shaped how she researches health inequities and attune her to the specific ways that structural racism and Whiteness foster health inequities.

Health and health inequities in the United States are the focus of our work, independently and collectively. Moreover, given intersectionality's historical roots in the United States and lineage in critical race theory, many of our examples focus on US populations and topics (e.g., structural racism, lack of universal health care, gender-affirming care bans) that may not be as relevant or salient in international contexts. We also made a conscious decision to highlight groups who are often intersectionally invisible (Purdie-Vaughns and Eibach 2008) in health equity scholarship and research—specifically, people with disabilities, Native Americans, and Alaska Natives—as well as issues of class and class exploitation (poverty, in particular), which are also underexamined in most US public health and intersectionality scholarship.

1.2 Does Intersectionality Apply Beyond Black and Other Racialized People in the United States?

Intersectionality's origins in the United States, combined with our positionality as US-based intersectionality scholars, inform the examples and data that form the foundation of this chapter. Intersectionality's reach as an analytic lens and tool for critical praxis, however, far transcends US borders. For example, Collins and Bilge (2020:1–36) discuss the deployment of intersectionality across a variety of global reproductive justice, climate change, sexual violence resistance, and human rights projects. As for the topic of intersectionality and health, systematic reviews highlight intersectionality's utility for health intervention research in high-income countries such as Canada, the United Kingdom, Australia, New Zealand, and Norway (as well as the United States) (Tinner et al. 2023) as well as in low- and middle-income countries such as India, Swaziland, Uganda, and Mexico (Larson et al. 2016), and Sub-Saharan Africa (Lynch et al. 2020).

There are valid questions to be raised about intersectionality's applicability beyond US borders. For example, in an incisive and insightful article, Nivedita Menon questions the universal validity of intersectionality with her assertion that the "the politics of *caste, religious community identity* and *sexuality* [italics in original]" (Moren 2015:38) in India long disrupted the notion of woman as a stable or homogeneous category. Menon rightly lambastes the timeworn assumption that concepts birthed in the Global North and White European West are universal, an assumption never granted to concepts that emerge from the Global South. Building on Nira Yuval-Davis's 2006 recounting of how Kimberlé Crenshaw introduced the concept of intersectionality in an invited talk at the 2001 World Conference against Racism in Durban South Africa, and the subsequent influx of funds for intersectional global projects, Menon also questions the influence of UN funds on the global uptake of intersectionality and how such funding flattens and depoliticizes intersectionality. There are also counternarratives about intersectionality's global travels, such as the amusing story that Patricia Hill Collins recounts in a conversation with a small group of Afro-Brazilian women scholars-activists who approached her after her keynote on US Black feminism and intersectionality in Brasilia, Brazil: "We thought intersectionality was for White feminists and that it had nothing to do with us" (Collins 2015:15).

In her wonderfully provocative article, "Re-thinking Intersectionality," Jennifer Nash encourages intersectionality and antiracist scholars to grapple with thorny questions such as "who is intersectional"; that is, "whether all identities are intersectional or whether only multiple marginalized subjects have an intersectional identity" (Nash 2008:9). Elsewhere, in advancement of the argument that intersectionality is primarily about intersecting power relations, not multiple intersecting identities, Bowleg (2023:105) asserts that "everyone has multiple intersecting identities," offering in essence a "both/and" answer to Nash's question. This attention to what Collins has called the "ebb and flow" of power and privilege is not new: "depending on the context, an individual may be an oppressor, a member of an oppressed group, or simultaneously oppressor and oppressed" (Collins 1990:225). Yuval-Davis's (2017) concept of situated intersectionality is also apt: "situated intersectionality analysis…[is] highly sensitive to the geographical, social and temporal locations of the particular individual or collective social actors examined by it, contested, shifting and multiple as they usually are" (Yuval-Davis 2017:5). Central to Yuval-Davis's situated intersectionality approach is the notion that intersectionality should be applied to all people, not just marginalized and racialized women, and that "our analytical intersectional gaze has to be directed also towards the powerful and not just the powerless" (Yuval-Davis 2015:638).

Other questions about intersectionality's applicability beyond the US context are more problematic when they question the epistemic validity of intersectionality or reveal epistemic resistance primarily *because* of its origins in US Black women's activism and knowledge production (Bilge 2014; Collins 1990, 2019; Settles et al. 2020). When concepts that White US or European scholars introduce are embraced with little or no concerns about their geographic boundedness—take, for example, the concept of the gender/sex entanglement—and the concepts that US Black women originate are scrutinized and interrogated for their applicability beyond their contexts, such questions implicitly and explicitly devalue and marginalize US Black women's intellectual contributions.

Questions about the applicability of intersectionality beyond the US context also hint at interest in or the legitimacy of a "whitened" intersectionality. "Whitening intersectionality" is the term that Sirma Bilge (2014) uses to describe "the ways of doing intersectionality that rearticulate it around Eurocentric epistemologies," one that eschews an emphasis on race and racism. Structural racism, however, is as relevant to those exposed to it as well as its beneficiaries. Thus, intersectionality's applicability as a critical theoretical framework transcends the United States and, consistent with its core themes—social inequality, intersecting power relations, social context, relationality, complexity, and social justice (Collins and Bilge 2020)—is ideally suited to analyze the complex intersections of marginalization and oppression, as well as power and privilege in the United States and beyond.

2 The Controversy: Gender/Sex Cannot Be Disentangled from Other Intersectional Positions

The primary controversy we address strikes at the heart of this Forum's single-axis focus on gender/sex and their entanglements. Informed by intersectionality and echoing critiques that gender-first, gender+, sex- and gender-based analysis (SGBA), and sex- and gender-based analysis plus (SGBA+) approaches reflect a fundamental misunderstanding of intersectionality (Anglin 1998; Hankivsky and Hunting 2022; Rotz et al. 2022), we reject the notion that gender/sex can (or should) be isolated as a primary axis of analysis. Single-axis approaches—such as those focused exclusively on gender/sex, gender-first, SGBA and SGBA+ (see, e.g., Government of Canada 2017)—violate a core tenet of intersectionality that gender/sex is mutually constituted with other intersectional positions, such that one cannot understand gender/sex absent its intersection with other historically marginalized positions (Collins 1990; Combahee River Collective 1977; Crenshaw 1989, 1991).

Even when they espouse intersectionality, gender-first or "plus" approaches reflect the primacy of the White racial frame (Feagin 2013) in shaping discourses about gender/sex, and increasingly intersectionality (Bilge 2014; Carbado 2013). The White racial frame describes "an overarching White worldview that encompasses a *broad and persisting set of racial stereotypes, prejudices, ideologies, images, interpretations and narratives, emotions and reactions to language accents, as well as racialized inclinations to discriminate* [italics in original]" (Feagin 2013:3).

Intersectionality disrupts the White racial frame that it is even possible to disentangle gender/sex as a unit of analysis; so does history. During slavery, for example, racism interlocked with gender/sex to justify the enslavement and brutal oppression of Black people. Racist notions about Black women's intrinsic inferiority and promiscuity justified White men's rape and sexual violence of Black girls and women. Sex/gendered racism was also the lifeline of the slavery repopulation project; Black women were forcibly (and legally) tasked with the role of reproducing the free labor source of enslaved people. Thus, in the context of slavery, it is simply not possible to talk about gender/sex and their entanglements—enslaved girls and women who bore children, for example—absent its intertwinement with racialized minority status and Black girls' and women's distinct and specific intersectional location in the "matrix of domination" of racism and sexism (Collins 1990:225).

The history of modern American gynecology further affirms the entanglements of racism with gender/sex. Starting in 1844, enslaved Black women in Alabama were subjected to "medical bondage"—gynecological "experiments" without anesthesia—by Dr. James Marion Sims, often called "the father of modern gynecology" (Owens 2017). Further underscoring the entanglements of racialization and racism with gender/sex, Irish women who were racialized as not-White were also subjected to medical experimentation (Owens 2017). From slavery onward, Black people, regardless of gender/sex, have been subjected to "medical apartheid" or medical experimentation (Washington 2006). Moreover, during and post-slavery, racist tropes about the innate hypersexuality of Black men legitimized their brutal and violent subjugation, in part to protect the putative innate purity and femininity of White women (Davis 1983a).

The historical vestiges of the intermingling of racialization, racism, and gender/sex endure. In her book *Fearing the Black Body: The Racial Origins of Fat Phobia* for example, Sabrina Strings (2019) describes Europeans' derision of Black people's fatness as evidence of African "savagery" and immorality, and the privileging of White people's leaner bodies as superior in the 18th century. Highlighting how racism, sexism, and classism intertwine in the lives of Black women, Strings argues that contemporary dialogues about obesity among poor Black women in the United States are historically rooted in racist and religious ideologies that "have been used to both degrade Black women *and* discipline White women" (Strings 2019:6). In more recent work, Strings (2023) uses pseudoscience in medical discourses about the body mass index to illuminate how racism is entangled with gender/sex with 19th and 20th century White supremacist notions about normal or standard bodies.

For a more contemporary take on the entanglement of racism with that of gender/sex, consider the case of Caster Semenya, the two-time Olympic South African gold medalist (see also Pape, this volume; Karkazis et al. 2012; Karkazis and Jordan-Young 2018). In her book, *The Race to Be Myself,* Semenya poignantly illustrates how racism and cisgenderism have intersected to oppress Black women and other women of color athletes, many of whom are poor, in professional sports (Semenya 2023). Born with what the International Association of Athletics Federation calls "differences of sexual development," women like Semenya who have higher than typical levels of testosterone are deemed to have an unfair advantage in women's sports. Semenya details the gynecological tests and other gender testing to which she's been subjected in order to compete. When one of these tests found that

Semenya had XY chromosomes, a team gynecologist recommended that Semenya have surgery to have her undescended testicles removed, a recommendation she rejected. In lieu of surgery, Semenya took birth control pills to boost her estrogen, a move that resulted in severe side effects. Underscoring how gender/sex are entangled not only with each other but, for Black women, with racism and sexism, Semenya makes a compelling argument for intersectional multiplicativity; that is, the distinct intersectional space that Black women occupy beyond the mere addition of racial group and gender/sex. Writing about the speculation that she would win her race at the 2016 Olympic Games because of her unfair genetic advantage, she reflects (Semenya 2023:251–252):

> [People] thought nothing of cheering on the seeming inevitability of wins by genetically gifted athletes like the sprinter Usain Bolt, who boasted a stride that was far longer than his peers. No one suggested Michael Phelps's dominance in the pool was unfair and he should have to take medication to ensure that he produced just as much lactic acid as his competitors or have surgery to fix his hypermobile joints. The swimmer Katie Ledecky was never accused of being a man because she smashed multiple world records and her ever-improving times in several events would qualify her for the men's Olympic trials. But they said such things about me because I represented something that was seen as abnormal.

In setting up these contrasts between Bolt (a Black man), Phelps (a White man), and Ledecky (a White woman), Semenya deftly illustrates the folly that it is possible to disentangle gender/sex from its intersections with racialized group. For Black women like Semenya, and other racialized and minoritized groups across the gender/sex continuum, racism interlocks with sexism in the case of cisgender women, misandry in the case of cisgender men, and cisgenderism in the case of transgender, nonbinary, and gender-expansive people to shape health and health inequities.

3   The Unsolved Problem: Language Limits Understanding about Intersectional Health Inequities

The language of gender/sex and their entanglements, particularly when it obscures entanglements with other intersectional positions, is not neutral. Language never is. Language offers critical insights into the role of power (Foucault 1978/1990; Freire 1970/2000; Richardson 2011). This is particularly so with the unsolved problem of the language of gender/sex and their entanglements with other intersectional positions, such as racialized minoritized status and sexual and gender minority status.

Albeit reflecting on the phenomenon of racial and sexual power dynamics in intimate relationships, not health and health inequities, couples therapist Ornal Guralnik deftly captured why language, or its absence, is an unsolved problem for the topic of gender/sex and intersectional health inequities (Guralnik 2023:39): "Language tends to evolve to better accommodate experiences of the dominant social group, leaving other experiences obscured from collective understanding, and thus silently perpetuating bias and harm. When these gaps are filled by new concepts, social change can follow."

Pursuant to Guralnik's argument, we examine the issue of intersectional language—both its presence and absence—about gender/sex and their entanglements with racialized and gender minority status. We begin with the terms *gendered racism* and *misogynoir* and what they illuminate about the intersectionally distinct health inequities that Black women experience. Thereafter, we explore how the relative void of intersectionally specific language about health inequities among Black transgender, nonbinary, and gender-expansive people and Black boys and men relates to issues of bias and structural violence.

3.1 Gendered Racism and Misogynoir

Gendered racism describes how sexism and racism "narrowly intertwine and combine under certain conditions into one hybrid phenomenon" (Essed 1991:31). Racism is also sexed, but we are not aware of any literature that has focused on sexed racism, sexed/gendered racism, or gendered/sex racism; our use of the latter two terms denotes the gender/sex or sex/gender-entangled version of racism. Although Black women in the United States have been the focus of most of the theory and research on gendered racism (see, e.g., Battle and Carty 2022; Jackson et al. 2001; Jones et al. 2022; Thomas et al. 2008), a smaller body of research has also applied the concept to Asian American men (Liu et al. 2018) and African American men (Hartfield et al. 2018; Ramseur II et al. 2024; Schwing et al. 2013). Misogynoir, a term coined by Bailey and Trudy in 2008, describes the "anti-Black racist misogyny that Black women experience" (Bailey and Trudy 2018:762). Although similar to gendered racism in meaning, misogynoir is not as commonly used within mainstream social and behavioral sciences and intersectional health equity research circles.

3.1.1 Black Women's Health Inequities

Intersectional neologisms such as gendered racism and misogynoir are important because they pinpoint Black women's intersectionally specific experiences with and exposure to interlocking racism and sexism. As Kimberlé Crenshaw (1989) explained in her trailblazing article on intersectionality, Black women share similar discrimination experiences with White women and Black men, yet they also often experience additive discrimination on the basis of their "race" and sex/gender. The crux of gendered racism and misogynoir is that "sometimes [Black women] experience discrimination as Black women—not the sum of race and sex [and gender] discrimination, but as Black women" (Crenshaw 1989:139).

The evidence for gendered racism and misogynoir in Black women's health inequities is stark. In 2021, for example, Black women in the United States, regardless of socioeconomic status (i.e., education, income), were three times more likely than non-Latino White women to die from pregnancy-related causes (Hoyert 2022). Similarly, according to the CDC, the preterm (i.e., babies born prematurely before the completion of 37 weeks of pregnancy) birth rate for Black women (14.4%) in 2020 was roughly 50% higher than that of White and Latina women (9.1% and 9.8%, respectively). The terms gendered racism and misogynoir deftly describe the multiplicative effects of Black women's experiences of interlocking structural oppression, thus highlighting why single-axis (i.e., gender/sex only) and additive (i.e., gender/sex plus "race") perspectives offer limited understanding of pregnancy-related health inequities (Burton Wanda 2022; Cazeau-Bandoo and Ho 2022; Jackson et al. 2001; Laster Pirtle and Wright 2021; Markin and Coleman 2023; Patterson et al. 2022; Rosenthal and Lobel 2020).

Gendered racism is the focus of a small and burgeoning area of research on US Black women's inequitable pregnancy outcomes (e.g., Jackson et al. 2001; Patterson et al. 2022; Rosenthal and Lobel 2020; Vedam et al. 2019). Underscoring their privileged status, the role of health-care providers in discriminating against Black women during pregnancy-related care has been woefully underexamined (for an exception, see Chambers et al. 2022). Yet, research documents the sexed/gendered racist stereotypes of health-care providers—who surveil and classify Black women as poor, having too many children, sexually promiscuous, or on drugs—and their disrespectful interactions with Black women (Mehra et al. 2020; Rosenthal and Lobel 2020). The resultant pregnancy-related stress serves as a potential pathway to Black women's adverse pregnancy-related outcomes.

Media reports of Black women celebrities' childbirth experiences (Chiu 2018; Williams 2022) and research such as the Giving Voice to Mothers Study (Vedam et al. 2019) document the unequal treatment and mistreatment that Black women routinely face during pregnancy-related care. Using patient-designed survey items to assess mistreatment during childbirth, Vedam et al. (2019) found that compared with White women, women of color and low-income women of color were more likely to report that health-care providers ignored or refused their requests and shouted at, scolded, or threatened them. The study also found that regardless of the woman's racial status, having a Black partner increased mistreatment. Further bolstering the need to focus on health-care providers' intersectional discrimination as a barrier to Black women's health equity, results from a Florida study found that death rates for newborns born to Black mothers—particularly for more complicated births and in hospitals that delivered more Black babies—were significantly higher when the physician who delivered and cared for the newborn was White (Greenwood et al. 2020).

Gendered racism, however, is not limited to pregnancy-related care. A classic 1999 study comparing referrals for cardiac catheterization for older Black women and men and White women and men found that Black women were the only group with the lowest rates of referral (Bowleg and Bauer 2016; Schulman et al. 1999). More recent research has found that Black women living with cancer had the lowest probability of being referred for psycho-oncology counseling (2%) compared with White women (10%), Black men (9%), and White men (5%) (Aburizik et al. 2023). When nursing caseloads decreased, referrals increased for all groups, except Black women. Other studies have linked gendered racism to Black women's psychological distress and depression (Jones et al. 2022; Thomas et al. 2008).

Albeit still a relatively small area of intersectional health equity research, the existence of the term gendered racism paves the way for broader conversations, research, and advocacy than might have happened absent the term. A case in point: in 2022, the US Centers for Disease Control and Prevention published, but has subsequently removed (status: 28.1.2025), a blog entitled *Gendered Racism Among*

Women of Color by S. Battle and D. Carty as part of its Conversations in Equity program. A March 2024 Google Scholar Search for the keywords "gendered racism and health" yielded 6,300 articles on the topic. Analogous to Gloria Steinem's wry observation, "When I was growing up in Toledo, there was no term for domestic violence; it was just called life" (National Network to End Domestic Violence 2016), intersectionally specific terms such as gendered racism and misogynoir are vital to health equity because they affirm the reality of Black women's shared experiences rendered invisible and silenced by the absence of language to name it. As these terms enter common parlance and academic disciplines, they will in turn catalyze research, advocacy, intervention, and policy and legislative action. Intersectionally specific language matters.

3.2 The Void of Intersectionally Specific Language for Black Men and Transgender People

Although the term gender has historically defined socially constructed norms, behaviors, and roles for cisgender girls and boys, women and men, Whiteness, as the eugenics and social Darwinism movements show, has historically been imbricated with gender/sex (Guthrie 2004; Owens 2017). In light of this racist history, it is hardly surprising to find—with the exception of the terms gendered racism and misogynoir, which are not yet common in mainstream scholarship and research—that language that centers the particular intersectional experiences of people of color is virtually nonexistent. Indeed, as we reviewed the literature on intersectional health inequities, we were struck by the relative void of language available to describe the distinct intersectional experiences of racism and misandry that Black, Latino, Native American, and other dark-skinned (i.e., colorism) men of color and transgender, nonbinary, and gender-expansive people disproportionately experience. Terms such as anti-Black misandry (Johnson 2022; Smith et al. 2020), anti-Black transmisogyny (in the case of Black transgender women and transfeminine people; Bailey and Trudy 2018; Krell 2017), and anti-Black transmisandry (in the case of Black transgender men and transmasculine people) are emergent and appear to be isolated to the humanities. None of these terms is common in the mainstream social and behavioral sciences or intersectional health equity literature.

3.2.1 Structural Violence: Black Transgender and Gender-Expansive People and Health Inequities

No intersectional neologism yet exists to describe the disproportionate health inequities that Black and Latino transgender, nonbinary, and gender-expansive people experience compared with their cisgender and White counterparts. Affirming the link between systems of interlocking structural oppression and health inequities are studies that show that transgender adults are more likely to be Black and low-income compared to their cisgender peers, and more likely to be uninsured (Koma et al. 2020). In line with our central argument that gender/sex cannot be disentangled from other intersectional positions and that health inequities are multiplicative, not simply additive, the *Injustice at Every Turn: A Report of the National Transgender Discrimination Survey* noted that transgender people of color reported lower access to health care overall, and that Black respondents "fared worse than all others" in many of the areas that the report examined, such as poverty, discrimination, and violence (Grant et al. 2011). Four years later, *The Report of the 2015 Transgender Survey* documented the "clear and disturbing" tenacity of intersectional inequity: "…transgender people of color experience deeper and broader patterns of discrimination than White respondents and the US population" (James et al. 2016:6). The absence of intersectionally specific language to describe the staggering and disproportionate health inequities that Black and other transgender people of color face—not as the addition of racism and cisgenderism but *because* of interlocking oppression based on being a Black transgender or other transgender person of color—is a form of structural violence.

Structural violence describes the myriad ways that "people are socially and culturally marginalized in ways that deny them the opportunity for emotional and physical well-being, or expose them to assault or rape, or subject them to hazards that can cause sickness and death" (Anglin 1998:145). The concept of structural violence is expansive in the sense that it encompasses how social and government policies sanction certain norms (e.g., White Christian cisgender heteronormative family values and norms; Dowland 2009) and castigate all others, provide financial resources and support for certain groups but not others, and fortify policies such as mass incarceration and police brutality for racialized and minoritized groups in service of maintaining order (Anglin 1998).

3.3 Police Brutality and Lethal Violence

In the United States, police shootings are a leading cause of death for Black boys and men, who are 2.5 times more likely to be killed by police than White boys and men (Edwards et al. 2019). This violence is often and correctly framed through the lens of structural racism, the "totality of ways in which societies foster racial discrimination throughout mutually reinforcing systems of housing, education, employment, earnings, benefit, credit, media, health care and criminal justice" (Bailey et al. 2017:1453). There is also an important need to foreground the issue as intersectional. Specifically, "being Black, but more distinctively, being a Black male in America seems to increase dramatically the chances that someone is likely to have an encounter with the police where the civilian ends up dead" (Hartfield et al. 2018:157). Beyond police violence, the distinct intersectional space that Black boys and men occupy is also "a robust marker of who is likely to experience more unfavorable and unfair outcomes in criminal justice and across other key sectors" (Hartfield et al. 2018:157) such as health care, education, employment, to name just a few. Viewing police violence against Black boys and men through the prism of anti-Black misandry is important because it spotlights how historically racialized and gendered stereotypes of Black men as violent, rapists, and threatening serves to rationalize state-sanctioned violence against them. This knowledge, in turn, is vital to inform public health and legal interventions that could mitigate this tragic and devastating injustice and inequity.

Although Black boys and men bear the disproportionate brunt of nonlethal and lethal police violence, Black cisgender and transgender girls and women are also explicitly targeted. In line with gendered racism, misogynoir, and transmisogynoir, they are also more likely to be victims of lethal and nonlethal police violence, including sexual violence, than their White counterparts. For example, Black women have a 1.4 greater risk over a lifetime of being killed by police compared with White women (Edwards et al. 2019). The framing of police violence against Black people as solely a structural racism problem, rather than as a structural intersectional problem (i.e., the intersection of racism and sexism, and/or anti-Black misandry, and/or transmisogynoir, and/or heterosexism, and/or cisgenderism) birthed the African American Policy Forum's #SayHerName (Crenshaw and African American Policy Forum 2023; Crenshaw and Ritchie 2015) campaign. This project was developed to raise awareness and document the numerous cases of police killings of Black cisgender and transgender girls and women, counter public silence, and advocate for an end to the violence.

In the United States, homicides of transgender people, particularly those who are Black transgender women, are an alarming example of structural violence at the intersections of racism, cisgenderism, and in the case of Black transgender women, anti-Black transmisogyny (Krell 2017). A 2024 report by Everytown for Gun Safety Support Fund, one of a handful of organizations tracking violence against transgender and gender-expansive people in the United States, documented 263 homicides of transgender, nonbinary, and gender-expansive people between 2017 (the year the organization started tracking the violence) and 2023. The data are both staggering and disturbing. Black transgender women accounted for 50% of the 35 gun homicides of transgender people in 2023 and 60% of the gun homicides of transgender people between 2017 and 2023.

With the notable exception of the #SayHerName project's work and the Hartfield et al. (2018) article on police violence against Black men as sexed/gendered racism, police violence against Black people is rarely framed in intersectional terms. It should be. More intersectionally specific language is needed to advance knowledge about the "specific and particular" (Crenshaw 1989:166) ways that sexism is racialized for Indigenous, Black, and Latina women who are disproportionately targets of violence in general (i.e., not just by police). Violence statistics for Indigenous women in the United States are stunning. For example, research from a 2018 Urban Indian Health Institute study found that Indigenous women were 10 times more likely to be murdered compared with women from other ethnic groups (Urban Indian Health Institute 2018).

The precedence of the application of the term *gendered racism* to men of color in research (Hartfield et al. 2018; Liu et al. 2018; Ramseur II et al. 2024; Schwing et al. 2013) affirms that researchers in the men and masculinities field deem the term to be gender neutral and sufficiently expansive to encompass violence against Black boys and men. This notwithstanding, we advocate for more intersectionally gender-precise language to denote the particular ways that racism is gendered and sexed for cisgender girls and women, cisgender boys and men, and transgender, nonbinary, gender-expansive, and intersex people. Two important issues are at stake. First, the term gendered racism was coined and has historically been used to refer to Black and other women of color (Essed 1991). Thus, the polysemous use for all genders is confusing and conceptually inconsistent. Second, having more intersectionally specific language, such as anti-Black misandry (or other intersectional neologisms), to describe the historical and stereotypical rationalizations that Black boys and men in the United States experience (e.g., stereotypes about Black boys and men as violent, dangerous, and sexually predatory), and the anti-Black transmisogyny that Black transgender women and other gender-expansive people experience, is necessary to galvanize research to document the scope of the violence, raise public awareness, and promote advocacy for interventions and legislative action to reduce these crimes and tragedies.

The absence of intersectionally specific language to describe social injustice and health inequities that disproportionately and distinctly affect Black boys and men, and Black transgender, nonbinary, and other gender-expansive people shares the same consequence of the "no data, no problem" (Krieger 1992) phenomenon. The absence of language to describe particular instantiations of racism and misandry, racism and cisgenderism (e.g., anti-Black transmisogny or transmisandry), and racism and heterosexism in the case of lesbian, gay, and bisexual people (who may also be transgender or gender expansive) is analogous to the absence of data problem because it impedes understanding, prevention (Stotzer 2017), and legislative or policy intervention. Pursuant to Anglin's definition of structural violence, the dearth of intersectionally specific language to describe health and other inequities that are racialized and (specifically) gendered, "engender[s] a kind of structural violence that is normalized and accepted as part of the 'status quo' but that is experienced as injustice and brutality at particular intersections of race, ethnicity, class, nationality, gender, and age" (Anglin 1998:146).

4 The Open Question: Why the Lag of Structural Intersectionality Research in the Field of Gender/Sex?

Traditional biomedical, psychosocial, and biobehavioral frameworks of health frame health primarily as a property of individuals (e.g., their cognitions, beliefs, and behaviors), their biology, or viruses or other disease-causing agents (Weber and Parra-Medina 2003). Furthermore, principles such as social justice and health equity are typically nonexistent in traditional approaches to health. Spurred by mounting criticism that conventional biomedical, psychosocial, biobehavioral (Bailey et al. 2021; Bailey et al. 2017; Bowleg 2012; Weber and Parra-Medina 2003), and social determinants of health (SDOH) frameworks (Yearby 2020) are inadequate for understanding intersectional health inequities among historically oppressed groups, attention is burgeoning toward the structural and political determinants of health and health inequities (Beckfield and Krieger 2009; Dawes 2020; Kickbusch 2015; Lynch and Perera 2017; Navarro 2004; Navarro and Shi 2001). Legal epidemiology—"the scientific study and deployment of law as a factor in the case, distribution, and prevention of disease and injury" (Burris et al. 2016:139)—is an emergent branch of epidemiology that offers important insights about the role of law in shaping health and health inequities. In his book, *The Political Determinants of Health*, Daniel Dawes (2020) asserts that SDOH have commanded the most attention, neglecting the factor with the greatest power "over all aspects of health" (Dawes 2020:1): political determinants.

In the United States, for example, the lack of universal health care is a fundamental structural and political determinant of health: one associated with disproportionate and excess morbidity and mortality for racialized and minoritized groups and/or people who are poor. Underscoring the entanglement of racialized and minoritized categories with gender/sex, no gender/sex is spared. Negative health outcomes and health inequities (e.g., HIV, diabetes, COVID-19, maternal morbidity and mortality, negative mental health, substance use) are worse for Black and Latino/a/x transgender and gender-expansive people (Grant et al. 2011; James et al. 2016), Black cisgender women (Homan et al. 2021), and Black and Native American cisgender men (Hill et al. 2023) compared with similarly gendered White counterparts.

Further bolstering the inextricable link between structural and political determinants of health and intersectional health inequities, the Everytown for Gun Safety Support Fund's (2024) report on gun homicides of transgender and gender-expansive people in the United States observed a relationship between the uptick in homicides in 2023 and weak gun safety laws, particularly in the South, combined with an increase in anti-LGBQ+ legislation that targets transgender people for discrimination, such as bathroom bans (i.e., the criminalization of using a gendered bathroom that does not correspond to the sex assigned at birth, regardless of gender identity or presentation). Thus, having argued that structural intersectionality is an essential antidote to the predominantly individualistic, biomedical, and psychosocial bent of much of the literature on health and health inequities, we are confronted with this open question: Why has structural intersectionality research lagged within the field of gender/sex and their entanglements?

In the United States, particularly in the wake of the murder of George Floyd, a Black man murdered by a White police officer in Minnesota in 2020 and witnessed by millions in a video that went viral, there has been burgeoning theory (Adkins-Jackson et al. 2021; Bailey et al. 2021; Bailey et al. 2017; Krieger 2019) and research (Agénor et al. 2021; Garcia et al. 2021; Hardeman et al. 2022; Selden and Berdahl 2020) on the effects of structural racism on health inequities. Theory and research on how to measure or empirically investigate health inequities in the context of forms of structural oppression, such as structural sexism (Homan 2021) and structural heterosexism (Sell and Krims 2021), are in their infancy and, in the case of structural ableism (Lundberg and Chen 2024) and structural cisgenderism (Zubizarreta et al. 2024), inchoate. By comparison, theory and research on structural intersectional health equity is virtually nonexistent.

The lag of structural intersectionality research can be traced to at least three factors. First, research on structural discrimination is relatively nascent; just 15 years old (Krieger 2019). Research on intersectional discrimination is younger still. Second, with the exception of researchers trained in disciplines that emphasize structure such as sociology, legal epidemiology, and political science, most social and behavioral science researchers hail from disciplines such as psychology and public health, where individual-level theories and methods dominate. Not surprisingly, many researchers from these disciplines gravitate toward the individual-level theories and methods (e.g., self-report measures of individual and interpersonal discrimination, experimental techniques) in which they were trained. Indeed, even researchers well-versed about intersectionality's attention to interlocking structural oppression and social-structural context tend to default to designing and conducting studies that examine individual-level intersectional "identities" [sic]. Finally, absent formal training in structural methodological approaches, it is likely more feasible for researchers to conceptualize individual-level studies, relying on variables that they can readily measure and statistically control, than the considerably more complicated task of measuring the effects of structural-level intersectionally discriminatory laws, policies, and rules. Structural intersectionality research is further compounded by an array of complex structural study design considerations such as comparison groups, temporal issues, and the availability and validity of exposure data (Krieger 2019). Adding structural intersectionality to the mix likely increases the complexity at least tenfold.

In one of the first explicitly structural intersectionality studies, sociologist Patricia Homan (2021) analyzed data from the US Behavioral Risk Factor Surveillance System to examine the effects of intersecting structural oppression (i.e., structural racism, structural sexism, and income inequality) on health inequities in the United States. Results showed that intersecting structural oppression (a) varied markedly across states, (b) intersected in multiple ways that did not strongly or positively covary, (c) highlighted individual and joint effects on health, and (d) were consistently linked to poor health outcomes for Black women. We are aware of just two other explicitly structural intersectional studies. A 2022 study of nationally representative US data found that non-Hispanic older (i.e., age 65 or older) Black people who lived in states with high indicators of structural racism and sexism (but not income inequality) had 60% higher odds of edentulism (i.e., missing one or more teeth or being wholly or partially toothless) (Bastos et al. 2022). A 2023 study using the same dataset as the 2022 study found that Black men who lived in states with high structural racism, sexism, and high economic inequality had the highest frequency of edentulism (Bastos et al. 2023).

Collectively, these studies pinpoint a dire need for more structural intersectionality research, in general, and for scholars who study sex and sex differences, gender in the context of gender studies, and the entanglements of gender/sex to become intersectionally structurally competent (Bowleg 2023). Empirical evidence from Homan et al.'s (2021) structural intersectionality research that structural intersectional inequities undermine population health lays important groundwork for structural interventions that have the potential to dismantle interlocking systems of oppression, not simply target oppression unidimensionally. Primarily because a structural intersectionality approach seeks to transform, disrupt, and dismantle inequitable systems (Fox et al. 2009) that seed intersectional health inequities in the first place, not just document them (Ford and Airhihenbuwa 2010), structural intersectionality research is a vital tool for the field of gender/sex and their entanglements, as well as for most other cross-disciplinary fields committed to social justice and health equity.

5 Conclusion

In her book, *Intersectionality as Critical Social Theory*, Patricia Hill Collins (2019) invites readers to engage with the concept of relationality, a core theme in intersectionality. Relationality disrupts Western knowledge logics of "oppositional difference" (i.e., either/or thinking; Collins and Bilge 2020:233) in favor of relational logics that acknowledge interconnectedness (i.e., both/and thinking). From a relationality perspective, when an additive framework such as intersectionality is added to a field such as gender/sex and their entanglements, it "disrupts taken-for-granted knowledge" (Collins 2019:227); namely, that gender/sex has meaning or can be understood absent its entanglements with other intersectional positions such as racialized minoritized or sexual or gender minority status, or interlocking systems of oppression such as structural racism, structural sexism, structural heterosexism, and structural cisgenderism.

There are likely countless other controversies, unsolved problems, and open questions about the topic of gender/sex and their entanglements beyond the ones we chose to address in this chapter. From an intersectionality and structural intersectionality perspective however, the notion that gender/sex can be disentangled from other intersectional positions such as racialized or sexual minority status is not one of them. It is simply impossible.

References

Aburizik A, Brindle M, Johnson E, et al. (2023) Black Women's Distress Matters: Examining Gendered Racial Disparities in Psycho-Oncology Referral Rates. Psychooncology 32 (6):933–941. https://doi.org/10.1002/pon.6134

Adkins-Jackson PB, Chantarat T, Bailey ZD, and Ponce NA (2021) Measuring Structural Racism: A Guide for Epidemiologists and Other Health Researchers. Am J Epidemiol 191 (4):539–547. https://doi.org/10.1093/aje/kwab239

Agénor M, Perkins C, Stamoulis C, et al. (2021) Developing a Database of Structural Racism–Related State Laws for Health Equity Research and Practice in the United States. Public Health Rep 136 (4):428–440. https://journals.sagepub.com/doi/full/10.1177/0033354920984168

Alvidrez J, Greenwood GL, Johnson TL, and Parker KL (2020) Intersectionality in Public Health Research: A View from the National Institutes of Health. Am J Public Health 111 (1):95–97. https://doi.org/10.2105/AJPH.2020.305986

Anglin MK (1998) Feminist Perspectives on Structural Violence. Identities 5 (2):145–151. https://doi.org/10.1080/1070289X.1998.9962613

Anzaldúa G (1987/2012) Borderlands/La Frontera: The New Mestiza. Fourth edn. Aunt Lute Books, San Francisco

——— (1990) Haciendo Caras, Una Entrada. In: Anzaldúa G (ed) Making Face, Making Soul: Haciendo Caras. Aunt Lute Books, San Francsisco, pp xv–xxviii

Bailey M, and Trudy (2018) On Misogynoir: Citation, Erasure, and Plagiarism. Fem Media Stud 18 (4):762–768. https://doi.org/10.1080/14680777.2018.1447395

Bailey ZD, Feldman JM, and Bassett MT (2021) How Structural Racism Works: Racist Policies as a Root Cause of US Racial Health Inequities. N Engl J Med 384 (8):768–773. https://doi.org/10.1056/NEJMms2025396

Bailey ZD, Krieger N, Agénor M, et al. (2017) Structural Racism and Health Inequities in the USA: Evidence and Interventions. Lancet 389 (10077):1453–1463. https://doi.org/10.1016/S0140-6736(17)30569-X

Bastos JL, Constante HM, Schuch HS, Haag DG, and Jamieson LM (2022) How Do State-Level Racism, Sexism, and Income Inequality Shape Edentulism-Related Racial Inequities in Contemporary United States? A Structural Intersectionality Approach to Population Oral Health. J Public Health Dent 82 (S1):16–27. https://doi.org/10.1111/jphd.12507

——— (2023) The Mouth as a Site of Compound Injustices: A Structural Intersectionality Approach to the Oral Health of Working-Age US Adults. Am J Epidemiol 192 (4):560–572. https://doi.org/10.1093/aje/kwac205

Beckfield J, and Krieger N (2009) Epi+ Demos+ Cracy: Linking Political Systems and Priorities to the Magnitude of Health Inequities—Evidence, Gaps, and a Research Agenda. Epidemiol Rev 31 (1):152–177. https://doi.org/10.1093/epirev/mxp002

Berger R (2015) Now I See It, Now I Don't: Researcher's Position and Reflexivity in Qualitative Research. Qual Res 15 (2):219–234. https://doi.org/10.1177/1468794112468475

Bilge S (2014) Whitening Intersectionality: Evanescence or Race in Intersectionality Scholarship. In: Hund WD, Lentin A (eds) Racism and Sociology. Lit. Verlag, Vienna, pp 175–205

Bowleg L (2012) The Problem with the Phrase "Women and Minorities": Intersectionality, an Important Theoretical Framework for Public Health. Am J Public Health 102 (7):1267–1273. https://doi.org/10.2105/AJPH.2012.300750

——— (2013) "Once You've Blended the Cake, You Can't Take the Parts Back to the Main Ingredients": Black Gay and Bisexual Men's Descriptions and Experiences of Intersectionality. Sex Roles 68 (11):754–767. https://doi.org/10.1007/s11199-012-0152-4

——— (2022) The Problem with Intersectional Stigma and HIV Equity Research. Am J Public Health 112 (S4):S344–S346. https://doi.org/10.2105/AJPH.2022.306729

——— (2023) Beyond Intersectional Identities: Ten Intersectional Structural Competencies for Critical Health Equity Research. In: Nash JC, Pinto S (eds) Routledge Companion to Intersectionalities. Routledge, London, pp 101–116

Bowleg L, and Bauer GR (2016) Invited Reflection: Quantifying Intersectionality. Psychol Women Q 40 (3):337–341. https://doi.org/10.1177/0361684316654282

Bowleg L, Burkholder GJ, Massie JS, et al. (2013) Racial Discrimination, Social Support, and Sexual HIV Risk among Black Heterosexual Men. AIDS Behav 17 (1):407–418. https://doi.org/10.1007/s10461-012-0179-0

Bowleg L, Huang J, Brooks K, Black A, and Burkholder GJ (2003) Triple Jeopardy and Beyond: Multiple Minority Stress and Resilience among Black Lesbians. J Lesbian Stud 7 (4):87–108. https://doi.org/10.1300/J155v07n04_06

Bowleg L, Malekzadeh AN, AuBuchon KE, Ghabrial MA, and Bauer GR (2023) Rare Exemplars and Missed Opportunities: Intersectionailty with in Current Sexual and Gender Diversity Research and Scholarship in Psychology. Curr Opin Psychol 49:101511. https://doi.org/10.1016/j.copsyc.2022.101511

Bowleg L, Malekzadeh AN, Mbaba M, and Boone CA (2022) Ending the HIV Epidemic for All, Not Just Some: Structural Racism as a Fundamental but Overlooked Social-Structural Determinant of the US HIV Epidemic. Curr Opin HIV AIDS 17 (2):40–45. https://doi.org/10.1097/COH.0000000000000724

Burris S, Ashe M, Levin D, Penn M, and Larkin M (2016) A Transdisciplinary Approach to Public Health Law: The Emerging Practice of Legal Epidemiology. Annu Rev Public Health 37 (1):135–148. https://doi.org/10.1146/annurev-publhealth-032315-021841

Burton Wanda M (2022) Gendered Racism: A Call for an Intersectional Approach. J Psychosoc Nurs Ment Health Serv 60 (12):3–4. https://doi.org/10.3928/02793695-20221109-01

Carbado DW (2013) Colorblind Intersectionality. Signs 38 (4):811–845. https://doi.org/10.1086/669666

Cazeau-Bandoo SIV, and Ho IK (2022) The Role of Structural Gendered Racism in Effective Healthcare Utilization among Black American Women with Herpes Simplex Virus. J Prev Health Promot 3 (1):3–29. https://doi.org/10.1177/26320770211049257

Chambers BD, Taylor B, Nelson T, et al. (2022) Clinicians' Perspectives on Racism and Black Women's Maternal Health. Womens Health Rep 4 (3):476–482. https://doi.org/10.1089/whr.2021.0148

Chiu A (2018) Beyoncé, Serena Williams Open up about Potentially Fatal Childbirths, a Problem Especially for Black Mothers. Washington Post, August 7, 2018

Cho S, Crenshaw KW, and McCall L (2013) Toward a Field of Intersectionality Studies: Theory, Applications, and Praxis. Signs 38 (4):785–810. https://doi.org/10.1086/669608

Coates T-N (2015) Between the World and Me. Spiegel & Grau, New York

Collins PH (1990) Black Feminist Thought: Knowledge Consciousness, and the Politics of Empowerment. Routledge, New York

——— (2015) Intersectionality's Definitional Dilemmas. Annu Rev Sociol 41 (1):1–20. https://doi.org/10.1146/annurev-soc-073014-112142

——— (2019) Intersectionality as Critical Social Theory. Duke Univ Press, Durham

Collins PH, and Bilge S (2020) Intersectionality: Key Concepts. 2nd edn. Polity Press, Cambridge

Combahee River Collective (1977) The Combahee River Collective Statement. In: Smith B (ed) Home Girls: A Black Feminist Anthology. Kitchen Table: Women of Color Press, New York, pp 272–282

Cooper AJ (1892) A Voice from the South: A Black Woman of the South. Aldine Printing Press, Xenia, OH

Crenshaw KW (1989) Demarginalizing the Intersection of Race and Sex: A Black Feminist Critique of Antidiscrimination Doctrine, Feminist Theory and Antiracist Politics. Univ Chic Leg Forum 1989 (1):8. https://chicagounbound.uchicago.edu/uclf/vol1989/iss1/8

——— (1991) Mapping the Margins: Intersectionality, Identity Politics, and Violence against Women of Color. Stanford Law Rev 43:61

Crenshaw KW, and African American Policy Forum (2023) #Sayhername: Black Women's Stories of Police Violence and Public Silence. Haymarket, Chicago

Crenshaw KW, and Ritchie AJ (2015) Say Her Name: Resisting Police Brutality against Black Women. African American Policy Forum, New York

Davis AY (1983a) Rape, Racism and the Myth of the Black Rapist. In: Women, Race and Class. Vintage, New York, pp 172–201

——— (1983b) Women, Race & Class. First Edition edn. Vintage, New York

Dawes DE (2020) The Political Determinants of Health. Johns Hopkins Univ Press, Baltimore

Dowland S (2009) "Family Values" and the Formation of a Christian Right Agenda. Church Hist 78 (3):606–631. https://doi.org/10.1017/S0009640709990448

Edwards F, Lee H, and Esposito M (2019) Risk of Being Killed by Police Use of Force in the United States by Age, Race–Ethnicity, and Sex. PNAS 116 (34):16793. https://doi.org/10.1073/pnas.1821204116

Essed P (1991) Understanding Everyday Racism: An Interdisciplinary Theory. Sage, Newbury Park

Everytown for Gun Safety Support Fund (2024) New Everytown Data on Transgender Homicides Reveals Concentration in the South. Press Release Feb 13, 2024

Feagin JR (2013) The White Racial Frame: Centuries of Racial Framing and Counter-Framing. 2nd edn. Routledge, New York

Finlay L (2002) "Outing" the Researcher: The Provenance, Process, and Practice of Reflexivity. Qual Health Res 12 (4):531–545. https://doi.org/10.1177/104973202129120052

Ford CL, and Airhihenbuwa CO (2010) Critical Race Theory, Race Equity, and Public Health: Toward Antiracism Praxis. Am J Public Health 100 (Suppl 1):S30–S35. https://doi.org/10.2105/ajph.2009.171058

Foucault M (1978/1990) The History of Sexuality, vol 1: An introduction. Vintage, New York

Fox D, Prilleltensky I, and Austin S (eds) (2009) Critical Psychology: An Introduction. Sage, London

Freire P (1970/2000) Pedagogy of the Oppressed. 30th Anniversary edn. Continuum, New York

Garcia MA, Homan PA, García C, and Brown TH (2021) The Color of COVID-19: Structural Racism and the Disproportionate Impact of the Pandemic on Older Black and Latinx Adults. J Gerontol B Psychol Sci Soc Sci 76 (3):e75–e80. https://doi.org/10.1093/geronb/gbaa114

Government of Canada (2017) Intersectionality Job Aid: A Guide to Applying an Intersectional Lens/Mindset to Your Gender-Based Analysis Plus (GBA Plus). https://women-gender-equality.canada.ca/gbaplus-course-cours-acsplus/assets/modules/job-aid-EN.pdf

Grant JM, Mottet LA, Tanis J, et al. (2011) Injustice at Every Turn: A Report of the National Transgender Discrimination Survey. http://www.thetaskforce.org/static_html/downloads/reports/reports/ntds_full.pdf

Greenwood BN, Hardeman RR, Huang L, and Sojourner A (2020) Physician–Patient Racial Concordance and Disparities in Birthing Mortality for Newborns. PNAS 117 (35):21194–21200. https://doi.org/10.1073/pnas.1913405117

Guralnik O (2023) Domestic Disturbance. May 21, 2023, Sunday Magazine, p 37, New York Times

Guthrie RV (2004) Even the Rat Was White: A Historical View of Psychology. 2nd edn. Pearson, Boston

Hankivsky O, and Hunting G (2022) From Gender Mainstreaming Towards Mainstreaming Intersectionality. In: Morrow M, Hankivsky O, Varcoe C (eds) Women's Health in Canada: Challenges of Intersectionality. 2nd edn. Univ Toronto Press, Toronto, pp 186–208

Hardeman RR, Homan PA, Chantarat T, Davis BA, and Brown TH (2022) Improving the Measurement of Structural Racism to Achieve Antiracist Health Policy. Health Aff 41 (2):179–186. https://doi.org/10.1377/hlthaff.2021.01489

Harper KC (2021) The Ethos of Black Motherhood in America: Only White Women Get Pregnant, Lexington. Rowman & Littlefield, Lanham

Hartfield JA, Griffith DM, and Bruce MA (2018) Gendered Racism Is a Key to Explaining and Addressing Police-Involved Shootings of Unarmed Black Men in America. In: Inequality, Crime, and Health among African American Males, vol 20. Research in Race and Ethnic Relations. Emerald Publ Ltd., Leeds, pp 155–170

Hill L, Ndugga M, and Artiga S (2023) Key Data on Health and Health Care by Race and Ethnicity. https://www.kff.org/racial-equity-and-health-policy/report/key-data-on-health-and-health-care-by-race-and-ethnicity/

Homan P (2021) Sexism and Health: Advancing Knowledge through Structural and Intersectional Approaches. Am J Public Health 111 (10):1725–1727. https://doi.org/10.2105/AJPH.2021.306480

Homan P, Brown TH, and King B (2021) Structural Intersectionality as a New Direction for Health Disparities Research. J Health Soc Behav 62 (3):350–370. https://doi.org/10.1177/00221465211032947

Hooks B (1981) Ain't I a Woman: Black Women and Feminism. South End Press, Boston

Hoyert DL (2022) Maternal Mortality Rates in the United States, 2020. https://dx.doi.org/10.15620/cdc:113967

Hull GT, Bell Scott P, and Smith B (eds) (1982) But Some of Us Are Brave: All the Women Are White, All the Blacks Are Men: Black Women's Studies. The Feminist Press, Old Westbury

Jackson FM, Phillips MT, Hogue CJ, and Curry-Owens TY (2001) Examining the Burdens of Gendered Racism: Implications for Pregnancy Outcomes among College-Educated African American Women. Matern Child Health J 5 (2):95–107. https://doi.org/10.1023/A:1011349115711

Jacobs HA (1861/2015) Incidents in the Life of a Slave Girl: An Autobiographical Account of an Escaped Slave and Abolitionist. Dover, Mineola

James SE, Herman JL, Rankin S, et al. (2016) The Report of the 2015 US Transgender Survey. National Center for Transgender Equality, Washington, DC

Johnson TH (2022) Is Anti-Black Misandry the New Racism? J Black Sex Relatsh 8 (4):77–107. https://doi.org/10.1353/bsr.2022.0006

Jones MK, Leath S, Settles IH, Doty D, and Conner K (2022) Gendered Racism and Depression among Black Women: Examining the Roles of Social Support and Identity. Cultur Divers Ethnic Minor Psychol 28 (1):39–48. https://doi:10.1037/cdp00004

Karkazis K, Jordan-Young R, Davis G, and Camporesi S (2012) Out of Bounds? A Critique of the New Policies on Hyperandrogenism in Elite Female Athletes. Am J Bioeth 12 (7):3–16. https://doi.org/10.1080/15265161.2012.680533

Karkazis K, and Jordan-Young RM (2018) The Powers of Testosterone: Obscuring Race and Regional Bias in the Regulation of Women Athletes. Fem Format 30 (2):1–39. http://doi.org/10.1353/ff.2018.0017

Kickbusch I (2015) The Political Determinants of Health: 10 Years on. Br Med J 350:h81. https://doi.org/10.1136/bmj.h81

Koma W, Rae M, Ramaswamy A, et al. (2020) Demographics, Insurance Coverage, and Access to Care among Transgender Adults. https://www.kff.org/health-reform/issue-brief/demographics-insurance-coverage-and-access-to-care-among-transgender-adults/

Krell EC (2017) Is Transmisogyny Killing Trans Women of Color?: Black Trans Feminisms and the Exigencies of White Femininity. Transgend Stud Q 4 (2):226–242. https://doi.org/10.1215/23289252-3815033

Krieger N (1992) The Making of Public Health Data: Paradigms, Politics, and Policy. J Public Health Policy 13 (4):412–427

———— (2019) Measures of Racism, Sexism, Heterosexism, and Gender Binarism for Health Equity Research: From Structural Injustice to Embodied Harm: An Ecosocial Analysis. Annu Rev Public Health 41:37–62. https://doi.org/10.1146/annurev-publhealth-040119-094017

Larson E, George A, Morgan R, and Poteat T (2016) 10 Best Resources on… Intersectionality with an Emphasis on Low- and Middle-Income Countries. Health Policy Plan 31 (8):964–969. https://doi.org/10.1093/heapol/czw020

Laster Pirtle WN, and Wright T (2021) Structural Gendered Racism Revealed in Pandemic Times: Intersectional Approaches to Understanding Race and Gender Health Inequities in COVID-19. Gend Soc 35 (2):168–179. https://psycnet.apa.org/doi/10.1177/08912432211001302

Liu T, Wong YJ, Maffini CS, Mitts NG, and Iwamoto DK (2018) Gendered Racism Scales for Asian American Men: Scale Development and Psychometric Properties. J Couns Psychol 65 (5):556–570. https://doi.org/10.1037/cou0000298

Lorde A (1984) Sister Outsider: Essays and Speeches. Crossing Press, Freedom, CA

Lundberg DJ, and Chen JA (2024) Structural Ableism in Public Health and Healthcare: A Definition and Conceptual Framework. Lancet Reg Health Am 30:100650. https://doi.org/10.1016/j.lana.2023.100650

Lynch I, Isaacs N, Fluks L, et al. (2020) Intersectionality in African Research: Findings from a Systematic Literature Review. http://hdl.handle.net/20.500.11910/15940

Lynch JF, and Perera IM (2017) Framing Health Equity: US Health Disparities in Comparative Perspective. J Health Polit Policy Law 42 (5):803–839. https://doi.org/10.1215/03616878-3940450

Markin RD, and Coleman MN (2023) Intersections of Gendered Racial Trauma and Childbirth Trauma: Clinical Interventions for Black Women. Psychotherapy 60 (1):27–38. https://doi.org/10.1037/pst0000403

May VM (2015) Pursuing Intersectionality, Unsettling Dominant Imaginaries. Routledge, New York

Mehra R, Boyd LM, Magriples U, et al. (2020) Black Pregnant Women "Get the Most Judgment": A Qualitative Study of the Experiences of Black Women at the Intersection of Race, Gender, and Pregnancy. Womens Health Issues 30 (6):484–492. https://doi.org/10.1016/j.whi.2020.08.001

Menon N (2015) Is Feminism about "Women"? A Critical View on Intersectionality from India. Econ Polit Wkly 50 (17):37–44

Moraga C, and Anzaldúa G (eds) (1981) This Bridge Called My Back: Writings by Radical Women of Color. Kitchen Table: Women of Color Press, New York

Nash JC (2008) Re-Thinking Intersectionality. Fem Rev 89:1–15. https://doi.org/10.1057/fr.2008.4

National Network to End Domestic Violence (2016) Looking Back, Moving Forward: An Exclusive Interview with Gloria Steinem. https://nnedv.org/latest_update/exclusive-interview-gloria-steinem/

Navarro V (2004) The Politics of Health Inequalities Research in the United States. Int J Health Serv 34 (1):87–99. https://doi.org/10.2190/0KT0-AQ1G-5MHA-9H7R

Navarro V, and Shi L (2001) The Political Context of Social Inequalities and Health. Int J Health Serv 31 (1):1–21. https://doi.org/10.2190/1GY8-V5QN-A1TA-A9KJ

Owens DC (2017) Medical Bondage: Race, Gender, and the Origins of American Gynecology. Univ Georgia Press, Athens

Owens DC, and Fett SM (2019) Black Maternal and Infant Health: Historical Legacies of Slavery. Am J Public Health 109 (10):1342–1345. https://doi.org/10.2105/ajph.2019.305243

Patterson EJ, Becker A, and Baluran DA (2022) Gendered Racism on the Body: An Intersectional Approach to Maternal Mortality in the United States. Popul Res Policy Rev 41:1261–1294. https://doi.org/10.1007/s11113-021-09691-2

PHAC (2022) How to Integrate Intersectionality Theory in Quantitative Health Equity Analysis? A Rapid Review and Checklist of Promising Practices. https://www.canada.ca/content/dam/phac-aspc/documents/services/publications/science-research-data/how-integrate-intersectionality-theory-quantitative-health-equity-analysis/phac-siithia-checklist.pdf (accessed Nov. 4, 2024)

Ramseur II K, Cattaneo LB, Stori S, and Adams L (2024) Black Men Need Friends: Social Support Moderates the Connection between Gendered Racism and Psychological Distress. Psychol Men Masc 25 (1):13–26. https://doi.org/10.1037/men0000449

Rausch D (2018) Promoting Reductions in Intersectional StigMa (PRISM) to Improve the HIV Prevention Continuum. NIMH

Richardson L (2011) Gender Stereotyping in the English Language. In: Rosenblum KE, Travis T-MC (eds) The Meaning of Difference: American Constructions of Race, Sex and Gender, Social Class, and Sexual Orientation. 3rd edn. McGraw Hill, New York, pp 509–516

Rosenthal L, and Lobel M (2020) Gendered Racism and the Sexual and Reproductive Health of Black and Latina Women. Ethn Health 25 (3):367–392. https://doi.org/10.1080/13557858.2018.1439896

Rotz S, Rose J, Masuda J, Lewis D, and Castleden H (2022) Toward Intersectional and Culturally Relevant Sex and Gender Analysis in Health Research. Soc Sci Med 292:114459. https://doi.org/10.1016/j.socscimed.2021.114459

Schulman KA, Berlin JA, Harless W, et al. (1999) The Effect of Race and Sex on Physicians' Recommendations for Cardiac Catheterization. N Engl J Med 340 (8):618–626. https://doi.org/10.1056/NEJM199902253400806

Schwing AE, Wong YJ, and Fann MD (2013) Development and Validation of the African American Men's Gendered Racism Stress Inventory. Psychol Men Masc 14 (1):16–24. https://doi.org/10.1037/a0028272

Selden TM, and Berdahl TA (2020) COVID-19 and Racial/Ethnic Disparities in Health Risk, Employment, and Household Composition. Health Aff 39 (9):1624–1632. https://doi.org/10.1377/hlthaff.2020.00897

Sell R, and Krims EI (2021) Structural Transphobia, Homophobia, and Biphobia in Public Health Practice: The Example of COVID-19 Surveillance. Am J Public Health 111 (9):1620–1626. https://doi.org/10.2105/AJPH.2021.306277

Semenya C (2023) The Race to Be Myself. W. W. Norton, New York

Settles IH, Warner LR, Buchanan NT, and Jones MK (2020) Understanding Psychology's Resistance to Intersectionality Theory Using a Framework of Epistemic Exclusion and Invisibility. J Soc Issues 76 (4):796–813. http://doi.org/10.1111/josi.12403

Smith WA, David R, and Stanton GS (2020) Racial Battle Fatigue: The Long-Term Effects of Racial Microaggressions on African American Boys and Men. In: Majors R, Carberry K, Ransaw TS (eds) The International Handbook of Black Community Mental Health. Emerald Publ Ltd., Leeds, pp 83–92

Stewart MW (1831) Religion and the Pure Principles of Morality: The Sure Foundation on Which We Must Build. https://teachingamericanhistory.org/document/religion-and-the-pure-principles-of-morality-the-sure-foundation-on-which-we-must-build/

Stotzer RL (2017) Data Sources Hinder Our Understanding of Transgender Murders. Am J Public Health 107 (9):1362–1363. https://doi.org/10.2105/AJPH.2017.303973

Strings S (2019) Fearing the Black Body: The Racial Origins of Fat Phobia. New York Univ Press, New York

———— (2023) How the Use of BMI Fetishizes White Embodiment and Racializes Fat Phobia. AMA J Ethics 25 (7):E535–539. https://doi.org/10.1001/amajethics.2023.535

Thomas AJ, Witherspoon KM, and Speight SL (2008) Gendered Racism, Psychological Distress, and Coping Styles of African American Women. Cultur Divers Ethnic Minor Psychol 14 (4):307–314. https://doi.org/10.1037/1099-9809.14.4.307

Tinner L, Holman D, Ejegi-Memeh S, and Laverty AA (2023) Use of Intersectionality Theory in Interventional Health Research in High-Income Countries: A Scoping Review. Int J Environ Res Public Health 20 (14):6370. https://doi.org/10.3390/ijerph20146370

Urban Indian Health Institute (2018) Missing and Murdered Indigenous Women and Girls: A Snapshot of Data from 71 Urban Cities in the United States. https://www.uihi.org/wp-content/uploads/2018/11/Missing-and-Murdered-Indigenous-Women-and-Girls-Report.pdf

Vedam S, Stoll K, Taiwo TK, et al. (2019) The Giving Voice to Mothers Study: Inequity and Mistreatment during Pregnancy and Childbirth in the United States. Reprod Health 16 (1):77. https://doi.org/10.1186/s12978-019-0729-2

Washington HA (2006) Medical Apartheid: The Dark History of Medical Experimentation on Black Americans from Colonial Times to the Present. Doubleday, New York

Weber L, and Parra-Medina D (2003) Intersectionality and Women's Health: Charting a Path to Eliminating Health Disparities. Adv Gend Res 7:181–230. https://doi.org/10.1016/S1529-2126(03)07006-1

Williams S (2022) How Serena Williams Saved Her Own Life. https://www.elle.com/life-love/a39586444/how-serena-williams-saved-her-own-life/

Yearby R (2020) Structural Racism and Health Disparities: Reconfiguring the Social Determinants of Health Framework to Include the Root Cause. J Law Med Ethics 48 (3):518–526. https://doi.org/10.1177/1073110520958876

Yuval-Davis N (2015) Situated Intersectionality: A Reflection on Ange-Marie Hancock's Forthcoming Book: Intersectionality—an Intellectual History. New Polit Sci 37 (4):637–642. https://doi.org/10.1080/07393148.2015.1089045

———— (2017) Situated Intersectionality and the Meanings of Culture. Europa Fortaleza. Fronteiras, Valados, Exilios, Migracións 18. https://consellodacultura.gal/mediateca/extras/Texto_Nira_maquetado.pdf

Zubizarreta D, Trinh M-H, and Reisner SL (2024) Quantitative Approaches to Measuring Structural Cisgenderism. Soc Sci Med 340:116437. https://doi.org/10.1016/j.socscimed.2023.116437

10

Sex and Gender Should Be Considered Continuous Variables in Cancer Research

Wei Yang, Jason Wong, and Joshua B. Rubin

Abstract Significant sex and gender differences exist in cancer mechanisms, incidence, and survival, yet it is difficult to translate these important differences because cancer phenotypes do not segregate into dichotomous male versus female or man versus woman categories. Instead, sex and gender are developmental and environmental forces that work together to establish the exquisite diversity of human phenotypes from imprints to death. Sex and gender effects are entangled in human phenotypes, which means that sex- or gender-specific cancer treatments are unrealistic. The translational goal of cancer research should be to establish the effects of gender-sex entanglement (GSE) on cancer protection at the cellular, tissue, and systems levels. This will be essential for the adaptation of therapy to the varying effects of GSE on cancer phenotypes. To take a step in this direction, this chapter examines similarities in 8,370 transcriptomes of 26 different adult and 4 different pediatric cancers. Individual transcriptomic phenotype is assumed to be a product of GSE, and each patient's transcriptome was allowed to cluster by similarity into naturally occurring local clusters that reflected XX or XY characteristics. A transcriptomic index (TI, ranging from 0 to 1) was then calculated using a metric based on the local enrichment of male- or female-specific characteristics, which was subsequently used to identify reference poles. Using TI values, patient-specific GSE effects on targetable pathways (e.g., cell cycle signaling and immunity) are described. This novel approach to patient-specific phenotyping can be used for more realistic GSE adaptations of precision cancer treatments.

Joshua B. Rubin (✉)
Depts. of Pediatrics and Neuroscience, Washington University School of Medicine,
St. Louis, MO 63110, USA
Email: rubin_j@wustl.edu

Wei Yang
Dept. of Genetics, Washington University School of Medicine, St. Louis, MO 63110, USA

Jason Wong
Dept. of Pediatrics, Washington University School of Medicine, St. Louis, MO 63110, USA

© Frankfurt Institute for Advanced Studies (FIAS) 2025
L. Z. DuBois et al. (eds.), *Sex and Gender*, Strüngmann Forum Reports,
https://doi.org/10.1007/978-3-031-91371-6_10

Keywords Sex/gender differences, cancer, cell cycle regulation, inflammation/ immunity, Bayesian analyses, personalized medicine

1 Introduction

Regardless of beliefs about gender identities, roles, and effects, an ever-increasing number of people worldwide are becoming familiar with the concept of nonbinary *genders*. In contrast, it seems that people are less accustomed to thinking about *sex* as nonbinary. Limiting sex effects to a mechanism for genetic diversity and the perpetuation of the species neglects to consider the effects of genetic diversification through sexual reproduction on mammalian sex specification; that is, XX, XY, and sufficiency of Y (SRY) in establishing testes (Eggers et al. 2014) as well as functional polymorphic variants in sex hormone production and response mechanisms which further individualize sex hormone effects in aging (Jiang and Huhtaniemi 2004), mood and cognition (Sundermann et al. 2010), and cancer (Jahandoost et al. 2017; Lillie et al. 2003; Schleutker 2012). It also fails to incorporate the unequal allelic diversification of the X and Y chromosomes as only small telomeric portions of Y can cross over during meiosis, and allelic diversity is more dependent upon genetic drift and is only relevant to Y-containing genotypes (Jobling and Tyler-Smith 2017). Consequently, a conceptual framework of chromosomal and gonadal sex that does not treat them as highly variable is simply not correct. While sex chromosomes are not fluid like gender across the lifespan, their diversity in sequence and number results in a spectrum of chromosomal and gonadal effects on phenotypes. Together, the continuously varying nature of sex effects and the fluidity of gender suggest that there are an enormous number of possible gender and sex entangled human phenotypes. Adoption of a gender-sex entanglement (GSE) framework for phenotypic variation is realistic and attractive because of its inclusivity. Individual GSE considerations could leapfrog the personalization of cancer treatments and, seemingly, almost any endeavor in human health, aging, and disease.

An obstacle in moving the continuous nature of sex and hormone effects on phenotype forward is a general lack of knowledge of the developmental processes by which transgenerational epigenetics (imprinting), sex chromosome haplotype, and varying sex hormone actions along with environmental effects function as variable determinants of individual GSE biology, from the cellular to systemic scales. Working together, these elements result in GSE differences that range in the population between what might be envisioned as extremes in phenotype, where one extreme could be the shortest in terms of human stature and the other the tallest (Figure 10.1). A frequency histogram of human heights from twenty different countries, representing most pillar ancestries, illustrates the continuous nature of normal human height as a variable in which the shortest are exclusively XX, while XY is similarly correlated with the tallest. The continuous nature of height can then be used to identify mechanisms underlying GSE effects on height. Importantly, while these extremes or poles exist, they only account for 10% of the population. The majority of the population exhibits continuous and overlapping variation that is skewed toward their respective sex specified poles.

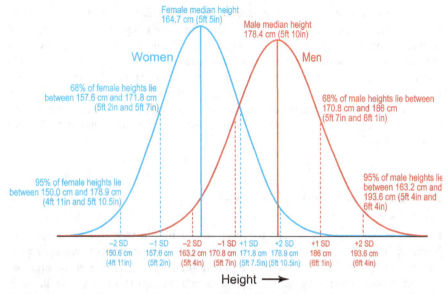

Figure 10.1 Distribution of male and female heights. Data shown are for adult heights of males and females across 20 North American, European, East Asian, and Australian populations. Source is Jelenkovic et al. (2016). Visualization is from OurWorldinData.org. Licensed under CC-BY by Cameron Appel (https://creativecommons.org/licenses/by/4.0/).

Failure to discern the spectrum of GSE effects results in a tendency to dichotomize sex and treat its related biology as a categorical variable. While many important biological traits aggregate around the XX/ovaries and XY/testes dichotomous poles, it is those poles that reveal the spectrum of phenotypes at a distance from them. It is important to note that even at the level of chromosomes and gonads, binary does not include everyone. Sex chromosome aneuploidies occur with a frequency of approximately one in every 1,500 births (Samango-Sprouse et al. 2016), and variable sensitivity to sex hormones contributes to variable phenotypes; for instance, longer or shorter polyglutamine repeats on androgen receptor functions have a substantial impact on neurological disease risk (Palazzolo et al. 2008), prostate cancer risk in males (Qin et al. 2017), and possibly ovarian cancer in females (Deng et al. 2017). In this chapter, we will use the aggregated traits around XX/ovaries and XY/testes to help illuminate the spectrum of varying GSE phenotypes.

The continuous nature of GSE warrants careful consideration of statistical approaches for identifying its significant effects and generating hypotheses regarding underlying mechanisms. An experiment in which measure the effects of two independent variables, GSE, and dose of a novel drug on the dependent variable activation of its molecular target in human cells, as measured by western blot, illustrates the point. We expect that dose will have a significant effect on response and are investigating if GSE does as well. We have chosen human cells that represent multiple GSE phenotypes across the spectrum and want to know if GSE (TI values) also effects the dose response. If it does, it can do so in two different ways: (a) GSE affects dose response in an additive manner with the dose effect and (b) GSE and dose interact such that the result is either greater or lesser than the sum of their effects. To evaluate both requires two-way ANOVA (analysis of variance) to generate p-values for the two independent variables separately and then a third p-value for their interaction (Rubin et al. 2020). Additive effects indicate the absence of interaction whereas synergy or antagonism indicate that the two independent variables are interacting with each other as determinants of dose response. This analysis impacts the kinds of mechanistic hypotheses that can be generated from the data. Even in inbred laboratory animals, where we assume there are no gender effects or genetic variation, we still see continuous variation in cancer-relevant phenotypes between XX/ovaries and XY/testes poles. This, too, demands statistics like two-way ANOVA.

2 Cancer Population-Level Analysis

Sexual reproduction has existed for more than 1 billion years (Goodenough and Heitman 2014). In all sexually reproducing species, fitness requirements for reproductive success differ in the alternate gamete producers, resulting in the array of sex differences we observe in biology (Hosken et al. 2019). In humans and other mammals, male reproductive success requires large and strong, ornamented bodies. In females, survival and longevity are paramount for reproductive success. These sex differences are relevant to cancer as increasing body size in humans is differently correlated with increased cancer risk in males and females (Nunney 2018). Longevity places a greater demand on lifelong protections against disease and decay, including tumor suppressor functions, metabolism, and immunity. All are important in cancer. Thus, we anticipated that through evolution there would be conserved mechanisms and effects of sex specification and differentiation that would be evident across cancer types in humans and shared between humans and other species, particularly with regard to growth regulation, metabolism, tumor suppresser function, and immunity (Rubin 2022; Rubin et al. 2020).

As in other sex differences, we expected there to be continuous phenotypic variation in these mechanisms and systems as a consequence of GSE, and that this variation could be visualized through whole transcriptome analysis. We reasoned that if we let whole transcriptomes cluster according to similarity, we would discover separation of cases into those that clustered similarly around a female extreme, or pole, and male cases that clustered around a male pole. Earlier, we observed (and published) this phenomenon when we performed unsupervised clustering of metabolite abundance in 44 male and 32 female patient-derived glioblastoma specimens. In that instance, there was continuous variation in female specimens to a second profile that was almost entirely comprised of male specimens as defined by clinical records (Sponagel et al. 2022). Once identified, poles such as these can inform us about the genes and pathways underlying the differences. To accomplish this for the transcriptomic analysis, we developed and applied a Bayesian-based analysis that captures the nature of GSE as a continuous variable. We developed and demonstrated the potential utility of this approach in an analysis of over 8,000 adult and pediatric cancer transcriptomes (Yang and Rubin 2024).

The first step was to identify the population distributions for a given parameter. As illustrated in Figure 10.1, this could be as simple as the population distributions of male or female height. Our parameter of interest was similarity between cancer transcriptomes; thus, we derived similarity scores and let them cluster by similarity on a graph known as a uniform manifold approximation and projection (UMAP) (Figure 10.2). To explore polarization by GSE on this map of transcriptome similarity, we applied a Bayesian Nearest Neighbor (BNN) algorithm to calculate the posterior probability of predicting the chromosomal status of an individual patient by their similarity with other individuals (Nuti 2019; Nuti et al. 2019). We refer to position along this polarization axis of Bayesian posterior probability as the transcriptomic index (TI), which ranges from 0 (female) to 1 (male). The closer an individual is to 0 or to 1, the greater the probability of correctly identifying them as female or male, respectively.

We can visualize differences in population data distributions using ridgeline plots in which histograms are transformed into smooth lines with shaded areas under the curves. Individual data points can be represented by tick marks along the X-axis. Ridgeline plots representing different components of the population are then vertically stacked to facilitate comparisons between them (Figure 10.3). Adrenocortical carcinoma illustrates the utility of ridgeline plots. Tick marks represent individual patient TI values. Female TI values range from close to zero to 0.5, with a small number of cases having higher TI values. In contrast, male TI values range from approximately 0.3 to nearly 1. The ridgeline plots indicate that female and male patients exhibit distinct but overlapping TI values. It is important to recognize that

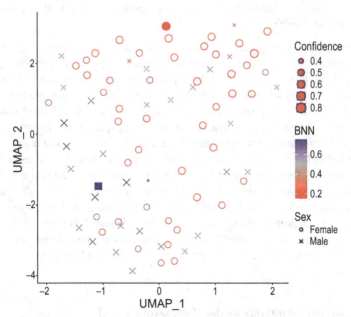

Figure 10.2 Transcriptomic similarity is skewed by sex in adrenocortical carcinoma (ACC). UMAP of 79 ACC transcriptomes (31 males, 48 females) clustered by similarity. Male (blue X's) and female (red circles) distribute throughout the transcriptional space. Local enrichments for male and female transcriptomes were recognized and quantified to define female (filled red circle) and male (filled blue square) poles of gene expression. The BNN value is color-coded and confidence in the posterior probability is indicated by symbol size. Adapted from Yang and Rubin (2023) and licensed under CC-BY (https://creativecommons.org/licenses/by/4.0/).

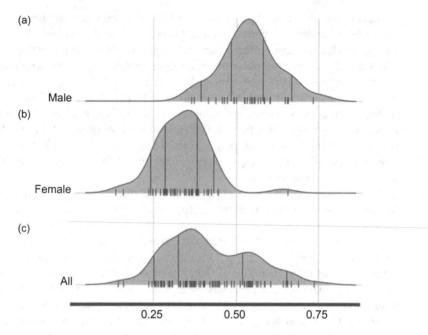

Figure 10.3 Ridgeline plots for TI population distributions for adrenocortical carcinoma. Distribution of TI values for male (top), female (middle), and all (bottom) patients with adrenocortical carcinoma. Adapted from Yang and Rubin (2023) and licensed under CC-BY (https://creativecommons.org/licenses/by/4.0/).

while we are aggregating many elements of our current understanding of biology around a binary sex chromosome framework, this is required to reveal the diversity of GSE phenotypes within each of these populations. As Dubois et al. (2021) have suggested, increased knowledge of biological mechanisms may make this unnecessary, and instead, analogous population analyses would fall out from analyses of metabolomics, proteomics, or disease metrics (e.g., incidence, severity, survival). We saw this in our analysis of glioblastoma metabolomics. When females and males have similar index values, for example, for height, this does not necessarily mean that they grew to the same height by the same mechanism. The TI approach allows you to ask a broader question: Do overlapping heights arise through female and male convergence in the biology of height determination, or through mechanisms that oppose their pole-defining biology, or both? These types of questions more completely address the nature of biased actions, such as those underlying GSE differences, and are therefore more likely to produce actionable results.

3 Downstream Analysis at the Population Level

Once TI values are established, they can be used to inform us about the transcriptional differences between the poles and cases near or far away from the poles. To this end, we performed pathway analysis and looked at which pathways distinguished the poles. We were reassured to find that immunity/inflammation pathways most frequently characterized the female pole in specific cancers. This served as an important gut check for the approach because important GSE differences in immunity are well described overall (Klein and Flanagan 2016) and in cancer (Hunt and Alspach 2024). Further, we were encouraged to find that cell cycle regulation was the most common feature of the male pole, as we had previously found cell cycle to characterize male cases in a prior application of the "joint and individual variance explained" (JIVE) analysis of transcriptome data from patients with glioblastoma (Yang et al. 2019).

Next, we used the TI values to identify pathways that were altered when female cases with midrange TI values were compared to female cases with TI values closer to zero. Similarly, we were able to identify pathways that were altered in male cases with midrange values compared to male cases with TI values closer to one. Importantly, the pathways that were altered in activity in females and males with midrange TI values were not the same. This suggests that GSE differences in mechanisms are evident even with similar TI values. This result highlights an important aspect of this approach: it is able to detect differences between females as well as differences between males, thus negating the categorical concept of "male" or "female." This could have important applications in the clinical setting because it could support GSE-informed stratifications for therapies that target specific pathways. At the very least, it provides a unique biological variable that could have important correlations with objective treatment response, toxicity, quality of life, and survival.

4 Detecting the Continuous Effect of Sex in the Lab

In the laboratory setting, we must consider how to establish TI values for cells and animal models. We assume that although rodent animal models do not have gendered phenotypes, their cells and tissues will exhibit chromosomal and gonadal sex effects on gene expression. Thus, when discussing rodent-derived *in vitro* and *in vivo* models, we will use the term *sex* to distinguish individual animals with XX/ovaries from those with XY/testes specification. It remains to be determined whether the BNN approach will have value when applied to inbred strains of laboratory animals.

In contrast, TI values for patient-derived cancer specimens reflect GSE-patterned phenotype and can be determined by placing an individual patient-derived transcriptome in the context of the appropriate cancer type-specific analysis that we performed as the reference range (Yang et al. 2019). Once a given cell line is located on the TI axis, it becomes possible to choose patient-derived specimens that represent the spectrum of GSE effects on phenotype. This will support greater fidelity in modeling GSE in cancer in the laboratory. GSE-informed modeling allows hypotheses to be generated regarding specific hallmark pathway contributions to biology, and the responsiveness of cells to *in vitro* treatments or xenograft sensitivity to treatment of tumor-bearing immunocompromised mice.

We validated an approach for capturing a range of variability in sex differences in mouse model systems. It is based primarily on the variable masculinization of female pups as a function of the male/female litter ratio and whether the female pup is positioned next to one or two male pups (Monclus and Blumstein 2012). To control for litter effects on biology, we separately pool male and female postnatal day 1 astrocytes from littermates as a single biological replicate. We routinely use astrocytes derived from up to five litters of pups for an n=5 biological replication to create a range of sex differences in experimental specimens. As this approach primarily captures *in utero* effects of testosterone, we see a greater range in phenotypic variance in female compared to male pups. This variance is well illustrated in Figure 10.4 across three very different assays: radiation-induced senescence, TNF-induced NF-κβ reporter assay, and doubling times in culture. In all cases, the range in experimental values is greater in female cells than in male cells.

To analyze murine data collected in this fashion, we performed parallel male and female linear regressions between one independent and one dependent variable. In the example shown in Figure 10.5, we regressed a range of radiation-induced changes in p21/Cdk2 protein ratios (independent variable), which is a critical determinant of cellular senescence, measured by senescence-associated β-galactosidase (dependent variable). With this analysis, we can compare consistency in the relation between the variables (r values) for males and females and differences in the slopes of their relations. In the analysis presented, female cellular senescence exhibits a higher correlation and steeper slope in relation to p21/Cdk2 than male cells. In fact, the male cells do not exhibit a significant correlation to p21/Cdk2. Thus, we can hypothesize that different control mechanisms are operative in radiation-induced senescence in male and female cells. This has immediate clinical significance as treatment-induced senescence can have anti-tumor or tumor-promoting effects (Schmitt et al. 2022). The availability of senolytic agents makes it possible to think about inducing senescence and the ablation of senescent cells in cancer therapies (Kirkland and Tchkonia 2020).

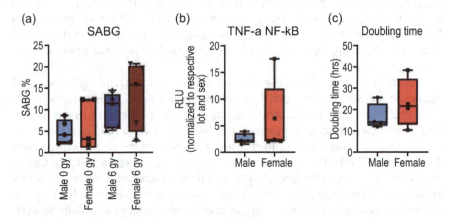

Figure 10.4 Male and female variance across litters. Litter effects on male and female variability is evident across different experimental measures. (a) Senescence-associated β-galactosidase (SABG) under basal conditions and following 6 Gy irradiation. (b) NF-B reporter assay following TNF treatment. (c) Doubling time for male and female cells in normal growth media.

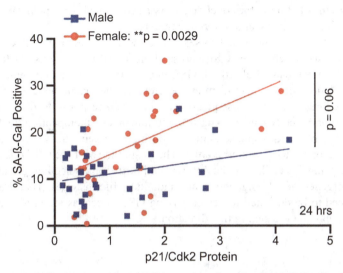

Figure 10.5 Correlation between the p21/Cdk2 protein ratio and radiation-induced cellular senescence. The p21 and CDK2 protein abundance was determined from western blot quantification
24 hours after irradiation with 0, 3, 6, or 8 Gy. Five days after irradiation, the percentage of SA-
β-Gal positive cells in male and female Nf1−/− DNp53 astrocytes was measured. Male correlation: r=0.30, p=0.0997. Female correlation: r=0.52, p=0.0029. Male slope vs. Female slope:
p=0.0627. Adapted from Broestl et al. (2022) and licensed under CC-BY (https://creativecom-
mons.org/licenses/by/4.0/).

5 Changing the Kinds of Questions to Ask

As discussed above, we can look for two different kinds of sex or GSE effects on
biology. In the first, one or both of two experimentally independent variables exert
a statistically significant effect on results, but they do not significantly interact with
each other (e.g., additive effects). In the second, the two independent variables exert
significant independent effects and significantly interact in producing dependent variable responses (e.g., synergy or antagonism). These differing effects on results can
be distinguished by two-way ANOVA. Evaluation of sex or GSE effects in this manner is more appropriate for the spectrum of biology that arises from their effects and
increases the likelihood of discovering their significance in an experimental setting.

Both dichotomous effects (significant and constant sex or GSE effects) and interaction effects (significant but varying magnitude of an interaction effect) have important mechanistic implications. The first, the categorical effect, suggests mechanisms that are more dichotomous, such as *XIST* expression, in those with multiple X
chromosomes or the dependency of gonad specification on the presence or absence
of a Y chromosome. The mechanistic link between sex and these effects is likely to
be more direct, like the necessary and sufficient actions of SRY in the formation of
testes (Eggers et al. 2014). The second, the interaction effect, suggests something
different: The effect of sex or GSE may not be direct; instead, sex or GSE may
exert a dynamic modulatory effect on the biology being studied. Not only does this
support looking into other known components of the pathway being studied for the
target interactor(s), it also suggests that important differences in mechanism must
exist between males and females. Therefore, looking at the sex or GSE differences

increases the likelihood for novel discovery as you can observe pathway function in two naturally occurring and different contexts.

Clinical studies must adapt to sex/gender as a continuous variable. We cannot expect consistent effects of GSE across clinical trial populations. At the simplest level, like age (Bisht et al. 2024), GSE impacts (a) systemic and cellular metabolism, including drug metabolism and excretion (pharmacokinetics); (b) gene and protein expression, and subsequently function, and response to drug-target interactions (pharmacodynamics); and (c) determinants of therapeutic effects, like the epigenetics of stress response pathways during treatment stress and immunity, including response to immunotherapies. The magnitude of these GSE effects on each of these pathways is likely to be different and further influenced by germline genetics, transgenerational epigenetics, and individual life histories. Therefore, clinical trials need to be designed to detect varying GSE effects as a continuous variable. The TI approach is a step in that direction.

6 Moving Forward

6.1 Clinical Studies

Demonstrating the clinical utility of the TI approach will ultimately rely on prospective evaluation in clinical trials of personalized treatment approaches for cancer patients. Prior to that, there are several kinds of retrospective studies that could be helpful in further developing the approach for clinical trial application. An important approach to retrospective studies would be the post hoc calculation of TI values for patients treated with targeted therapeutics during clinical trials. TI values are correlated with gene expression and pathway activation. As gene expression and assessments of target pathway activation are used for designing individualized treatment plans in the future, they could be utilized as a stratification tool in precision treatment approaches. The utility of this approach could be evaluated retrospectively by using post hoc TI values in analyses of response to targeted data. We published a use case for immune checkpoint inhibition (ICI) in lung cancer (Yang and Rubin 2024). Male patients are known to exhibit a greater response to ICI than female patients (Conforti et al. 2019, 2021; Santoni et al. 2022). Immunity and inflammation pathways were highly polarized in our lung cancer analysis, suggesting that females furthest away from their pole and males nearest to their pole would be the most likely subset of patients of each sex to exhibit a strong response to ICI. This and other evaluations like it could be readily performed retrospectively.

Another retrospective kind of analysis with great potential impact would be to examine the consistency of TI values across multisector tumor biopsies. We hypothesize they will be very similar, as we expect that the TI value reflects fundamental aspects of sexual differentiation for that patient. Similarly, it will be important to determine whether TI values are the same or different in primary versus recurrent tumor biopsies/resections from individual patients. Again, our hypothesis is that although there may be some variance between the samples, the effect size of that variability will be small, and TI values will remain relatively consistent.

6.2 Laboratory Studies

As described above, TI values may be very useful in laboratory experiments, especially in the design of experiments involving patient-derived specimens to ensure that the specimens represent a spectrum of TI values. Whether TI values can be similarly useful in animal models and cells remains to be determined. Key to determining the utility of this application will be establishing TI values across cancer models to determine whether poles can be defined and if meaningful distinctions can be made between the genes and pathways that characterize the female and male poles and the midrange values. The large amount of available RNA-seq data would support this kind of new analysis.

It may also be possible that specific pathways can be analyzed in patient data by generating a list of pathway-involved genes that can be analyzed from initial UMAP through histograms of TI values in the population. This would support identifying GSE differences in the expression of specific pathway component genes. These results should generate new hypotheses regarding mechanisms underlying GSE differences in the specific pathway. This could then be used to perform experiments to establish sufficiency and requirement for the specific gene for cancer cell fitness under basal and therapy-induced stress. The incorporation of this approach in preclinical studies could improve the development of GSE-informed cancer treatments. In parallel with the application of TI values in stratification of patients for pathway-directed treatments, the use of this approach in preclinical development would likely provide additional guidance for the use of TI values in the clinical setting.

7 Conclusion

Discussion of gender and sex as variables in research has garnered enough attention that the onus is on those of us in the biomedical research community to set standards for how the entanglement of sex and gender should be evaluated. It is not enough to state that equal numbers of males and females will be used in experiments. A plan that addresses GSE as a biological variable should include how the interaction between GSE and other variables will be evaluated, including required experimental numbers to assess this interaction. With this chapter, we hope to inspire further discussion of this important issue.

Acknowledgments Work on sex differences in the Rubin lab is supported by the National Cancer Institute R01 CA174737-06, P01CA245705, Joshua's Great Things, Siteman Investment Program, Barnard Research Fund, St. Louis Children's Hospital Foundation, Taylor Rozier's Hope for a Cure Foundation, Prayers from Maria Foundation, and the Haubrich and Griffiths family foundations.

References

Bisht S, Mao Y, and Easwaran H (2024) Epigenetic Dynamics of Aging and Cancer Development: Current Concepts from Studies Mapping Aging and Cancer Epigenomes. Curr Opin Oncol 36 (2):82–92. https://doi.org/10.1097/CCO.0000000000001020

Broestl L, Warrington NM, Grandison L, et al. (2022) Gonadal Sex Patterns p21-Induced Cellular Senescence in Mouse and Human Glioblastoma. Commun Biol 5 (1):781. https://doi.org/10.1038/s42003-022-03743-9

Conforti F, Pala L, Bagnardi V, et al. (2019) Sex-Based Heterogeneity in Response to Lung Cancer Immunotherapy: A Systematic Review and Meta-Analysis. J Natl Cancer Inst 111 (8):772–781. https://doi.org/10.1093/jnci/djz094

Conforti F, Pala L, Pagan E, et al. (2021) Sex-Based Dimorphism of Anticancer Immune Response and Molecular Mechanisms of Immune Evasion. Clin Cancer Res 27 (15):4311–4324. https://doi.org/10.1158/1078-0432.CCR-21-0136

Deng Y, Wang J, Wang L, and Du Y (2017) Androgen Receptor Gene CAG Repeat Polymorphism and Ovarian Cancer Risk: A Meta-Analysis. Biosci Trends 11 (2):193–201. https://doi.org/10.5582/bst.2016.01229

DuBois LZ, Gibb JK, Juster RP, and Powers SI (2021) Biocultural Approaches to Transgender and Gender Diverse Experience and Health: Integrating Biomarkers and Advancing Gender/Sex Research. Am J Hum Biol 33 (1):e23555. https://doi.org/10.1002/ajhb.23555

Eggers S, Ohnesorg T, and Sinclair A (2014) Genetic Regulation of Mammalian Gonad Development. Nat Rev Endocrinol 10 (11):673–683. https://doi.org/10.1038/nrendo.2014.163

Goodenough U, and Heitman J (2014) Origins of Eukaryotic Sexual Reproduction. Cold Spring Harb Perspect Biol 6 (3):a016154. https://doi.org/10.1101/cshperspect.a016154

Hosken DJ, Archer CR, House CM, and Wedell N (2019) Penis Evolution across Species: Divergence and Diversity. Nat Rev Urol 16 (2):98–106. https://doi.org/10.1038/s41585-018-0112-z

Hunt KS, and Alspach E (2024) Battle Within the Sexes: Differences in Male and Female Immunity and the Impact on Antitumor Responses. Cancer Immunol Res 12 (1):17–25. https://doi.org/10.1158/2326-6066.CIR-23-0005

Jahandoost S, Farhanghian P, and Abbasi S (2017) The Effects of Sex Protein Receptors and Sex Steroid Hormone Gene Polymorphisms on Breast Cancer Risk. J Natl Med Assoc 109 (2):126–138. https://doi.org/10.1016/j.jnma.2017.02.003

Jelenkovic A, Hur Y-M, Sund R, et al. (2016) Genetic and Environmental Influences on Adult Human Height across Birth Cohorts from 1886 to 1994. eLife 5:e20320. https://doi.org/10.7554/elife.20320

Jiang M, and Huhtaniemi I (2004) Polymorphisms in Androgen and Estrogen Receptor Genes: Effects on Male Aging. Exp Gerontol 39 (11–12):1603–1611. https://doi.org/10.1016/j.exger.2004.06.017

Jobling M, and Tyler-Smith C (2017) Human Y-Chromosome Variation in the Genome-Sequencing Era. Nat Rev Genet 18:485–497. https://doi.org/10.1038/nrg.2017.36

Kirkland JL, and Tchkonia T (2020) Senolytic Drugs: From Discovery to Translation. J Intern Med 288 (5):518–536. https://doi.org/10.1111/joim.13141

Klein SL, and Flanagan KL (2016) Sex Differences in Immune Responses. Nat Rev Immunol 16 (10):626–638. https://doi.org/10.1038/nri.2016.90

Lillie EO, Bernstein L, and Ursin G (2003) The Role of Androgens and Polymorphisms in the Androgen Receptor in the Epidemiology of Breast Cancer. Breast Cancer Res 5:164–173. https://doi.org/10.1186/bcr593

Monclus R, and Blumstein DT (2012) Litter Sex Composition Affects Life-History Traits in Yellow-Bellied Marmots. J Anim Ecol 81 (1):80–86. https://doi.org/10.1111/j.1365-2656.2011.01888.x

Nunney L (2018) Size Matters: Height, Cell Number and a Person's Risk of Cancer. Proc R Soc Lond B Biol Sci 285 (1889):20181743. https://doi.org/10.1098/rspb.2018.1743

Nuti G (2019) An Efficient Algorithm for Bayesian Nearest Neighbours. Methodol Comput Appl Probab 21 (4):1251–1258. https://doi.org/10.1007/s11009-018-9670-z

Nuti G, Jiménez Rugama LA, and Cross I (2019) A Bayesian Decision Tree Algorithm. https://arxiv.org/abs/1901.03214

Palazzolo I, Gliozzi A, Rusmini P, et al. (2008) The Role of the Polyglutamine Tract in Androgen Receptor. J Steroid Biochem Mol Biol 108 (3–5):245–253. https://doi.org/10.1016/j.jsbmb.2007.09.016

Qin Z, Li X, Han P, et al. (2017) Association between Polymorphic CAG Repeat Lengths in the Androgen Receptor Gene and Susceptibility to Prostate Cancer: A Systematic Review and Meta-Analysis. Medicine (Baltimore) 96 (25):e7258. https://doi.org/10.1097/MD.0000000000007258

Rubin JB (2022) The Spectrum of Sex Differences in Cancer. Trends Cancer 8 (4):303–315. https://doi.org/10.1016/j.trecan.2022.01.013

Rubin JB, Lagas JS, Broestl L, et al. (2020) Sex Differences in Cancer Mechanisms. Biol Sex Differ 11 (1):17. https://doi.org/10.1186/s13293-020-00291-x

Samango-Sprouse C, Kirkizlar E, Hall MP, et al. (2016) Incidence of X and Y Chromosomal Aneuploidy in a Large Child Bearing Population. PLoS One 11 (8):e0161045. 10.1371/journal.pone.0161045

Santoni M, Rizzo A, Mollica V, et al. (2022) The Impact of Gender on the Efficacy of Immune Checkpoint Inhibitors in Cancer Patients: The MOUSEION-01 Study. Crit Rev Oncol Hematol 170:103596. https://doi.org/10.1016/j.critrevonc.2022.103596

Schleutker J (2012) Polymorphisms in Androgen Signaling Pathway Predisposing to Prostate Cancer. Mol Cell Endocrinol 360 (1–2):25–37. https://doi.org/10.1016/j.mce.2011.07.007

Schmitt CA, Wang B, and Demaria M (2022) Senescence and Cancer: Role and Therapeutic Opportunities. Nat Rev Clin Oncol 19 (10):619–636. https://doi.org/10.1038/s41571-022-00668-4

Sponagel J, Jones JK, Frankfater C, et al. (2022) Sex Differences in Brain Tumor Glutamine Metabolism Reveal Sex-Specific Vulnerabilities to Treatment. Med 3 (11):792–811.e712. https://doi.org/10.1016/j.medj.2022.08.005

Sundermann EE, Maki PM, and Bishop JR (2010) A Review of Estrogen Receptor Alpha Gene (ESR1) Polymorphisms, Mood, and Cognition. Menopause 17 (4):874–886. https://doi.org/10.1097/gme.0b013e3181df4a19

Yang W, and Rubin JB (2023) Preprint: Accounting for Sex Differences as a Continuous Variable in Cancer Treatments. Res Sq Jun 29. https://doi.org/10.21203/rs.3.rs-3120372/v1
——— (2024) Treating Sex and Gender Differences as a Continuous Variable Can Improve Precision Cancer Treatments. Biol Sex Differ 15 (1):35. https://doi.org/10.1186/s13293-024-00607-1

Yang W, Warrington NM, Taylor SJ, et al. (2019) Sex Differences in GBM Revealed by Analysis of Patient Imaging, Transcriptome, and Survival Data. Sci Transl Med 11 (473):eaao5253. https://doi.org/10.1126/scitranslmed.aao5253

11

Gender, Sex, and Gender/Sex Entanglement in Transgender Health Equity Research

Tonia Poteat and Lu Ciccia

Abstract This chapter considers conceptual understandings and usage of the terms *gender* and *sex* as well as the complexity of *gender/sex entanglement* in the context of health equity research focused on transgender, nonbinary, and gender-diverse people and populations. The aim is to outline current challenges and barriers to transgender health equity, to provide a broad overview of research frameworks and approaches in this area, and to address pressing questions about how best to define and operationalize gender, sex, and their entanglement. These terms and concepts require consideration because of their impacts on the design, implementation, and interpretation of research that seeks to advance health justice for transgender people. The frameworks and associated measures most frequently employed rely on distinguishing sex from gender identity to meaningfully include transgender people in research and identify populations for tailored interventions. Attending to gender, sex, and their entanglement in transgender health equity research exposes how biological essentialism is used to oppress transgender people; it also challenges widely accepted approaches to measuring transgender health disparities. It concludes with implications of shifting the paradigm of transgender health equity research and the questions that must be answered to ensure that such a shift is done in a way that promotes health equity and justice.

Keywords transgender health equity, gender/sex entanglement policy, public health

Tonia Poteat (✉)
School of Nursing and Co-Director of the Duke SGM Health Program, Duke University, Durham, NC 27708, USA
Email: tonia.poteat@duke.edu

Lu Ciccia
Gender in Science, Technology, and Innovation, Center for Research and Gender Studies, Universidad Nacional Autónoma de México, México City 04510, México

© Frankfurt Institute for Advanced Studies (FIAS) 2025   219
L. Z. DuBois et al. (eds.), *Sex and Gender*, Strüngmann Forum Reports,
https://doi.org/10.1007/978-3-031-91371-6_11

1 Introduction

Our aim in this chapter is to summarize key issues related to gender, sex, and their entanglement in transgender health equity research. As other chapters in this volume address gender/sex entanglement in research design, data collection, and measures, we will focus specifically on the issues raised by gender/sex entanglement in health research that seeks to promote health equity for transgender, nonbinary, and gender-diverse populations. Drawing on existing scientific literature and published best practices, we highlight how researchers may need to rethink fundamental frameworks, study designs, measurement, and interpretations to truly incorporate the complexity implicit in disrupting the notion of an asocial biology of sex and a strictly social construction of gender.

2 Definitional Challenges

Consistent definitions provide a basis for shared understanding. Scientific inquiry relies on shared definitions for effective communication and comparison of findings across research studies and over time. However, definitions of sex and gender-related terms (and how they may be entangled) can vary widely. How these terms are understood not only presents challenges in research, but has also been weaponized by state actors seeking to restrict the rights of transgender people (Perez-Brumer et al. 2024). For example, more than 600 pieces of anti-transgender legislation were proposed in the United States in 2023 and over 500 were proposed by June 2024 (ACLU 2024). While most bills do not use the term "transgender," many set legal definitions for sex in binary biological terms and explicitly equate sex with gender. For example, South Carolina Bill S0276 proposed an amendment to the state constitution "to provide that a person's biological sex at birth constitutes that person's gender" (Trans Legislation Tracker 2024). The bill goes on to define sex as male or female based on "reproductive potential or capacity, such as sex chromosomes, naturally occurring sex hormones, gonads, and nonambiguous internal and external genitalia present at birth, without regard to an individual's psychological, chosen, or subjective experience of gender" (South Carolina 2023). In effect, this bill erases the very existence of transgender people.

For the purposes of this chapter, which focuses on transgender health equity research, we make a conceptual distinction between sex and gender; that is, while sex and gender are entangled, they are not equivalent. We understand the notion of sex as biological parameters linked to reproduction and gender as the sociocultural expectations ascribed to bodies according to their external genitalia (Krieger 2003). This conceptual distinction between sex and gender, as categories, allows room for current definitions of the term transgender (DuBois and Shattuck-Heidorn 2021). The American Psychological Association (APA) uses the following definition: "transgender is an umbrella term for persons whose gender identity, gender expression or behavior does not conform to that typically associated with the sex to which they were assigned at birth" (APA 2023). The

National Center for Transgender Equality defines transgender as "a broad term that can be used to describe people whose gender identity is different from the gender they were thought to be when they were born" (NCTE 2023). Both definitions were published online a few months apart in 2023, yet reflect subtle differences in understandings of gender/sex entanglement. Both definitions rely on a difference between initial perceptions at birth and current self-identity. However, the APA definition denotes this difference as one that relies on assigned sex at birth while the NCTE definition does not reference sex at all. The reason for the difference in definitions is not readily apparent and raises meaningful questions about how the terms gender and sex, upon which these definitions rely, are understood and operationalized by these sources.

As the visibility of nonbinary sex and gender has increased (Risman et al. 2022), some researchers have sought to advance the terminology we use to describe sex and gender relationships. Ashley and colleagues have called for a distinction between gender identity and a concept known as "gender modality" (Ashley 2022; Ashley et al. 2024) and note: "a person's gender identity is their sense of gender at any given time. By contrast, gender modality refers to how a person's gender identity relates to the gender they were assigned at birth. It is a mode or way of being one's gender." According to Ashley, gender modalities go beyond cisgender (i.e., sex assigned at birth and gender identity align) and transgender (i.e., sex assigned at birth and gender identity differ) to include agender modalities for people who do not identify with any gender, detrans or retrans modalities for people who have stopped or changed a gender transition, and other potential modalities.

Here, we use the term transgender in its broadest sense, to include those who identify as transgender as well as agender, nonbinary, and gender-diverse people who may not embrace a transgender identity, acknowledging that the inequities experienced may differ according to the specificities of each identity. For example, health inequities may vary between transgender men, transgender women, and nonbinary people. We also include transgender people in their diverse embodiments, regardless of whether they legally affirm their gender (e.g., via changing identity documents to align with their correct gender), socially affirm their gender (e.g., via pronoun and name changes), and/or access gender-affirming care (e.g., access gender-affirming hormonal therapies or surgeries).

We understand gender and sex to be entangled, in that the aspects of biological embodiment that we call sex are influenced by aspects of social practices and experience that we call gender, and vice versa. At the same time, we note that no practice and experience of gender is an inherent consequence of, nor caused by, sex. Instead, the correlation observed today between genitalia and certain practices/experiences results from gender (Butler 1999; Ciccia 2022). Cisheteronormativity (Kinitz and Salway 2022)—the naturalization of a society's expected correlation between sex and gender, which implies that cisgender heterosexuality is *normal*, as is the pathologization and violence experienced by those who do not conform to it—can lead to specific health inequities for transgender populations, several of which we describe below.

3 Health Equity Research

The World Health Organization defines equity as the "absence of unfair, avoidable or remediable differences among groups of people, whether those groups are defined socially, economically, demographically, or geographically or by other dimensions of inequality" (WHO 2024). The US National Institutes of Health (NIH) recognizes health equity as a process of ensuring all populations have the opportunity to attain the best possible health by removing barriers to good health and allocating resources proportional to their needs (NIMHD 2023). As the largest funder of health research in the world by far (Viergever and Hendriks 2016), NIH exerts enormous influence in this arena. Therefore, how they understand and apply principles of health equity is fundamental to which research studies get funded. Prior to 2016, transgender people were not mentioned in the list of priority populations for the National Institute on Minority Health and Health Disparities (NIMHD); therefore, research on transgender health was not fundable based on gender identity-related disparities. In 2016, however, the director of NIMHD announced the addition of sexual and gender minority populations (including transgender people, broadly defined) to the groups designated as health disparities populations (Pérez-Stable 2016). This change took place a few months after the establishment of the NIH Sexual and Gender Minority Research Office and led to a substantial change in direct funding for transgender health research that has been reflected by an exponential increase in the number of published studies focused on transgender populations (Reisner et al. 2016a; Scheim et al. 2024).

Across every country in which data are available, transgender people experience health-related inequities (Scheim et al. 2024). In 2020, the National Academies of Science, Engineering, and Medicine (NASEM) published a consensus research report on the well-being of sexual and gender minority populations. The report identified inequities in a wide range of health behaviors and outcomes including, but not limited to, smoking, cardiovascular disease, HIV, sexually transmitted infections, cancer screening, mental health, substance use, and access to health care (NASEM 2021). However, a systematic review published as recently as 2024 found that the preponderance of transgender health data remain heavily skewed toward mental health and HIV, respectively, with less attention to other health topics and with the reaffirmation of stereotypes that link being transgender with psychiatric disorders and sexually transmitted diseases (Scheim et al. 2024).

The preponderance of research on mental health is likely related to two factors. First, persistent data indicate high rates of psychological distress and suicidality among transgender adolescents and adults (Pinna et al. 2022). For example, a US population-based study of transgender adults found an alarming prevalence of lifetime suicidal ideation (81%) and attempts (42%), with odds more than four- and sixfold higher when compared with their cisgender counterparts, after adjusting for age, race, ethnicity, income, and education (Kidd et al. 2023). Another factor is the historical labeling of transgender identities as "disordered" by the APA's *Diagnostic and Statistical Manual of Mental Disorders* (DSM) and the WHO's *International Classification of Diseases* (ICD) (Dakić 2020; Dora et al. 2021; Scheim et al. 2022). In brief, prior to 2013, "gender identity disorder" was classified as a mental disorder in the DSM and ICD. However, the DSM-5 replaced this diagnosis with "gender dysphoria" (APA 2013), and ICD-11 moved its reconceptualized "gender incongruence" diagnosis out of the chapter on mental disorders (Reed et al. 2016). These changes were pivotal for shifting the focus of medical interventions away from treating gender identity itself as disordered, toward addressing the clinically relevant distress that can accompany a difference between gender identity and sex assigned at birth (Reed et al. 2016; Scheim et al. 2022). Some researchers have hypothesized that gender dysphoria may be a product of societal stigma against transgender people rather than an innate response to gender incongruence (Galupo et al. 2020; Lindley and Galupo 2020; Riggs et al. 2015). A global systematic review has linked the high prevalence of mood and anxiety disorders among transgender people with anti-transgender societal stigma (Pinna et al. 2022).

HIV is the second most commonly researched health issue among transgender people (Scheim et al. 2024). Data are consistent that transgender people bear a disproportionate burden of HIV across the globe, though data from transgender men and gender nonbinary individuals are much more limited (CDC 2024; Stutterheim et al. 2021). Based on a meta-analysis of 98 studies from across the world, the prevalence of HIV is 19.9% among transgender women and 2.6% among transgender men, with odds of living with HIV being 6.6- and 6.8-fold higher, respectively, compared to the general population (Stutterheim et al. 2021). Attention to these notable disparities is clearly warranted. At the same time, transgender communities have called for health research that goes beyond a focus on HIV (Logie et al. 2022) and have encouraged greater attention to noninfectious chronic conditions (e.g., cardiovascular health), reproductive health and fertility, social determinants of health, and resilience factors (Feldman et al. 2016; LeBlanc et al. 2022; Veale et al. 2022).

Regardless of the specific health condition, without context and a conceptual framework, the persistence of health inequities among transgender people could be misconstrued to be a result of some deficit inherent in being transgender. However, an accurate, justice-focused approach requires that we acknowledge and examine the stigma and structural drivers that are at the root of health inequities. Health justice research that utilizes and tests conceptual frameworks to identify and name intersecting power relations as the drivers of health equities enable disruption of the status quo of existing oppressive power relations and center the embodied knowledge of transgender people who experience these health inequities (Wesp et al. 2019). Current transgender health research frameworks provide models for understanding how existing oppressive and stigmatizing social structures drive transgender health inequities. A few common frameworks are described below.

4 Research Frameworks for Understanding Transgender Health Inequities

4.1 Minority Stress

For decades, scientists have studied how psychosocial stressors influence mental and physical health (McEwen 1998; McEwen and Stellar 1993; Schneiderman et al. 2005; Yaribeygi et al. 2017). The term *minority stress* has been used to describe the excess psychosocial stressors experienced by people with minoritized identities, such as racism experienced by African Americans and prejudice experienced by lesbian, gay, and bisexual people (Clark et al. 1999; Meyer 2003). Virginia Brooks first described a sexual minority stress model in relation to drivers of mental health disparities among lesbian women in her book published in 1981 (Brooks 1981). Ilan Meyer drew from this model in his seminal paper on sexual minority stress among gay men in 1995 (Meyer 1995).

By 2003, the minority stress models described by Brooks and Meyer had been expanded into an explanatory theory designed to explicate structural, social, and psychological factors leading to mental health inequities experienced by sexual minority populations, inclusive of lesbian, gay, and bisexual people (Meyer 2003). The expanded model outlines how people with minoritized sexual orientations experience distal (i.e., discrimination) and proximal (i.e., expectations of rejection, concealment, and internalized homophobia) stressors which lead to increased risk for negative mental health outcomes. Positive coping and social support are theorized to operate as resilience factors that mitigate the impact of these negative stressors on health. Since its development, the minority stress model has gained widespread use in health equity research (Frost and Meyer 2023) and has been applied to physical as well as mental health inequities (Flentje et al. 2020).

In 2015, Testa et al. (2015) developed the gender minority stress and resilience (GMSR) measure based on Meyer's minority stress model and adapted it to reflect the experiences of transgender populations. GMSR assesses the following constructs as minority stressors: gender-related discrimination, gender-related rejection, gender-related victimization, non-affirmation of gender identity, internalized transphobia, negative expectations for future events, and nondisclosure of gender identity. Resilience constructs include transgender pride and transgender community connectedness as forms of coping and social support, respectively. Multiple studies have applied the gender minority stress model to explain associations between anti-transgender stigma and health across populations that are diverse along lines of age, race, gender identities, and geographic locations (Delozier et al. 2020; DuBois and Juster 2022; Puckett et al. 2024; Rich et al. 2020). For example, Puckett et al. (2024) found enacted and internalized stigma (i.e., gender-related discrimination and internalized transphobia) to have mediating effects between perceptions of local sociopolitical context and mental health outcomes among transgender adults in four different US states. A growing body of research goes further to examine how minority stress experienced by transgender people can be directly embodied as physiological dysregulation (Cohen et al. 2021; Dubois 2012; DuBois et al. 2024; Juster et al. 2019; Kumaraguru et al. 2023), demonstrating the biomaterialization of cisheteronormativity and its direct negative impact on health.

Minority stress models and their related psychometric measures demonstrate how cisheteronormativity provides the frame of reference from which both transgender and cisgender people develop their subjectivity. For example, the distal stress caused by discrimination requires perpetrators to identify that another individual is transgender, either by disclosure or presumption. Proximal stress caused by internalized transphobia requires someone to understand themselves as transgender. Mitigating factors in minority stress models require that transgender people feel a connection with others who have a shared transgender identity either internally (i.e., pride in being transgender) or externally (e.g., a sense of connection to transgender community). To recruit transgender participants to participate in studies using these measures and to interpret study findings, researchers must implement eligibility criteria that *ipso facto* define who is or is not transgender. In addition to these inherent limitations, some researchers have noted that the minority stress model fails to account for intersectional experiences of multiply marginalized transgender people who experience stigma and discrimination along more than one axes, such as racism, ableism, and classism (Tan et al. 2020).

4.2 Gender Affirmation

The Gender Affirmation Framework, developed by Jae Sevelius in 2013, elucidates pathways that lead to the extraordinary burden of HIV among transgender women of color. Sevelius defines gender affirmation as "an interpersonal, interactive process whereby a person receives social recognition and support for their gender identity and expression" (Sevelius 2013:676). The framework outlines a pathway to elevated HIV risk that originates with intersectional stigma, which drives both social oppression (e.g., transphobia, victimization) and psychological distress (e.g., internalized transphobia, depression/anxiety). Social oppression reduces access to gender affirmation, while psychological distress increases the need for gender affirmation. The gap between the increased need for gender affirmation and the decreased access to it creates an unmet need that leads to high-risk contexts and behaviors that make transgender women of color, in particular, more vulnerable to HIV than their cisgender peers.

The Gender Affirmation Framework has been applied to research on engagement in HIV care among transgender women (Rosen et al. 2019) and has been extended to HIV among transgender men (Reisner et al. 2016b). It has also been applied to other health outcomes, such as psychological well-being (Glynn et al. 2016). Measures of the framework's key constructs have been developed and validated (Sevelius et al. 2020, 2021a, b). Since the initial framing over a decade ago, the concept of "gender affirmation" has been expanded to include social, medical, and legal dimensions (Hughto et al. 2020; King and Gamarel 2020; Reisner et al. 2016a).

Similar to minority stress models, the Gender Affirmation Framework raises questions about who should determine, and how, transgender status for the purposes of research when gender/sex are entangled. This question becomes particularly salient when the concept of gender affirmation is expanded to include medical interventions that change sex-linked anatomy and physiology (e.g., exogenous hormone treatments, genital reconstruction surgeries) as well as legal interventions that alter the sociocultural and political landscape in which gender/sex are debated and boundaries adjusted and/or enforced. Regardless of whether there are surgical and/or hormonal interventions that accompany gender affirmation, said affirmation undoubtedly involves mental processes, which are simultaneously experienced in the body (Lakoff and Johnson 1999; Van der Kolk 1994). These are processes that occur as part of the entanglement between gender and sex. The Gender Affirmation Framework enables us to explore how the symbolic dimension of living as a transgender person can be embodied in cisheteronormative cultures.

4.3 Emerging Frameworks

While minority stress and gender affirmation are the most commonly used frameworks in transgender health equity research, new frameworks continue to emerge and existing frameworks are being expanded. These efforts are often, although not always, led by transgender researchers (Rosenberg and Tilley 2020; Santora 2021; Streed et al. 2023). Matsuno and colleagues have adapted the gender minority stress model to incorporate resilience strategies used by transgender people. This model, called the transgender resilience intervention model (TRIM), includes resilience promoting factors at both group (e.g., activism) and individual (e.g., self-acceptance) levels (Matsuno and Israel 2018). Another effort to counter deficit-based approaches to transgender health research (Holloway 2023; Shuster and Westbrook 2024) is the development and testing of the Minority Strengths Model, which provides a framework for how personal and collective strengths in minoritized populations (including transgender people) create resilience and lead to positive health outcomes (Perrin et al. 2020). Restar et al. (2021) have outlined an expanded gender equity continuum model that provides a transgender-inclusive framework for addressing gender-based health inequities. Gender euphoria is a term increasingly used in transgender communities (Austin et al. 2022), and ongoing research explores how to measure it (Blacklock et al. 2023) as well as how it is linked to health outcomes (Reisner et al. 2023).

5 Approaches to Research on Transgender Health Inequities

Each of the aforementioned research frameworks, and the ability to identify and measure population-specific health inequities itself, rely on the stable existence of a defined/definable transgender population and the ability to assess their health outcomes consistently over time and across contexts. In other words, two central questions arise for transgender health research: First, who and how is it decided whether or not one is part of the transgender population? This question is particularly salient in medical settings where cisheteronormative values prevent healthcare professionals from accepting first person narratives of living a transgender experience, even without surgical and/or hormonal interventions. Second, how do we investigate the negative health effects of living as a transgender person in cisheteronormative societies, considering the entanglement between sex and gender? These are more than semantic questions since interventions addressing the impacts of stigma and stress rely on amplifying factors that mitigate those impacts. Intervening on those factors will require that we know who we are intervening with and how we are operationalizing this shared understanding of gender/sex to apply to transgender population health. In a similar fashion, measuring stigma, stressors, resilience, and strengths related to transgender status requires an implicit, if not explicit, understanding of who is transgender. The entanglement and fluidity of gender/sex creates challenges for measuring gender- and sex-linked health outcomes and potential inequities in these outcomes. Below, we provide examples of how these challenges manifest when tracking health outcomes and then conclude with an examination of measurement in transgender health research and the implications of gender/sex entanglement.

6 Tracking Health Outcomes

6.1 Gender/Sex-Linked Health Conditions

Most public health surveillance systems and disease registries that track health outcomes do so based on binary sex categories and assumptions about the anatomy and physiology associated with them (Danielsen et al. 2022; Richardson 2022; Shattuck-Heidorn et al. 2021). For example, cervical cancer is tracked among individuals designated as female while prostate cancer is tracked among males. However, this conflation of binary sex categories with presumed reproductive anatomy (Fausto-Sterling 2021; Pape et al. 2024) not only excludes and/or renders invisible transgender people whose anatomy may not match assumptions associated with their designated sex, it can also lead to broader misinterpretations of surveillance data. For example, in a study of racial inequities in cervical cancer among cisgender women, researchers found that prior failure to remove females who had a hysterectomy from the denominator led to an underestimation of racialized inequities in cervical cancer incidence and mortality (Beavis et al. 2017). When estimating cervical cancer incidence, researchers divided the number of people at risk for cervical cancer by the number of people who developed cervical cancer. This inadvertently included women who no longer had a cervix in the pool of people at risk (i.e., the denominator), even though it is not possible to develop cervical cancer without a cervix. Since Black women were more likely to have had a hysterectomy than White women, this error inflated the denominator for Black women, thereby underestimating the incidence of cervical cancer in this group. In sum, the assumption that the category of female was an appropriate proxy for a person with a cervix led to inaccurate findings and hid the magnitude of racial disparities in cervical cancer.

The assumption of binary sex categories poses obstacles to the production of knowledge itself, since it feeds the idea of qualitative differences that also negatively impact the health of the cisgender population. Thus, prostate cancer in cisgender women (Slopnick et al. 2022) and breast cancer in cisgender men (Rudd et al. 2023) are underappreciated. We do not know the incidence of these cancers for certain, since diagnoses are biased by our dimorphic interpretation of sex. In addition, we fail to explore how gender practices can be connected with variables linked to sex to explain the incidence of prostate and breast cancers in cisgender women and men, respectively (Ciccia 2023).

An intersectional perspective (described in detail by Bowleg et al., Chapter 9) requires consideration of other mutually constituted identities equally relevant to understanding the complexity of gender/sex (Bowleg 2021; Collins and Bilge 2020; Crenshaw 2013). Bowleg (2012:267) defines intersectionality as the ways in which "multiple social categories (e.g., race, ethnicity, gender, sexual orientation, socioeconomic status) intersect at the micro level of individual experience to reflect multiple interlocking systems of privilege and oppression at the macro, social-structural level (e.g., racism, sexism, heterosexism)." For example, Abesamis (2022) has outlined how transgender health inequities in the Philippines are produced by

coexisting and interacting discriminatory legal, educational, and medical institutions, where the historical experiences of colonization, the hegemony of cisgenderism, and the impact of capitalism create gaps in the Philippine health-care system,
which perpetuate the lack of gender-affirming health care and the limited transgender competence of health-care providers. In the United States, Lett et al. (2020)
have applied an intersectional framework to a national dataset and demonstrated
how health outcomes among Black American transgender people differ from those
of both cisgender Black Americans and transgender White Americans.

Current research with transgender people who have a cervix has identified structural (e.g., gendered office settings) and interpersonal (e.g., health-care providers
without transgender-care competence) barriers to cervical cancer screening that
may ultimately result in inequitable distribution of cancer incidence and mortality
(Peitzmeier et al. 2020). However, tracking transgender people in population-based
surveillance data and disease-specific registries is complicated by low numbers
of transgender people identified in these databases, likely due to a failure to collect nuanced gender/sex data (Gomez et al. 2019; Kaplan-Marans et al. 2023).
Emerging attempts to identify transgender people in large population-based datasets has required complex and imperfect algorithms, with low sensitivity and/or
specificity, that rely on inadequate inclusion criteria such as transgender-related
diagnosis codes (e.g., gender incongruence) and/or reported genders that do not
align with anticipated anatomy (e.g., males with cervical cancer) (Alpert et al.
2021; Gomez et al. 2019).

6.2 Gender/Sex-Linked Clinical Tools

Health-care professionals use a variety of clinical tools to estimate disease risk and
to make clinical diagnoses. These tools often use the terms sex and gender interchangeably without providing a definition for either (Mohottige et al. 2024; Poteat
et al. 2023a, b). They require clinicians to provide data about the patient's gender/
sex to calculate an actionable result. For example, the most commonly used cardiovascular disease risk calculator requires the clinician to designate the patient as
either male or female; whether the calculator names the query sex or gender, the
response options provided are only "male" or "female" (Goff et al. 2014). The new-
est cardiovascular risk calculator has removed the social category of race from the
calculation but has retained gender/sex (Khan et al. 2024). Similarly, reported sex is
used in clinical calculations that estimate, among others, renal function and respira-
tory function (Haynes and Stumbo 2018; Mohottige et al. 2024).

Even for directly measured (i.e., not calculated) laboratory measures, many use
sex-based reference ranges to determine whether a result is abnormal, including
values for hemoglobin, creatinine, alkaline phosphatase, alanine aminotransferase,
aspartate aminotransferase, and high-density lipoprotein (Krasowski et al. 2024). It
is unclear if the use of gender/sex in clinical calculations and sex in reference ranges
is intended to represent something about the patient's hormonal status, anatomy,
genetic makeup, social experiences, or something else. Clinicians who care for
transgender people are left with no guidance for what to enter and how to interpret

the results of these gender/sex-based algorithms or the sex-based laboratory tests (Krasowski et al. 2024; Poteat et al. 2023b).

Behavioral health screening tools, such as the Alcohol Use Disorders Identification Test-Consumption (AUDIT-C), ask patients a series of questions to assess for hazardous drinking (Simon et al. 2024). However, the scores are interpreted differently based on binary gender, with a cut-off score of three or higher indicating an alcohol use disorder for women and a score of four or higher for men. Similar to other screening tools, it is unclear how to use this tool with transgender people since the rationale for different interpretation by gender is unclear and no guidance is provided for individuals with a difference between assigned sex at birth and current gender.

7 Trans Inclusive Gender/Sex Measures

7.1 Existing Measures

Terminology used to describe transgender people is dynamic and context dependent, varies based on language and geographic location, has certainly changed over time, and continues to evolve as communities find more precise and affirming ways to describe their identities and lived experiences (Poteat et al. 2019). The dynamic and evolving nature of this terminology presents a challenge to scientific efforts to create valid, standardized, and reliable quantitative measures of who is transgender. That stated, efforts to measure gender/sex in health research in transgender-inclusive ways are quite recent. Using the term "informational erasure," Bauer et al. (2009) note that most health research had not allowed for identification of transgender participants nor addressed questions relevant to transgender communities. Instead, research participants were presumed to be cisgender; transgender participants were systematically erased and their experiences rendered invisible. While an increasing number of survey instruments have begun to include transgender-inclusive measures (NIH 2023), the conflation of sex and gender in many studies continues to create gaps in knowledge about transgender health (Morrison et al. 2021; Poteat et al. 2021; Tordoff et al. 2022).

In 2022, NASEM published a report entitled, *Measuring Sex, Gender Identity, and Sexual Orientation* (NASEM et al. 2022), wherein researchers are encouraged to make a consistent distinction between measures of gender and measures of sex for precision and construct validity. They recommend a two-step method of assessing gender and sex that asks a participant's sex assigned at birth in one question and their current gender identity in another (NASEM et al. 2022). For analysis and interpretation, participants would be categorized as transgender if they either self-identify as transgender in the current gender question or if they report a current gender that differs from the sex they were assigned at birth. While this straightforward approach has its strengths, in reality, efforts to simplify and standardize definitions and measures of transgender status are complicated by the reality of gender/sex entanglement and challenged by individuals' desires to claim agency in defining and labeling their own sexed and gendered identities outside of boxes provided to them by researchers.

Some researchers have called for the use of more complex measures to allow for greater transgender visibility in survey research. Comparing the two-step method recommended by NASEM with multidimensional measures developed in Canada, Bauer et al. (2017) found agreement between the two measures to be very high (K = 0.9081) in assessment of gender identity. However, gender identity was a poor proxy for other dimensions of sex or gender among transgender participants. Therefore, they recommend consideration of a new "Multidimensional Sex/Gender Measure," which includes three questions that assess sex assigned at birth, current gender identity, and current lived gender (Bauer et al. 2017). More recently, Bauer (2023) has called for sex and gender multidimensionality in epidemiologic research. Using case examples, Bauer demonstrates how conflation of dimensions within and between sex and gender presents a validity issue wherein proxy measures are substituted for dimensions of interest (e.g., "woman" for "person with a uterus") often without explicit acknowledgement and with the potential to generate erroneous findings. Recommendations are made to clearly specify dimensions of "biological sex and/or social gender" relevant to the study and how they are measured, including when proxy measures are used and their limitations (Bauer 2023:128). Importantly, Bauer recommends acknowledgement of the ways sexed biology shapes and is shaped by gendered social and behavioral factors.

Bauer's call to acknowledge the entanglement of gender and sex addresses the way that two-step measures alone, influenced by cisheteronormative values, are based on implicit conceptualizations of sex as something material that exists prior to gender (i.e., at birth) and that defines our way of being in the world. This perspective may explain the heavy focus on mental health in transgender health research. Researchers assume that the main health consequence experienced by transgender people is in the mental sphere based on two related beliefs: (a) our biological configurations are independent of (i.e., impermeable to) the experience of gender and (b) mental health is understood as something separable from physical health. This perpetuates a mind-body dichotomy, which explains why gendered mental states are not considered with respect to how they can be embodied and affect variables that we associate with sex.

These beliefs run counter to the well-established phenomena of plasticity and epigenetics, which defy notions of biology as something fixed and impermeable to social practices. Instead, these phenomena require transcendence of false dichotomies. They acknowledge that lived experiences are embodied molecularly, thus blurring the boundaries between sex and gender. At the same time, mental states are not separable from biological states (Ciccia 2022; Jordan-Young and Karkazis 2019; Kaiser 2016; Rippon 2019). In short, being transgender in cisheteronormative cultures involves specific biomaterialization.

A growing body of scientific literature describes how gender and sex are neither binary nor independent of each other (DuBois and Shattuck-Heidorn 2021). In a recent conceptual review, Fausto-Sterling (2019:533) notes: "Gendered structures change biological function and structure. At the same time, biological structure and function affect gender, gender identity, and gender role at both individual and cultural levels." Further, she uses transdisciplinary dynamic systems theory to explain how sociocultural differences become bodily difference, debunking dichotomies between biology and culture. Therefore, it is impossible to consider sex without our gender practices, which involve occupations, interests, and desires. At the same time, when we investigate gender practices, there is always a biological dimension tied to these practices.

7.2 Measurement Gaps and Implications

In the field of health equity research, measurement of nonbinary and multifaceted gender/sex is in its infancy. We are far from determining how best to integrate what science and the lived experience of individuals tell us about gender/sex entanglement with how we go about measuring gender/sex for the purposes of research. We know that the two-step method of asking about sex assigned at birth separately from current gender identity is an improvement over prior one-item approaches of simply asking if a participant was male, female, or transgender (Bauer et al. 2017; NASEM et al. 2022). The two-step measurement approach has allowed for visibility of transgender people in survey data while not requiring that gender-diverse people self-identify as transgender if that label is not how they see themselves. At the same time, there are limitations. First, this type of measurement presents sex dimorphically (i.e., according to two forms defined by reproductive possibilities) and mutually exclusive (Joel 2012; Richardson 2022). This renders invisible the intersex population who do not conform to this dimorphic sex norm (Ashley et al. 2024; Bauer et al. 2017). At the same time, it implies a biased interpretation of the endosex (i.e., not intersex) population (Richardson 2022). In this sense, it has been emphasized that the parameters directly linked to reproduction, such as testosterone concentrations, are not genetically defined. Rather, they vary with highly gendered social practices (van Anders 2024). Furthermore, parameters not directly linked to reproduction, but which we associate with it, are not defined by sex chromosomes and hormonal concentrations. For example, hepatic metabolization rates widely purported to be based on sex (Della Torre 2021) actually vary by factors such as weight and height (not defined by sex) and also by gendered social practices, such as the frequency of physical activity and the consumption of bioactive components (Ciccia 2022; DiMarco et al. 2022).

The second limitation of the two-step approach is the potential to reify a dichotomy, rather than entanglement, between gender and sex. As exemplified above in the case of liver metabolization, this dichotomy does not reflect our biological reality (Pape et al. 2024; Richardson 2022). In short, this type of measurement implies the existence of dimorphic differences in the cisgender population, due to roles in reproduction, without considering the gender factor in this reading of the differences, thereby projecting this bias to the transgender population. Two-step measures may obscure gender/sex differences *within* transgender populations, and they provide no information on physical embodiment, leaving unanswerable questions about the effects of hormonal milieu, genetic makeup, and anatomical structures on health inequities experienced by transgender people.

Even though limited, two-step measures provide an important starting place from which to advance transgender health equity research. Future efforts to improve gender/sex measures must retain the two-step measures' strengths of creating transgender visibility in survey data, while also allowing for incorporation of gen-der/sex entanglement. This may raise questions about when and how transgender health equity researchers should go beyond centering sex assigned at birth and in-dividualized self-reported gender identity in research, especially in sociopolitical contexts that seek to erase transgender existence (Perez-Brumer et al. 2024). The recent introduction of new concepts, such as gender modality (Ashley 2022, 2023; Ashley et al. 2024), provide an example of how to think about transgender visibility beyond self-identity. Bauer (2023) recommends specifying which aspects of sex biology (e.g., genes, hormones, anatomy) and/or social gender (modality, identity, roles, norms) are relevant to a particular research question based on anticipated explanatory pathways, including the most valid available measures of those specific dimensions in one's study, analyzing the data with attention to multidimensional ef-fects, and interpreting the results with an understanding that biology and physiology both shape and are shaped by social context and behavior. Such an approach moves beyond identity alone and *requires* engagement with the entanglement of sex and gender. If gender/sex data are consistently operationalized in this way, it may allow us to account better for their dynamic nature over time and across contexts in our longitudinal research.

8 The Way Forward

Novel transgender-led and/or transgender-inclusive explanatory frameworks for identifying and leveraging transgender community strengths advance the science of health equity (Blacklock et al. 2023; Matsuno and Israel 2018; Perrin et al. 2020; Restar et al. 2021). Likewise, generating more nuanced, multidimensional measures are important steps for moving transgender health equity research forward (Ashley et al. 2024; Bauer 2023). However, there is more to be done to attend to the en-tanglement of sex and gender in research with transgender populations. Remaining questions include, but are not limited to:

- How do we develop a research implementation (vs. explanatory) framework to guide incorporation of gender/sex entanglement into research designs in ways that account for the complex nature and social reality of gender, sex, and their entanglement?
- How do we ensure that researchers make well-explicated, transparent deci-sions about why and how they selected certain measures and strategies to study transgender health inequities?
- What analytic approaches are most appropriate for health equity studies that incorporate multidimensional, potentially interacting, measures of gender/sex?

One approach for attending to these issues of complexity and transparency is col-laboration across disciplinary silos, such as biological and social sciences, to facili-tate generation of gender/sex measures and analytic plans that render transgender communities visible in the data and relevant to specific research questions. In ad-dition, because ways of operationalizing and interpreting the categories of sex and gender and their relationship create biases—not only for transgender populations, but also for cisgender populations—we encourage rejection of essentialist conceptualizations that permeate our current interpretation of biological differences as the basis for understanding the prevalence, development, and treatment of diseases. Beginning to disaggregate the variables of interest for a given study, including all the corporealities that embody such variables, can be a starting point to embrace the notions of plasticity and epigenetics and update our interpretations from such notions, foregrounding the biological dynamism that characterizes the human experience.

We also encourage researchers to consider, when and where possible, replacing dichotomies with continuums that do not assume qualitative differences between bodies. Instead, we suggest acknowledging that the phenomenon of plasticity requires recognition of the complexity of our genetics and disrupt the nature-culture dichotomy. Our corporealities are not the sum of different behaviors and/or biological variables. Instead, in an intersectional way, we biologically embody the multidimensionality of our practices and experiences.

In sum, we emphasize that reducing existing complexity when conducting gender/sex related research is not a simplification. Rather, reductionist approaches to gender/sex measurement applies essentialist biases to our understanding of ourselves as biomaterial beings. Thus, it is necessary to consider gender/sex in its multidimensionality (DuBois et al. 2021). A gender/sex binary, contingent and historically situated, assumes cisheteronormative values that are embodied based on our social locations of relative privilege and oppression. Exploring the dimension of these values in the field of health equity research requires imagination to develop new questions that consider the plasticity that characterizes us as a species, and the impact that this can have on the biological differences that we observe today between cisgender women and men as well as among transgender populations. In short, we believe that rigorous approaches to understanding the health of transgender populations entail, at the same time, changing our understanding of gender/sex in cisgender populations. Of course, this change involves questioning the profound dichotomous lenses on which our current interpretation of bodies rests.

References

Abesamis LEA (2022) Intersectionality and the Invisibility of Transgender Health in the Philippines. Glob Health Res Policy 7 (1):35. https://doi.org/10.1186/s41256-022-00269-9

ACLU (2024) Mapping Attacks on LGBTQ Rights in U.S. State Legislatures in 2024. (June 28, 2024). https://www.aclu.org/legislative-attacks-on-lgbtq-rights-2024

Alpert AB, Komatsoulis GA, Meersman SC, et al. (2021) Identification of Transgender People with Cancer in Electronic Health Records: Recommendations Based on CancerLinQ Observations. JCO Oncol Pract 17 (3):e336–e342. https://doi.org/10.1200/OP.20.00634

APA (2013) Diagnostic and Statistical Manual of Mental Disorders. American Psychiatric Association, Washington, DC

APA (2023) Understanding Transgender People, Gender Identity and Gender Expression. https://www.apa.org/topics/lgbtq/transgender-people-gender-identity-gender-expression

Ashley F (2022) "Trans" Is My Gender Modality: A Modest Terminological Proposal. In: Erickson-Schroth L (ed) Trans Bodies, Trans Selves, 2nd edition. Oxford Univ Press, Oxford, p 22
——— (2023) What Is It Like to Have a Gender Identity? Mind 132 (528):1053–1073. https://doi.org/10.1093/mind/fzac071
Ashley F, Brightly-Brown S, and Rider GN (2024) Beyond the Trans/Cis Binary: Introducing New Terms Will Enrich Gender Research. Nature 630 (8016):293–295. https://doi.org/10.1038/d41586-024-01719-9
Austin A, Papciak R, and Lovins L (2022) Gender Euphoria: A Grounded Theory Exploration of Experiencing Gender Affirmation. Psychol Sexual 13 (5):1406–1426. https://doi.org/10.1080/19419899.2022.2049632
Bauer GR (2023) Sex and Gender Multidimensionality in Epidemiologic Research. Am J Epidemiol 192 (1):122–132. https://doi.org/10.1093/aje/kwac173
Bauer GR, Braimoh J, Scheim AI, and Dharma C (2017) Transgender-Inclusive Measures of Sex/Gender for Population Surveys: Mixed-Methods Evaluation and Recommendations. PLoS One 12 (5):e0178043. https://doi.org/10.1371/journal.pone.0178043
Bauer GR, Hammond R, Travers R, et al. (2009) "I Don't Think This Is Theoretical; This Is Our Lives": How Erasure Impacts Health Care for Transgender People. J Assoc Nurses AIDS Care 20 (5):348–361. https://doi.org/10.1016/j.jana.2009.07.004
Beavis AL, Gravitt PE, and Rositch AF (2017) Hysterectomy-Corrected Cervical Cancer Mortality Rates Reveal a Larger Racial Disparity in the United States: Corrected Cervix Cancer Mortality Rates. Cancer 123 (6):1044–1050. https://doi.org/10.1002/cncr.30507
Blacklock CA, Tollit MA, Pace CC, et al. (2023) The Gender Euphoria Scale (GES): A Protocol for Developing and Validating a Tool to Measure Gender Euphoria in Transgender and Gender Diverse Individuals. Front Psychol 14:1284991. https://doi.org/10.3389/fpsyg.2023.1284991
Bowleg L (2012) The Problem with the Phrase "Women and Minorities": Intersectionality, an Important Theoretical Framework for Public Health. Am J Public Health 102 (7):1267–1273. https://doi.org/10.2105/AJPH.2012.300750
——— (2021) Evolving Intersectionality Within Public Health: From Analysis to Action. Am J Public Health 111 (1):88–90. https://doi.org/10.2105/ajph.2020.306031
Brooks VR (1981) Minority Stress and Lesbian Women. Lexington Books, Lexington
Butler J (1999) Gender Trouble: Tenth Anniversary, Second Edition. Routledge, New York
CDC (2024) Diagnoses, Deaths, and Prevalence of HIV in the United States and 6 Territories and Freely Associated States, 2022. https://stacks.cdc.gov/view/cdc/156509 (accessed Nov. 4, 2024)
Ciccia L (2022) La Invención de Los Sexos: Cómo la Ciencia Puso el Binarismo en Nuestros Cerebros y Cómo Los Feminismos Pueden Ayudarnos a Salir de Ahí. Siglo XXI Editores, Madrid
——— (2023) (Epi)Genealogía del Cuerpo Generizado. Diánoia 68 (91):113–145. https://doi.org/10.22201/iifs.18704913e.2023.91.1993
Clark R, Anderson NB, Clark VR, and Williams DR (1999) Racism as a Stressor for African Americans: A Biopsychosocial Model. Am Psychol 54 (10):805–816. https://doi.org/10.1037//0003-066x.54.10.805
Cohen M, Karrington B, Trachtman H, and Salas-Humara C (2021) Allostatic Stress and Inflammatory Biomarkers in Transgender and Gender Expansive Youth: Protocol for a Pilot Cohort Study. JMIR Res Protoc 10 (5):e24100. https://doi.org/10.2196/24100
Collins PH, and Bilge S (2020) Intersectionality: Key Concepts. 2nd edn. Polity Press, Cambridge
Crenshaw KW (2013) Mapping the Margins: Intersectionality, Identity Politics, and Violence against Women of Color. In: The Public Nature of Private Violence. Routledge, New York, pp 93–118

Dakić T (2020) New Perspectives on Transgender Health in the Forthcoming 11th Revision of the International Statistical Classification of Diseases and Related Health Problems: An Overview of Gender-Depathologization, Considerations and Recommendations for Practitioners. Psychiatr Danub 32 (2):145–150. https://doi.org/10.24869/psyd.2020.145

Danielsen AC, Lee KM, Boulicault M, et al. (2022) Sex Disparities in COVID-19 Outcomes in the United States: Quantifying and Contextualizing Variation. Soc Sci Med 294:114716. https://doi.org/10.1016/j.socscimed.2022.114716

Della Torre S (2021) Beyond the X Factor: Relevance of Sex Hormones in NAFLD Pathophysiology. Cells 10 (9):2502. https://doi.org/10.3390/cells10092502

Delozier AM, Kamody RC, Rodgers S, and Chen D (2020) Health Disparities in Transgender and Gender Expansive Adolescents: A Topical Review from a Minority Stress Framework. J Pediatr Psychol 45 (8):842–847. https://doi.org/10.1093/jpepsy/jsaa040

DiMarco M, Zhao H, Boulicault M, and Richardson SS (2022) Why "Sex as a Biological Variable" Conflicts with Precision Medicine Initiatives. Cell Rep Med 3 (4):100550. https://doi.org/https://doi.org/10.1016/j.xcrm.2022.100550

Dora M, Grabski B, and Dobroczyński B (2021) Gender Dysphoria, Gender Incongruence and Gender Nonconformity in Adolescence–Changes and Challenges in Diagnosis. Psychiatr Pol 55 (1):23–37. https://doi.org/10.12740/pp/onlinefirst/113009

Dubois LZ (2012) Associations between Transition-Specific Stress Experience, Nocturnal Decline in Ambulatory Blood Pressure, and C-Reactive Protein Levels among Transgender Men. Am J Hum Biol 24 (1):52–61. https://doi.org/10.1002/ajhb.22203

DuBois LZ, Gibb JK, Juster RP, and Powers SI (2021) Biocultural Approaches to Transgender and Gender Diverse Experience and Health: Integrating Biomarkers and Advancing Gender/Sex Research. Am J Hum Biol 33 (1):e23555. https://doi.org/10.1002/ajhb.23555

DuBois LZ, and Juster R-P (2022) Lived Experience and Allostatic Load among Transmasculine People Living in the United States. Psychoneuroendocrinology 143:143:105849. https://doi.org/10.1016/j.psyneuen.2022.105849

DuBois LZ, Puckett JA, Jolly D, et al. (2024) Gender Minority Stress and Diurnal Cortisol Profiles among Transgender and Gender Diverse People in the United States. Horm Behav 159:105473. https://doi.org/10.1016/j.yhbeh.2023.105473

DuBois LZ, and Shattuck-Heidorn H (2021) Challenging the Binary: Gender/Sex and the Bio-Logics of Normalcy. Am J Hum Biol 33 (5):e23623. https://doi.org/10.1002/ajhb.23623

Fausto-Sterling A (2019) Gender/Sex, Sexual Orientation, and Identity Are in the Body: How Did They Get There? J Sex Res 56 (4-5):529–555. https://doi.org/10.1080/00224499.2019.1581883

——— (2021) A Dynamic Systems Framework for Gender/Sex Development: From Sensory Input in Infancy to Subjective Certainty in Toddlerhood. Front Hum Neurosci 15:613789. https://doi.org/10.3389/fnhum.2021.613789

Feldman J, Brown GR, Deutsch MB, et al. (2016) Priorities for Transgender Medical and Healthcare Research. Curr Opin Endocrinol Diabetes Obes 23 (2):180–187. https://doi.org/10.1097/med.0000000000000231

Flentje A, Heck NC, Brennan JM, and Meyer IH (2020) The Relationship between Minority Stress and Biological Outcomes: A Systematic Review. J Behav Med 43 (5):673–694. https://doi.org/10.1007/s10865-019-00120-6

Frost DM, and Meyer IH (2023) Minority Stress Theory: Application, Critique, and Continued Relevance. Curr Opin Psychol 51:101579. https://doi.org/10.1016/j.copsyc.2023.101579

Galupo MP, Pulice-Farrow L, and Lindley L (2020) "Every Time I Get Gendered Male, I Feel a Pain in My Chest": Understanding the Social Context for Gender Dysphoria. Stigma Health 5 (2):199. https://awspntest.apa.org/doi/10.1037/sah0000189

Glynn TR, Gamarel KE, Kahler CW, et al. (2016) The Role of Gender Affirmation in Psychological Well-Being among Transgender Women. Psychol Sex Orientat Gend Divers 3 (3):336–344. https://doi.org/10.1037/sgd0000171

Goff DC, Jr., Lloyd-Jones DM, Bennett G, et al. (2014) 2013 ACC/AHA Guideline on the Assessment of Cardiovascular Risk: A Report of the American College of Cardiology/ American Heart Association Task Force on Practice Guidelines. Circulation 129 (25):49–73. https://doi.org/10.1161/01.cir.0000437741.48606.98

Gomez SL, Duffy C, Griggs JJ, and John EM (2019) Surveillance of Cancer among Sexual and Gender Minority Populations: Where Are We and Where Do We Need to Go? Cancer 125 (24):4360–4362. https://doi.org/10.1002/cncr.32384

Haynes JM, and Stumbo RW (2018) The Impact of Using Non-Birth Sex on the Interpretation of Spirometry Data in Subjects with Air-Flow Obstruction. Respir Care 63 (2):215–218. https://doi.org/10.4187/respcare.05586

Holloway BT (2023) Highlighting Trans Joy: A Call to Practitioners, Researchers, and Educators. Health Promot Pract. Health Promot Pract 24 (4):612–614. https://doi.org/10.1177/15248399231152468

Hughto JMW, Gunn HA, Rood BA, and Pantalone DW (2020) Social and Medical Gender Affirmation Experiences Are Inversely Associated with Mental Health Problems in a U.S. Non-Probability Sample of Transgender Adults. Arch Sex Behav 49 (7):2635–2647. https://doi.org/10.1007/s10508-020-01655-5

Joel D (2012) Genetic-Gonadal-Genitals Sex (3G-Sex) and the Misconception of Brain and Gender, or, Why 3G-Males and 3G-Females Have Intersex Brain and Intersex Gender. Biol Sex Differ 3 (27):1–6. https://doi.org/10.1186/2042-6410-3-27

Jordan-Young RM, and Karkazis K (2019) Testosterone: An Unauthorized Biography. Harvard Univ Press, Cambridge, MA

Juster RP, de Torre MB, Kerr P, et al. (2019) Sex Differences and Gender Diversity in Stress Responses and Allostatic Load among Workers and LGBT People. Curr Psychiatry Rep 21 (11):110. https://doi.org/10.1007/s11920-019-1104-2

Kaiser A (2016) Sex/Gender Matters and Sex/Gender Materialities in the Brain. In: Pitts-Taylor V (ed) Mattering: Feminism, Science, and Materialism. Biopolitics. New York Univ Press, New York, pp 122–139

Kaplan-Marans E, Zhang TR, Zhao LC, and Hu JC (2023) Transgender Women with Prostate Cancer Are Under-Represented in National Cancer Registries. Nat Rev Urol 20 (4):195–196. https://doi.org/10.1038/s41585-022-00688-w

Khan SS, Matsushita K, Sang Y, et al. (2024) Development and Validation of the American Heart Association's PREVENT Equations. Circulation 149 (6):430–449. https://doi.org/10.1161/CIRCULATIONAHA.123.067626

Kidd JD, Tettamanti NA, Kaczmarkiewicz R, et al. (2023) Prevalence of Substance Use and Mental Health Problems among Transgender and Cisgender U.S. Adults: Results from a National Probability Sample. Psychiatry Res 326:115339. https://doi.org/10.1016/j.psychres.2023.115339

King WM, and Gamarel KE (2020) A Scoping Review Examining Social and Legal Gender Affirmation and Health among Transgender Populations. Transgend Health 6 (1):5–22. https://doi.org/10.1089/trgh.2020.0025

Kinitz DJ, and Salway T (2022) Cisheteronormativity, Conversion Therapy, and Identity among Sexual and Gender Minority People: A Narrative Inquiry and Creative Non-Fiction. Qual Health Res 32 (13):1965–1978. https://doi.org/10.1177/10497323221126536

Krasowski MD, Hines NG, Imborek KL, and Greene DN (2024) Impact of Sex Used for Assignment of Reference Intervals in a Population of Patients Taking Gender-Affirming Hormones. J Clin Transl Endocrinol 36:100350. https://doi.org/10.1016/j.jcte.2024.100350

Krieger N (2003) Genders, Sexes, and Health: What Are the Connections—and Why Does It Matter? Int J Epidemiol 32 (4):652–657. https://doi.org/10.1093/ije/dyg156

Kumaraguru M, Chellappa LR, I MA, and Jayaraman S (2023) Association between Perceived Stress and Salivary Biomarkers of Allostatic Load among Gender Minorities in Chennai: An Observational Cross-Sectional Study. Cureus 15 (9):e46065. https://doi.org/10.7759/cureus.46065

Lakoff G, and Johnson ML (1999) Philosophy in the Flesh: The Embodied Mind and Its Challenge to Western Thought. Basic Books, New York

LeBlanc M, Radix A, Sava L, et al. (2022) Focus More on What's Right Instead of What's Wrong: Research Priorities. BMC Public Health 22 (1):1741. https://doi.org/10.1186/s12889-022-14139-z

Lett E, Dowshen NL, and Baker KE (2020) Intersectionality and Health Inequities for Gender Minority Blacks in the U.S. Am J Prev Med 59 (5):639–647. https://doi.org/10.1016/j.amepre.2020.04.013

Lindley L, and Galupo MP (2020) Gender Dysphoria and Minority Stress: Support for Inclusion of Gender Dysphoria as a Proximal Stressor. Psychol Sex Orientat Gend Divers 7 (3):265. https://psycnet.apa.org/doi/10.1037/sgd0000439

Logie CH, Kinitz DJ, Gittings L, et al. (2022) Eliciting Critical Hope in Community-Based HIV Research with Transgender Women in Toronto, Canada: Methodological Insights. Health Promot Int 37 (Suppl 2):ii37–ii47. https://doi.org/10.1093/heapro/daac017

Matsuno E, and Israel T (2018) Psychological Interventions Promoting Resilience among Transgender Individuals: Transgender Resilience Intervention Model (TRIM). Counsel Psychol 46 (5):632–655. https://psycnet.apa.org/doi/10.1177/0011000018787261

McEwen BS (1998) Protective and Damaging Effects of Stress Mediators. N Engl J Med 338 (3):171–179. https://doi.org/10.1056/NEJM199801153380307

McEwen BS, and Stellar E (1993) Stress and the Individual: Mechanisms Leading to Disease. Arch Intern Med 153 (18):2093–2101. https://doi.org/10.1001/archinte.1993.00410180039004

Meyer IH (1995) Minority Stress and Mental Health in Gay Men. J Health Soc Behav 36 (1):38–56. https://doi.org/10.2307/2137286

——— (2003) Prejudice, Social Stress, and Mental Health in Lesbian, Gay, and Bisexual Populations: Conceptual Issues and Research Evidence. Psychol Bull 129 (5):674–697. https://doi.org/10.1037/0033-2909.129.5.674

Mohottige D, Farouk S, Poteat T, Radix A, and Witchel SF (2024) Considerations of Sex as a Binary Variable in Clinical Algorithms. Nat Rev Nephrol 20 (6):347–348. https://doi.org/10.1038/s41581-024-00840-2

Morrison T, Dinno A, and Salmon T (2021) The Erasure of Intersex, Transgender, Nonbinary, and Agender Experiences through Misuse of Sex and Gender in Health Research. Am J Epidemiol 190 (12):2712–2717. https://doi.org/10.1093/aje/kwab221

NASEM (2021) Understanding the Well-Being of LGBTQI+ Populations. National Academies Press, Washington, DC, https://doi:10.17226/25877

NASEM, Becker T, Chin M, and Bates N (eds) (2022) Measuring Sex, Gender Identity, and Sexual Orientation. National Academies Press, Washington, DC

NIH (2023) Sexual & Gender Minority Measurement & Data. https://dpcpsi.nih.gov/sgmro/measurement-and-data/surveys-and-measures (accessed Nov. 4, 2024)

NIMHD (2023) Minority Health and Health Disparities Definitions. https://www.nimhd.nih.gov/about/strategic-plan/nih-strategic-plan-definitions-and-parameters.html (accessed Nov. 4, 2024)

NCTE (2023) Understanding Transgender People: The Basics. https://transequality.org/issues/resources/understanding-transgender-people-the-basics (accessed Nov. 4, 2024)

Pape M, Miyagi M, Ritz SA, et al. (2024) Sex Contextualism in Laboratory Research: Enhancing Rigor and Precision in the Study of Sex-Related Variables. Cell 187 (6):1316–1326. https://doi.org/10.1016/j.cell.2024.02.008

Peitzmeier SM, Bernstein IM, McDowell MJ, et al. (2020) Enacting Power and Constructing Gender in Cervical Cancer Screening Encounters between Transmasculine Patients and Health Care Providers. Cult Health Sex 22 (12):1315–1332. https://doi.org/10.1080/13691058.2019.1677942

Perez-Brumer A, Valdez N, and Scheim AI (2024) The Anti-Gender Threat: An Ethical, Democratic, and Scientific Imperative for NIH Research/Ers. Soc Sci Med 351 (Suppl 1):116349. https://doi.org/10.1016/j.socscimed.2023.116349

Pérez-Stable EJ (2016) Sexual and Gender Minorities Formally Designated as a Health Disparity Population for Research Purposes. 2023. https://www.nimhd.nih.gov/about/directors-corner/messages/message_10-06-16.html

Perrin PB, Sutter ME, Trujillo MA, Henry RS, and Pugh M, Jr. (2020) The Minority Strengths Model: Development and Initial Path Analytic Validation in Racially/Ethnically Diverse LGBTQ Individuals. J Clin Psychol 76 (1):118–136. https://doi.org/10.1002/jclp.22850

Pinna F, Paribello P, Somaini G, et al. (2022) Mental Health in Transgender Individuals: A Systematic Review. Int Rev Psychiatry 34 (3–4):292–359. https://doi.org/10.1080/09540261.2022.2093629

Poteat T, Rachlin K, Lare S, Janssen A, and Devor AH (2019) History and Prevalence of Gender Dysphoria. In: Transgender Medicine: A Multidisciplinary Approach. Humana Press, Cham, pp 1–24

Poteat TC, Lett E, Rich AJ, et al. (2023a) Effects of Race and Gender Classifications on Atherosclerotic Cardiovascular Disease Risk Estimates for Clinical Decision-Making in a Cohort of Black Transgender Women. Health Equity 7 (1):803–808. https://doi.org/10.1089/heq.2023.0066

Poteat TC, Rich AJ, Jiang H, et al. (2023b) Cardiovascular Disease Risk Estimation for Transgender and Gender-Diverse Patients: Cross-Sectional Analysis of Baseline Data from the Lite Plus Cohort Study. AJPM Focus 2 (3):100096. https://doi.org/10.1016/j.focus.2023.100096

Poteat TC, van der Merwe LLA, Sevelius J, and Keatley J (2021) Inclusion as Illusion: Erasing Transgender Women in Research with MSM. J Int AIDS Soc 24 (1):e25661. https://doi.org/10.1002/jia2.25661

Puckett JA, Huit TZ, Hope DA, et al. (2024) Transgender and Gender-Diverse People's Experiences of Minority Stress, Mental Health, and Resilience in Relation to Perceptions of Sociopolitical Contexts. Transgend Health 9 (1):14–23. https://doi.org/10.1089/trgh.2022.0047

Reed GM, Drescher J, Krueger RB, et al. (2016) Disorders Related to Sexuality and Gender Identity in the ICD-11: Revising the ICD-10 Classification Based on Current Scientific Evidence, Best Clinical Practices, and Human Rights Considerations. World Psychiatry 15 (3):205–221. https://doi.org/10.1002/wps.20354

Reisner SL, Pletta DR, Harris A, et al. (2023) Exploring Gender Euphoria in a Sample of Transgender and Gender Diverse Patients at Two U.S. Urban Community Health Centers. Psychiatry Res 329:115541. https://doi.org/10.1016/j.psychres.2023.115541

Reisner SL, Poteat T, Keatley J, et al. (2016a) Global Health Burden and Needs of Transgender Populations: A Review. Lancet 388 (10042):412–436. https://doi.org/10.1016/s0140-6736(16)00684-x

Reisner SL, White Hughto JM, Pardee D, and Sevelius J (2016b) Syndemics and Gender Affirmation: HIV Sexual Risk in Female-to-Male Trans Masculine Adults Reporting Sexual Contact with Cisgender Males. Int J STD AIDS 27 (11):955–966. https://doi.org/10.1177/0956462415602418

Restar AJ, Sherwood J, Edeza A, Collins C, and Operario D (2021) Expanding Gender-Based Health Equity Framework for Transgender Populations. Transgend Health 6 (1):1–4. https://doi.org/10.1089/trgh.2020.0026

Rich AJ, Williams J, Malik M, et al. (2020) Biopsychosocial Mechanisms Linking Gender Minority Stress to HIV Comorbidities among Black and Latina Transgender Women (LITE Plus): Protocol for a Mixed Methods Longitudinal Study. JMIR Res Protoc 9 (4):e17076. https://doi.org/10.2196/17076

Richardson SS (2022) Sex Contextualism. PTPBio 14:2. https://doi.org/10.3998/ptpbio.2096

Riggs D, Ansara G, and Treharne GJ (2015) An Evidence-Based Model for Understanding the Mental Health Experiences of Transgender Australians. Aust Psychol 50 (1):32–39. https://doi.org/10.1111/ap.12088

Rippon G (2019) The Gendered Brain: The New Neuroscience That Shatters the Myth of the Female Brain. Random, New York

Risman BJ, Travers, and Fleming C (2022) Category X: What Does the Visibility of People Who Reject the Gender Binary Mean for the Gender Structure? Int J Gend Stud 11 (21):1–34. https://doi.org/10.15167/2279-5057/AG2022.11.21.2005

Rosen JG, Malik M, Cooney EE, et al. (2019) Antiretroviral Treatment Interruptions among Black and Latina Transgender Women Living with HIV: Characterizing Co-Occurring, Multilevel Factors Using the Gender Affirmation Framework. AIDS Behav 23 (9):2588–2599. https://doi.org/10.1007/s10461-019-02581-x

Rosenberg S, and Tilley PJM (2020) A Point of Reference: The Insider/Outsider Research Staircase and Transgender People's Experiences of Participating in Trans-Led Research. Qual Res 21 (6):923–938. https://doi.org/10.1177/1468794120965371

Rudd É, Fortin J, and Brunet A (2023) Men's Stories of Living with Breast Cancer: A Systematic Review, Metasynthesis, and Proposed Framework. Psychol Men Masc 24 (4):291–310. https://psycnet.apa.org/doi/10.1037/men0000438

Santora T (2021) How Four Transgender Researchers Are Improving the Health of Their Communities. Nat Med 27 (12):2074–2077. https://doi.org/10.1038/s41591-021-01597-y

Scheim AI, Baker KE, Restar AJ, and Sell RL (2022) Health and Health Care among Transgender Adults in the United States. Annu Rev Public Health 43:503–523. https://doi.org/10.1146/annurev-publhealth-052620-100313

Scheim AI, Rich AJ, Zubizarreta D, et al. (2024) Health Status of Transgender People Globally: A Systematic Review of Research on Disease Burden and Correlates. PLoS One 19 (3):e0299373. https://doi.org/10.1371/journal.pone.0299373

Schneiderman N, Ironson G, and Siegel SD (2005) Stress and Health: Psychological, Behavioral, and Biological Determinants. Annu Rev Clin Psychol 1:607–628. https://doi.org/10.1146/annurev.clinpsy.1.102803.144141

Sevelius JM (2013) Gender Affirmation: A Framework for Conceptualizing Risk Behavior among Transgender Women of Color. Sex Roles 68 (11–12):675–689. https://doi.org/10.1007/s11199-012-0216-5

Sevelius JM, Chakravarty D, Dilworth SE, Rebchook G, and Neilands TB (2020) Gender Affirmation through Correct Pronoun Usage: Development and Validation of the Transgender Women's Importance of Pronouns (TW-IP) Scale. Int J Environ Res Public Health 17 (24):9525. https://doi.org/10.3390/ijerph17249525

——— (2021a) Measuring Satisfaction and Comfort with Gender Identity and Gender Expression among Transgender Women: Development and Validation of the Psychological Gender Affirmation Scale. Int J Environ Res Public Health 18 (6):3298. https://doi.org/10.3390/ijerph18063298

Sevelius JM, Chakravarty D, Neilands TB, et al. (2021b) Evidence for the Model of Gender Affirmation: The Role of Gender Affirmation and Healthcare Empowerment in Viral Suppression among Transgender Women of Color Living with HIV. AIDS Behav 25 (Suppl 1):64–71. https://doi.org/10.1007/s10461-019-02544-2

Shattuck-Heidorn H, Danielsen AC, Gompers A, et al. (2021) A Finding of Sex Similarities Rather Than Differences in COVID-19 Outcomes. Nature 597 (7877):E7–E9. https://doi.org/10.1038/s41586-021-03644-7

Shuster SM, and Westbrook L (2024) Reducing the Joy Deficit in Sociology: A Study of Transgender Joy. Soc Probl 71 (3):791–809. https://doi.org/10.1093/socpro/spac034

Simon CB, McCabe CJ, Matson TE, et al. (2024) High Test-Retest Reliability of the Alcohol Use Disorders Identification Test-Consumption (AUDIT-C) Questionnaire Completed by Primary Care Patients in Routine Care. Alcohol Clin Exp Res 48 (2):302–308. https://doi.org/10.1111/acer.15245

Slopnick EA, Bagby C, Mahran A, et al. (2022) Skene's Gland Malignancy: A Case Report and Systematic Review. Urology 165:36–43. https://doi.org/10.1016/j.urology.2022.02.004

South Carolina (2023) Biological Sex Constitutional Amendment, January 10, 2023. S.276. https://www.scstatehouse.gov/sess125_2023-2024/bills/276.htm

Streed CG, Jr., Perlson JE, Abrams MP, and Lett E (2023) On, with, by-Advancing Transgender Health Research and Clinical Practice. Health Equity 7 (1):161–165. https://doi.org/10.1089/heq.2022.0146

Stutterheim SE, van Dijk M, Wang H, and Jonas KJ (2021) The Worldwide Burden of HIV in Transgender Individuals: An Updated Systematic Review and Meta-Analysis. PLoS One 16 (12):e0260063. https://doi.org/10.1371/journal.pone.0260063

Tan KKH, Treharne GJ, Ellis SJ, Schmidt JM, and Veale JF (2020) Gender Minority Stress: A Critical Review. J Homosex 67 (10):1471–1489. https://doi.org/10.1080/00918369.2019.1591789

Testa RJ, Habarth J, Peta J, Balsam K, and Bockting W (2015) Development of the Gender Minority Stress and Resilience Measure. Psychol Sex Orientat Gend Divers 2 (1):65–77. https://doi.org/10.1037/sgd0000081

Tordoff DM, Minalga B, Gross BB, et al. (2022) Erasure and Health Equity Implications of Using Binary Male/Female Categories in Sexual Health Research and Human Immunodeficiency Virus/Sexually Transmitted Infection Surveillance: Recommendations for Transgender-Inclusive Data Collection and Reporting. Sex Transm Dis 49 (2):e45–e49. https://doi.org/10.1097/olq.0000000000001533

Trans Legislation Tracker (2024) Trans Legislation Tracker. https://translegislation.com/bills/2024/SC/S0276 (accessed Nov. 4, 2024)

van Anders SM (2024) Gender/sex/ual Diversity and Biobehavioral Research. Psychol Sex Orientat Gend Divers 11 (3):471–487. https://doi.org/10.1037/sgd0000609

Van der Kolk BA (1994) The Body Keeps the Score: Memory and the Evolving Psychobiology of Posttraumatic Stress. Harvard Rev Psychiatry 1 (5):253–265. https://doi.org/10.3109/10673229409017088

Veale JF, Deutsch MB, Devor AH, et al. (2022) Setting a Research Agenda in Trans Health: An Expert Assessment of Priorities and Issues by Trans and Nonbinary Researchers. Int J Transgend Health 23 (4):392–408. https://doi.org/10.1080/26895269.2022.2044425

Viergever RF, and Hendriks TC (2016) The 10 Largest Public and Philanthropic Funders of Health Research in the World: What They Fund and How They Distribute Their Funds. Health Res Policy Syst 14:12. https://doi.org/10.1186/s12961-015-0074-z

Wesp LM, Malcoe LH, Elliott A, and Poteat T (2019) Intersectionality Research for Transgender Health Justice: A Theory-Driven Conceptual Framework for Structural Analysis of Transgender Health Inequities. Transgend Health 4 (1):287–296. https://doi.org/10.1089/trgh.2019.0039

WHO (2024) Health Equity. https://www.who.int/health-topics/health-equity (accessed Nov. 4, 2024)

Yaribeygi H, Panahi Y, Sahraei H, Johnston TP, and Sahebkar A (2017) The Impact of Stress on Body Function: A Review. EXCLI J 16:1057–1072. https://doi.org/10.17179/excli2017-480

12

Gender, Sex, and Their Entanglement

From Scientific Research to Policy and Practice

Alexandra Brewis, Paisley Currah, L. Zachary DuBois, Lorraine Greaves, Katharina Hoppe, Katrina Karkazis, Madeleine Pape, Paula-Irene Villa, and Amber Wutich

Abstract   Scientific findings on gender, sex, and their entanglement should and do extend to policy and practice. Through both theorizing and exemplifying, this chapter identifies how the application of sex and gender as discrete, disentangled constructs harms individuals and groups in policy application (such as via discrimination). It also discusses how the application of entangled gender/sex can support more robust and just policy as well as sciences

Keywords   gender, sex, entanglement, binary notions of sex, knowledge translation, transgender health, policy, public health

1 Introduction

Science impacts policy that addresses issues involving gender, sex and their entanglement. In addition to advancing scientific knowledge into the complexities of sex and gender co-constitution, there is an ongoing need to improve the translation of scientific knowledge to policy. Historically, legislators and other policy makers have relied on conventional ideas about sex and gender as reflecting entirely separate concepts. Currently, policy makers invoke the "authority" of science to underwrite simplistic and rigid definitions of sex, even as scientific research on sex and gender reveals their complexity. For example, a leaked memo from the US Department of Health and Human Services (DHHS) proposed establishing a federal definition of

Alexandra Brewis (✉)
School of Human Evolution and Social Change, Arizona State University,
Tempe, AZ 85287-2402, USA
Email: Alex.Brewis@asu.edu

Affiliations for the coauthors are available in the List of Contributors

© Frankfurt Institute for Advanced Studies (FIAS) 2025   241
L. Z. DuBois et al. (eds.), *Sex and Gender*, Strüngmann Forum Reports,
https://doi.org/10.1007/978-3-031-91371-6_12

Group photos (top left to bottom right)   Alexandra Brewis, Zachary DuBois, Paisley Currah, Paula-Irene Villa, Amber Wutich, Lorraine Greaves, Katrina Karkazis, Katharina Hoppe, Madeleine Pape, Zachary DuBois, Alexandra Brewis, Paisley Currah, Paula-Irene Villa, Katrina Karkazis, Amber Wutich, Lorraine Greaves, Madeleine Pape, Katharina Hoppe. Photos by Norbert Miguletz.

sex as immutable and rigidly binary (i.e., as "male or female, unchangeable, and determined by the genitals that a person is born with") and claimed its definition was rooted in "a biological basis that is clear, grounded in science, objective, and administrable" (Green et al. 2018). Despite the importance of having precise definitions and operationalizing them in both science and policy, the terms *gender* and *sex* are often undefined, used interchangeably, or conflated in problematic ways (Tadiri et al. 2021) and can profoundly impact the lives of individuals and their legal rights (Sudai et al. 2022).

Current science is clear: biological factors often associated with sex and gender do not map perfectly onto static, binary categories of female and male (DuBois and Shattuck-Heidorn 2021; Karkazis 2019; Ritz and Greaves 2022). Moreover, when gender- and sex-related factors or sex-linked attributes are left undefined or are not clearly operationalized, it becomes difficult to identify which of the many factors associated with gender and/or sex are pertinent to any particular study or policy. It is thus crucial that we engage with their *entanglements* because in reality, gender interacts with sex, and sex with gender, to shape each individual and their social experience (Greaves and Ritz 2022; Ritz and Greaves 2022).

Combining the terms sex and gender as *sex/gender* (Fausto-Sterling 2000, 2012; Kaiser 2012; Krieger 2003; Springer et al. 2012) or *gender/sex* (van Anders 2015) offers ways to engage with the dynamics of their entanglement. Here in this chapter, we adopt the use of gender/sex; for an elaboration of these terms and their history, see Chapters 1 and 5. These neologisms, however, are currently being deployed in only a very small sector of scientific research and have yet to appear in policy making. Thus, the work of advancing rigorously granular precision in the study of factors associated with sex and gender is still at an early stage.

Distinguishing between sex and gender has often been proposed as an initial step in advancing precision (Clayton 2016) and has been characteristic for much research over the last decades. Now, however, what had been a useful but inaccurate heuristic—namely, that sex is a biological binary that exists separately from gender as a less binary sociocultural formation—is becoming solidified in potentially harmful laws and policies. For example, the state of Montana recently enacted legislation (Montana State 2023, Section 1-1-201) that states:

> In human beings, there are exactly two sexes, male and female, with two corresponding types of gametes. The sexes are determined by the biological and genetic indication of male or female, including sex chromosomes, naturally occurring sex chromosomes, gonads, and unambiguous internal and external genitalia present at birth, without regard to an individual's psychological, behavioral, social, chosen, or subjective experience of gender.

Given the potential for harm, it is crucial that researchers and policy makers be informed about the entanglement of sex/gender; that is, about the irreducible co-constitution and interwovenness of sex and gender as a biosocial fact. As will be elaborated below, we refer to entanglement as the complex and dynamic interplay of gender and sex; it is effectively impossible to *dis*entangle these concepts meaningfully or to isolate pristine effects of one or the other on a given outcome.

This chapter summarizes emergent understandings from our discussions at the Ernst Strüngmann Forum in Frankfurt, Germany. Members of our working group represented a range of expertise across the social and biological sciences:

biocultural and cultural anthropologists who study stigma, health disparities, trans-gender health, feminist science, and science and technology studies; a global health policy expert; a political scientist who explicitly considers implications of gender/sex within policy; a medical sociologist who focuses on women's health and sex and gender; and a sociologist engaged in gender and biopolitics research. Although we did not agree on every point, our interactions were productive, enabling us to reach consensus in many areas and to further dialogue in others. Our discussion was guided by the following key questions:

- What are the consequences of disentangled versus entangled definitions of gender and sex when used in policy and practice?
- How can entanglements of gender and sex be leveraged to foster science that can be translated into more effective and equitable policies and better practice?
- How can policy support the pursuit of scientific knowledge about gender/sex entanglement?

Here, we provide a general map of cases, theories, and implications that researchers, policy makers, and practitioners can draw on as they consider the entanglement of sex and gender in their own areas of interest. The selected cases reflect our topical expertise (i.e., policies and practices regarding sports, health and health care as well as legal discrimination) and include examples of policies for inclusion and exclusion, rights and access to resources, and scientific and biomedical practice. These examples are embedded within the scope of the Global North or Anglophone international organizations, since this is where our expertise lies. This brings to the fore a major gap in the current literature: the need for research in more diverse geo-political contexts where gender and sex terminologies and schema are defined and practiced in highly varied ways. Such research is currently impeded by publication and subscription costs, language hegemony, and the lengthy review processes that generate structural disadvantages for researchers located beyond the Global North (e.g., Naidu et al. 2024).

We discuss challenges and opportunities that the notion of entanglement offers and draw attention to existing challenges faced in potential implementation of these ideas in practice. One key point that guided our discussions concerns how policy and practice can be viewed at many analytic levels. A concrete example of how gender/sex entanglement impacts both sports policy and practice at many different levels is seen in the case of anterior cruciate ligament (ACL) rupture.

The ACL—a tiny band of connective tissue in the knee—can rupture as a result of high-speed landings and/or pivoting. It is a common, painful, and debilitating injury in many sports and can be devasting to an athlete's career, causing extended rehabilitation or even retirement. The risk of rupture in women athletes is at least double (and perhaps as much as eight times) compared to men (Devana et al. 2022), a statistic of concern to major sporting bodies and women's sporting leagues. The proximate scientific explanations invoked in policy discussions and practice deci-sions related to ACL focus on "sex differences" in women's versus men's anatomy (e.g., decreased intercondylar notch width), physiology (e.g., being preovulatory), and biomechanics (e.g., knee abduction on landing). Injury prevention science thus advocates, for example, better training of women athletes on how to land (Renstrom et al. 2008). Issues related to access to care and bias among health providers have also been highlighted. Combined, these differences are a downstream gendered effect of the ongoing relative lack of institutional investment in trainers, training, or training surfaces for women athletes. Solutions to gender disparities in ACL injury, rooted in an entanglement approach, are best framed in gender transformative sports policy (see Chapter 14) and would create equitable frameworks of specific sports' training and financing.

This example shows how the absence of an entanglement approach can lead to an understanding of injury as seemingly *only* biologically based and thus *only* sex-related, without considering gender and gendered effects. Through this and other cases, we aim to make clear how policies and practices affect and can harm differently gendered bodies, and to show how policies and practices *produce* individual bodies as gendered and at risk at multiple analytic levels.

2 A Vocabulary for Entanglement

The concepts of *sex* and *gender* are dynamic and increasingly recognized as complex and varied in their use and definition (see Chapters 1 and 5). There is a lack of consensus in their definitions, mirroring wider debates and ongoing dialogues within the social sciences and allied fields; put simply, it is challenging to separate analytically that which in practice are deeply entangled concepts (Pape et al. 2024; Richardson 2022). Given that gender is a system of social hierarchies, societal debates around sex and gender definitions easily emerge as sites of political contestation (e.g., Fausto-Sterling 2000; Martin 1991). Political and policy-based criteria for determining who is male or a man and who is female or a woman vary greatly, and often those differences invariably reflect the purpose for which they are being defined (Currah 2022). Yet categorization reflects a choice, and thus the choice of definitions has important implications for policy making. Below, we elaborate on our discussion of these definitions and concepts.

2.1 Gender

As a term, gender is used in different ways in different disciplinary contexts. One useful way to conceptualize gender is as a system of hierarchical differences reproduced at the micro, meso, and macro levels of social life (Ridgeway and Correll 2004; Risman 2004). The micro level encompasses individual gender expression and identities as well as the reproduction of gender through socialization and interactions with others. The meso level is often taken to refer to gendered organizational practices, such as couple and family relations, workplace hierarchies, and gender discrepancies in hiring (Acker 1990). The macro level refers to the reproduction of gender through major social institutions, such as prevailing expectations around female leadership in family caregiving or male dominance of the military (Connell 1987). In practice, these three levels operate concurrently and can be mutually reinforcing.

It is important to note that gender is not reducible to individual expression or choice: it is an ideology of hierarchical difference that is deeply entangled with relations of inequality and power (Connell 1987). This understanding of gender as an overarching structure is perhaps the most challenging to translate to biomedical research, where gender is often conceptualized primarily in terms of individual identity, gender expression, and sometimes gendered practices and behaviors (Nielsen et al. 2021). Concepts of gender also integrate recognition of socially prescribed roles and opportunities, differentiated social institutions and structures, and context-based definitions of masculinity and femininity (e.g., Johnson et al. 2007, 2009; Wade and Marx-Ferre 2015). Importantly, in many social contexts, gender and sex are also racialized in ways that amplify morbidity, mortality, and other suffering, discussed further below (Crenshaw 1989; Frankenberg 1993; Hernandez 2000).

Whether it is regarded as a social structure or as prescribed roles, health researchers have demonstrated how gender can have negative consequences for health outcomes (Krieger 2003). Homan (2019), for example, has shown that gender inequality at the meso and macro levels (e.g., wage inequality within households or the percentage of men in elected office) correlate with poorer health outcomes in women. Similarly, in their study of transgender health in the United States, DuBois and Juster (2022) found gender minority stressors and stigma at the micro, meso, and macro levels to be associated with negative mental health effects and increased physiological stress as measured through allostatic load.

2.2 Sex

Sex is a similarly complex concept. It is a system of classification, often conceptualized as categories in a female/male binary (as exemplified in the DHHS memo and Montana legislation discussed above), that is taken to refer mainly to traits associated with reproductive anatomy and physiology. Recent research, however, has greatly expanded the range of variables associated with sex, and several scholars now argue that the concept of sex should instead be recognized as a categorical proxy reflecting what is actually a range of complex and covarying biocultural factors (Ainsworth 2015; DuBois and Shattuck-Heidorn 2021; Maney 2016; Springer et al. 2012). In most contexts, individual sex categories are assigned at birth based on assessment of external genitalia. However, assigned sex categories fail to provide researchers with a precise account of which sex-related variables and covariates are pertinent within a given research context (e.g., Richardson 2022). In addition, binary assigned categories do not sufficiently capture how factors vary across individuals (Karkazis 2019; Pape et al. 2024). Fausto-Sterling (2018) has urged moving beyond static, simplistic definitions of sex: "those looking to biology for an easy-to-administer definition of sex and gender can derive little comfort from the most important of these [research] findings."

2.3 What Do We Mean by "Sex Difference?"

An emerging body of work calls into question conventional approaches to the analysis of sex-related variation within biomedical research. Very often, such approaches seek to establish the existence of a difference between two groups, typically classified as female and male, relying on a comparison of means using classic parametric tests (e.g., ANOVA) to do so (Sanchis-Segura and Wilcox 2024). In the context of mandates requiring researchers to consider sex, there is a tendency to conclude that a sex difference exists without having rigorously tested for one (Maney 2016). A larger question is whether even rigorous comparisons of group means are sufficient to capture the complexity of variation within and across sex-classified groups. These are rarely dimorphic, including in animal models. Sanchis-Segura and Wilcox (2024) argue that biomedical researchers should develop both a new vocabulary and a statistical toolbox that emphasizes a description of complex distributions rather than a reliance on single data points to draw conclusions about sex-related variation. This would provide an opportunity to go beyond assigned categories to a more complex, context-specific focus on variables and covariates, and would reconceptualize how "difference" is understood in the study of sex.

2.4 Gender/Sex Entanglement

Building on all of these conceptual considerations and informed by scholarly work that has historically focused on North America and Europe, we use the concept of entanglement to refer to the complex and dynamic interplay of gender and sex: it is effectively impossible to disentangle one from the other in a meaningful way or to isolate pristine effects on a given outcome. The notion of entanglement, hence, points to the co-constitutive, co-evolving, and co-structuring character of what we will from here on refer to as gender/sex (see Barad 2007; Haraway 1991; Villa 2019).

Using the ACL injury discussed above as an example of gender/sex entanglement in sports policy and practice, we can observe that little attention has been given to the gendered environmental factors that could be primary contributors to injury disparities between women and men in sport settings (Fox et al. 2020; Parsons et al. 2021). It is well documented that the practice of sport is riddled with gender differences and disparities, which often amount to different developmental opportunities and pathways for men versus women athletes. A gender/sex entanglement approach offers researchers a framework for conceptualizing how these gendered practices and disparities are embodied as physical harm (Krieger 2005), encouraging investigation of potential gendered factors and their constitutive interactions with the anatomy, physiology, and biomechanics of bodies (Fausto-Sterling 2000). Resulting recommendations for injury prevention and treatment could also reflect a broader—and potentially more pertinent—range of underlying causes.

Gender/sex entanglement is an unavoidable condition of the production of scientific knowledge about "biological sex." Feminist scholars of science have interrogated how gender ideologies shape the very production of scientific knowledge about sex itself and the myriad practices that reinforce the binary in the face of contrary evidence (Bluhm 2012; Fausto-Sterling 1985, 2000; Fine 2013; Karkazis 2008; Martin 1991; Richardson 2013).

A crucial issue for both research and policy is the need for transparency about how gender and/or sex are being measured (e.g., via self-reported identity or gametes). Operationalizing sex or gender as unitary concepts is often unhelpful and lacking in specificity for science. Instead, defining *sex-related factors* (e.g., gonads and other aspects of anatomy or genetics) and *gender-related factors* (e.g., gender expression or gender roles and norms or opportunities) in ways specific to the research context could enable a more comprehensive and accurate understanding of gender/sex entanglements (Richardson 2022). Using these terms addresses the components of sex and/or gender of interest and importance; it also diffuses tendencies to adopt binary definitions and categories, as these factors can be overlapping in bodies, regardless of whether a person is assigned male or female at birth (Ritz and Greaves 2022).

2.5 Intersectionality

The emphasis on the co-constitution of gender/sex invites scholars to consider other forms of structures of inequality that are themselves entangled with gender and sex, such as class relations, sexuality, and race (see Chapters 7 and 9). As early as 1977, activists (Combahee River Collective 1977/1997) and then scholars (Crenshaw 1989; West and Fenstermaker 1995) pointed out that neither sex nor gender (and certainly not their entanglement) can be understood as isolated parameters but are constitutively interwoven with other sociohistorically specific differences, such as sexual orientation, race, age, or class relations. Not only do other axes of differences contribute to inequality, discrimination, and domination, what is constituted as a sexed body itself is a racialized and class-related process. The complex intersections of differences, which generate constitutive power, is what interests scholars deploying an intersectional framework. Many health-related examples show that categorization of bodies as female/male may not make the decisive difference, for example, when testing for drug effects or for specific health-related risks. It may make more sense to attend to the intersectional effects of gender, class, race, age, sexuality, (dis)ability, and further biosocial differences and experiences to assess precise risks, outcomes, and effects (Hankivsky 2012; Mena et al. 2019).

Frameworks that focus on different axes of oppression are particularly relevant for recognizing differential access to power and the impacts of structural violence and identity. Identity characteristics reside in the context of social practices (e.g., discriminatory attitudinal stances), nestled in larger ideologically based systemic forces (Figure 12.1). These theories recognize, albeit with differential emphases, that the biological and social co-constitute each other. *Entanglement is always present.*

Notions of both sex and gender continue to be embedded in historical notions of racial inferiority, and scholars explain that the binary itself was constructed as one dimension of upholding whiteness (e.g., Snorton 2017). Prior to the nineteenth century, bodies were often theorized in European thought as more alike than different (Laqueur 1990; Russett 1989). By the mid-nineteenth century, a new logic

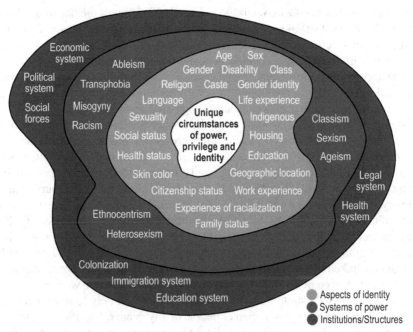

Figure 12.1 Graphical representation of intersectionality by CRIAW-ICREF, redrawn and used with permission.

of sex differentiation emerged that deeply imbricated race and gender within hierarchies that suited the political project of empire; some argue the very idea of sex differences emerged from racializing bodies (Markowitz 2001; Schiebinger 2004; Schuller 2018). Markowitz, for example, showed that categories of and arguments about gender/sex difference were "saturated with racial meaning for centuries" (Markowitz 2001:389). Looking specifically at how ideas of female pelvis sizes were used in conceptualizing racist frameworks of sex difference, she finds that scientists argued that the "more advanced" the race, the greater the gender/sex difference. The category of sex, she notes, rests "not on a simple binary opposition between male and female but rather on a scale of racially coded degrees of sex/gender difference" (Markowitz 2001:391). While such overtly racialized understandings of this dimorphism may have faded, they remain persistent, salient, and readily identifiable (Andersen and Collins 2013; Davis 1983; Lugones 2007, 2010; Mendoza 2015; Oyěwùmí 1997).

Accordingly, the concepts of gender and sex cannot be addressed adequately without considering their intersections with other forms of difference-making and inequality. We illustrate this point through two examples.

The first, a case from Australia, highlights entanglement of gender, sex, and indigeneity in men's health disparities: Aboriginal and Torres Strait Islander men face the worst health outcomes of any social group, including higher rates of suicide, self-harm, and other challenges related to mental health (Adams and Danks 2007; Brown et al. 2013). These men are also more likely than other Australians to develop cardiovascular disease and type 2 diabetes, as well as to experience drug- and alcohol-related illnesses (Brown 2012). Neither indigeneity (i.e., discrimination against Indigenous people in Australia) nor gender (i.e., men's relationship to health and the health system) alone are sufficient to explain observed health disparities. Rather, an intersectional and situated understanding of disease and illness pathways is needed to understand underlying causes and potential solutions.

The second illustrates intersectional entanglement through the case of the Black maternal health crisis in the United States: In the United States, as many as eight times more women die during pregnancy compared to similarly wealthy countries (Lister et al. 2019). Black women carry the greater burden of this disparity, being two to three times more likely to die from pregnancy-related complications than White women (Hoyert 2022). Black women are also more likely than White women to experience life-threatening conditions and other pregnancy-related complications (Winny and Bervell 2023). A United Nations report (UNFPA 2023) found that the primary causes lie at the intersection of systemic racism and sexism, which not only influence the general health of Black women but have consequences for how they are treated within the US health system while pregnant. Thus, in researching maternal health, any approach that seeks to separate sex-linked factors from gender, and gender-linked factors from race, cannot account for how these processes come together in the lived maternity experiences of differently situated women.

Both examples illustrate how intersections matter, not only to women's health but to the health experiences of men or people of any gender. A challenge is to resist the simple addition of gender, race, and indigeneity to research that begins with the study of sex-linked factors, but instead to create new methodologies that center entanglements from the outset. If research is mobilized for the public good, utilizing a gender transformative framework (Greaves et al. 2014; Pederson et al. 2015)—one that attends to improving gender equity at the same time as resolving a policy or practice issue—will enable improvements to be achieved in both domains.

2.6 Knowledge Mobilization

In the process of connecting science to policy through the lens of gender/sex entanglement, *knowledge mobilization* is an especially useful term. It refers to the conveying of research results, evidence, and/or information and data to a range of audiences who may make use of them. Knowledge mobilization is an umbrella term that encompasses knowledge translation, transfer, exchange, and brokerage along with translational and implementation science (e.g., Bennet et al. 2007). Knowledge mobilization can invoke a range of methods (e.g., communities of inquiry and practice, collaborative engagement models) and result in a range of products (e.g., policy briefs, evidence-based or informed backgrounders, clinical guidelines, infographics, opinion papers, commentaries, health advice, videos, science writing, and journalism). Content decisions as well as the processes of mobilization can invoke considerations of numerous sex/gender entanglements, dependent on a range of social and political factors. Indeed, knowledge mobilization is itself a messy process, dependent on politics, social attitudes, cultural context, ideological influences, opportunity, crises, readiness, public opinion, and polling, among other considerations.

3 Putting It into Practice: Defining and Devising Effective Policy

An assumption that underpinned many discussions at this Forum was that science could be improved if gender and sex were recognized as being fundamentally entangled, and that addressing these concepts in this way could lead to "better science" and hence "better policy." Science is never detached from society, with its values, norms, and (power) structures (e.g., Latour 1993). This means that science is also not free from imposing presuppositions about gender and sex (Alcoff and Potter 1993; Haraway 1989). Addressing this requires thoughtful and refined methodology and reflexive theorizing. For nonsocial sciences that rely on experimental, quantifiable methods, this might seem easy to achieve. Social sciences, by definition, address the very social processes inherent to human existence, including those that affect the development and uptake of knowledge, epistemology, and which use a wide range of methods including qualitative and quantitative, mixed methods and surveys. The humanities also utilize a wide range of methods (e.g., document, text, object, artifact analyses) and approaches (e.g., historical, semiotics, phenomenological). None of them, however, are "outside" this world, including its structuring by power relations and meaningful differences. "Better" science recognizes the involvement and contributions from all disciplines. "Better" policy is also an elusive concept, dependent on audience, culture, issue, context, and ideology. From the perspectives of social science, effective policy is based on replicable evidence but is also ethical and just; it interrogates how science is being constructed, deployed, and recognizes that all scientists necessarily view the world through their own positional lenses.

Sciences are sometimes referred to in ways (e.g., "hard" and "soft") that reflect hierarchies of thought, methods, and knowledge, thus mirroring broader contexts of power and domination rooted in the patriarchal dualism still persistent in wealthy industrialized societies or those that model themselves on them. These paradigms are relevant to the links between science and policy making: hierarchies of evidence are cemented in methods for assessing evidence (e.g., systematic reviews) and underpin practice approaches, such as evidence-based medicine (Al-Almaie and Al-Baghli 2003), where randomized control trials are seen as "gold standard" evidence on which to base clinical decisions (Rosenberg and Donald 1995). This dualism of "hard" and "soft" science corresponds to the distinction of natural and social sciences and reifies a mindset that systematically devalues nature, the body, and so on (Plumwood 1993:41–44). Along with these devaluations come those of qualitative methods and community-based research, storytelling, and narrative explorations, which are often associated with the feminine, women's ways of knowing and Indigenous approaches. In short: "soft" science or methods.

To counteract the danger of false binaries and oversimplification, gender/sex entanglement approaches try to make the complexity and co-constitution of sex and gender more visible. This invites approaches that reflect such complexity, with the hope of building ethical and useful science and its application. Ideally, science aims to inform practice in some way (e.g., by providing ideas or parameters for policy) but does not necessarily provide clear, singular, or immediate solutions. It refrains from the technocratic idea that it can supply ready-made solutions and instead sees its role in critiquing and raising new questions. While this does not diminish the importance of science as a source of advice for policy makers, it means that even the best science rarely translates neatly into policy or practice solutions, given the dynamics involved and the fluid nature of political agendas.

Policy making is the process by which government bodies, professional societies, nongovernmental organizations, and other institutions seek to address social, environmental, or economic issues that affect society. Given the numerous types of stakeholders (e.g., politicians, interest groups, researchers, interested citizens), the policy-making process is complex. It can also be controversial, as when stakeholders differ on how to frame or prioritize concerns or problems, how best to address them, what evidence counts, and to what extent values about the common good, competing interests, gender equality, universalism, freedom, or justice are to be shared. Scientific research is generally assumed to be an important component of policy making, with many researchers and policy makers calling for evidence-informed policy making (EIPM), which is distinct from evidence-based policy (Oxman et al. 2009). EIPM approaches have been adopted and formalized by bodies such as the World Health Organization and include both science-created evidence as well as tacit or colloquial evidence in the calculus for making more effective policy (WHO 2021). The OECD (2020) has also adopted EIPM but recognizes that science is but one (and not usually the predominant) type of evidence relevant to policy making, and that both intermediaries and cognitive biases affect the policy-making process.

Thus, the relationship between policy making and science is multifaceted, indirect, and complex, reflecting both science and policy as social practices that are marked by inherent biases and vested interests. Science and its evidence may form part of the platform for policy making, among many other factors, as reflected in the term "evidence-informed." Integrating science into practice emerges in many forms, including guidelines, codes, standards, competencies, regulatory processes, and laws. These mechanisms cross all fields of human society, such as medicine, health, clinical guidance, labor codes, building codes, electoral processes, industrial practices, consumer protection, and educational standards, among many others. This process of integration relies on science and synthesized reviews of evidence to guide practical decisions about standards and practices.

Consider, for example, the regulation of prescription drugs. Federal agencies make independent decisions on drug authorization and set guidance for marketing, clinical and consumer information, and post-market vigilance. This regulatory process implicates gender/sex entanglement in that both sex- and gender-related factors affect the licensing, processing, prescribing, and ingesting of drugs. However, these processes only inconsistently include gender- and sex-related information and evidence, or monitor and communicate about the drugs accordingly. This results in unknown risks being taken by consumers who are disproportionately female, pediatric, pregnant, and from minority and marginalized populations, reflecting uneven testing practices and gender/sex "blind" regulations (Greaves et al. 2023).

In theory, basing policy on scientific evidence is thought to be straightforward: scientists produce evidence that policy makers then use to make their decisions. In practice, however, multiple factors affect policy making as the goals of science and policy are quite different. Moreover, the complexities of policy making include varied vested interests, an oversupply of evidence, public interest, and active intermediaries. This means that policy makers may consider and evaluate much more than the scientific evidence on any given issue.

There are many other cases where gender/sex entanglements are pertinent and misinterpretation and confusion can occur. During the COVID-19 pandemic, for example, numerous scientific studies suggested fairly clear evidence of gender/sex (labeled as "sex disparities") in mortality rates (e.g., Danielsen et al. 2022). This reveals how evidence-informed policies need to be critical, systematically addressing the biosocial entanglement of epidemiological dynamics.

4 A Road Map Forward

How can more nuanced understandings of gender/sex be generated to encourage the integration of gender/sex evidence in policy and practice? Put simply, there are no formulaic answers, given the multiple factors that affect both science and policy, and the indirectness and partiality of knowledge mobilization. To assist in recognizing challenges inherent in creating and deploying knowledge relevant to gender/ sex entanglement, we offer a framework to support the development of effective and just policy and practice. In it, challenges are identified that reflect our expertise, but this is hardly a definitive list. Going forward, we hope that it will spur further dialogue and that other scholars will expand it by using very different challenges.

4.1 Reduce Harm Through Critical Reflection on "Settled Science"

Ideally, any policy issue should begin with a clear question and a large body of knowledge that points to how an issue should be addressed. In practice, however, science is often unable to provide a clear or simple answer. Those designing or implementing policy through law often think that decisions should be based on "settled science" (i.e., a knowledge consensus among experts). Yet in any complex human domain (like gender/sex), this is simply not possible. For example, to determine who qualified as "white" for purposes of US naturalization policy in the early part of the twentieth century, the US Supreme Court turned away from science and relied instead on "common sense" because scientists could not make "race" do the work the courts required; that is, to exclude certain types of immigrants (López 2006). As a result, the court's decision failed to recognize a vast, emerging set of understandings about racialization and set into place policies (now clearly recognized) that perpetuated and recreated racism.

4.2 Recognize Gender/Sex as Co-Constituted and Entangled

Many public policy issues are complex and may potentially involve biological, psychosocial, social, and ecological dimensions. They are further shaped by historical, political, and economic considerations. Only a rigorous scientific understanding of the cumulative interplay of these factors will allow us to grasp and respond to complex health or social policy issues. For example, understanding menopause from solely a biological perspective would miss how individuals in different socio-economic positions have varying exposures, susceptibilities, and resistance (both biological and political) through development and growth or in relation to disease (e.g., Sievert et al. 2021), and how this creates embodied inequality (Krieger 2012). Questions of scale of analysis and frame for interpretation are critical to using the available data and determining relevance for a particular issue.

In medical practice, one can clearly see the relevance of the co-constitution perspective in the example of breast implants: medical devices used for cosmetic breast augmentation (80%) and reconstruction (20%) (Pederson and Tweed 2004). Although the process of approval and licensing is regulated and rigorous in higher-income countries, numerous scandals, catastrophic illnesses, death, implant-related cancer, and class action suits have resulted from the use of these devices (Keith et al. 2017; Lampert et al. 2012). This is due to the assumption of "blind" regulatory processes, lack of sex and gender science, lack of funding for women's health research, and lack of gender/sex segregated post-market vigilance data, tracking, and reporting. As a result, consumer and clinician publications have been unable to process complete risk analyses and provide adequate warnings. Gender transformative regulatory efforts to improve the experience of women with breast reconstructions might focus on the development of knowledge products that directly address motivations and desires across all gender groups for augmentation and reconstruction, linking those to reduction of risks (e.g., Schall and Moses 2023).

4.3 Recognize That Knowledge Prioritization Happens in the Context of Hierarchies

Production of knowledge about gender/sex and its mobilization occurs in the context of a knowledge hierarchy: certain forms of knowledge and underpinning methodologies are valued over others, thus limiting the forms of knowledge that are accepted as relevant. This means that critical perspectives, lived experiences, qualitative data, and community-based knowledge are often ignored or deemed irrelevant in policy discussions. In contexts of contested science, policy makers, lobby groups, and other actors may incorrectly, incompletely, disingenuously, opportunistically, and/or ignorantly look to certain areas of knowledge production about sex and gender to justify a particular policy outcome, while dismissing the harms of those knowledge claims on the affected communities. The case of sex testing in sport exemplifies this challenge. In a 2015 case brought by Indian sprinter Dutee Chand before the Court of Arbitration for Sport (CAS), arbitrators dismissed critical perspectives on the contested scientific basis of sex- and testosterone-related eligibility regulations, and the resulting harms of such rules, as "sociological opinion, which does not equate to scientific and clinical knowledge and evidence" (CAS 2015:134).

Once in the world, results about scientific findings can take on a much larger significance when the media and policy makers read those findings through the lenses of binary sex. Scientists must be transparent in their methodologies and very clear about the claims they are making (and are not making) as well as their own relevant positionality. Current practices are insufficient to meet this need.

A study of language usage in articles published in *Science* found a trend toward exaggerated knowledge claims over recent decades (Brainard 2023). Similarly, gender considerations of audience receptivity and modes of learning vis-à-vis how evidence might be interpreted and applied also needs to be advanced as a core value and practice (Tannenbaum et al. 2016).

4.4 Be Transparent, Operationalize Context, and Recognize Use of Categorical Proxies

Research relevant to gender/sex entanglement should be precise in how it conceptualizes the *mechanisms at work* and, consequently, the *interventions needed*. Contrary to mandates like the Sex as a Biological Variable (SABV) policy in the United States, sex is not in and of itself a simple "variable." Rather, as described above, sex is a category, typically assigned at birth, which often serves as a proxy for a range of mechanisms and sex-related factors, such as hormone levels, gametes, and chromosomes. As such, there is no effect of sex in itself, but rather, actions of sex-linked factors or mechanisms.

As Karkazis (2008:13) has written: "If one postulates bodies (including genitals, gonads, chromosomes, and hormones), what more does the word sex buy us?....The body as a material fact is given, but sex is not." The key challenge that we identified, however, is how gender/sex can be brought into the knowledge production process to correct these historical tendencies *without* creating new problems, since the inclusion of sex and/or gender alone is insufficient to prevent inaccurate and inequitable knowledge claims (e.g., Ritz and Greaves 2022).

Sex category-specific dosing for zolpidem (Ambien®) further illustrates the potential pitfalls of this approach. When a lower dose for women was proposed by the US Federal Drug Agency in 2013, it was based on a finding that next-day drowsiness appeared to be more common in women than men. This recommended dosing was based on gender category despite there being no difference in residual concentrations of active drug or driving impairment. Greenblatt et al. (2019:189) concluded that not only was gender-specific dosing unwarranted, but "may in fact lead to underdosing and the consequent hazard of inadequately treated insomnia" in women. The resulting two-size-fits-all approach did not attend to overlapping distributions and lacked a clear understanding of the mechanism(s) underlying the disparity. Indeed, in the case of zolpidem, sex-specific dosing was recommended even in the absence of statistically significant differences between women and men. In practice, this means that some women are likely being deprived of adequate insomnia treatment while some men are being overtreated (Greenblatt et al. 2019; Zhao et al. 2023).

Strengthening the study of sex-linked factors to ensure complex variation and mechanisms is at the core of such research and is not the same as studying gender/sex entanglement. The concept of gender/sex entanglement moves closer to precision by resisting the temptation to separate out mechanisms associated with "sex" from those associated with "gender" and instead developing the methods and language to investigate and explain how these two concepts, which have been imposed on the body, are in fact entangled from the outset. The concept of gender/ sex entanglement may offer a more accurate and equitable starting or foundational concept for health research. The opportunity is to push toward more precise and clinically relevant knowledge by centering the concept of entanglement from the outset. This may work best in ideal circumstances when data and evidence are collected in more equitable ways, inclusive of sex, gender, and intersecting race/ ethnicity, class, and age-related factors, and engaging all relevant populations. As yet, this is not the case.

4.5 Assume Knowledge Mobilization Is Not Neutral

Many social science fields, including science and technology studies and feminist policy studies, hold that the use of knowledge in informing the development and implementation of policy is not a neutral or objective process. Policy makers do not simply "follow the science." They use forms of knowledge and expertise that have been integrated into a policy process mediated by a wide range of important factors (e.g., who is considered a relevant "knower," what forms of data or evidence are taken to be relevant, and the relative authority afforded to different forms of knowledge and scientific disciplines).

Policy concerning gender-affirming care for youth in some jurisdictions illustrates how disregard or selective interpretation of science can happen (Park et al. 2021). For example, some bans in the United States that exempt gender-affirming care for transgender youth allow it nonetheless for cisgender youth (Schall and Moses 2023). With respect to mental health, a positive correlation has been shown between access to hormone therapy for transgender teens and increased quality of life, decreased depression, and decreased anxiety (Baker et al. 2021; Turban et al. 2022). Nonetheless, in rationalizing their bans on gender-affirming care, some US states simply hold that there is no evidence of improved mental health outcomes for youth receiving gender-affirming care (e.g., Alabama State 2022).

Policy makers participate in a politics of expertise when they assert which forms of knowledge (and which knowledge makers) are relevant, which are excluded, and which among those included will be deemed most authoritative. It is not a given that certain forms of (scientific) knowledge are "naturally" the most authoritative; this depends on the policy context and the decisions actively taken by policy makers when affording authority to some forms of knowledge (and knowledge makers) and not others. It is also not given that policy makers will seek out the knowledge makers and forms of knowledge deemed by other actors and stakeholders to be the most pertinent; again, this depends on the particularities of the policy process at hand. Knowledge mobilization is never neutral. It requires greater transparency, more communication effort, and a careful localization of any knowledge production, including recognition of the limitations of the studies at hand.

4.6 Identify Inappropriate Attribution of Gender/Sex to Understand or Explain Issues

Not all "good" (e.g., rigorous, precise, effective, ethical) policies require scientific evidence, and identifying when this is the case can prove a challenge in knowledge mobilization. For example, consider the issue of mandating that individuals must use the restroom or toilet that corresponds to their sex assigned at birth. Here, science is not the best knowledge set to call on to devise a solution. Instead, one anchored in human rights is better suited to inform policy decisions. In another example concerning legal sex reclassification, "sex" may not even be an object until policy declares it to be so (see Chapter 15). This also coincides in real ways with other processes. In Germany, for example, the government liberalized the sex reclassification system in August 2023 by passing a self-determination law. A month later, it proposed wartime restrictions to the policy, which would, in the context of military conscription, define as male those individuals who were assigned male at birth but changed their gender in the preceding two months.

4.7 Embedding Gender/Sex Entanglement in Scientific Practice

The rich work of theorists of science has shown how gender/sex processes and their implications are not external to universities and research; they shape what is pursued and understood in science as knowledge (Barad 2007; Tuana 1989). Numerous examples of the bases and implications of this are provided throughout this volume. Many efforts to integrate gender/sex into research and policy can be linked to efforts to address legacies of racism and androcentrism, most notably in clinical trials, which in most countries have historically centered on (White, middle-class) privileged men (Epstein 2007). Today, mandates are slowly being reviewed and revised in several countries to increase gender balance and racial diversity in clinical trials.

More recent national policies of the European Commission, the Canadian Institutes of Health Research, and the US National Institutes of Health require SABV consideration in animal research, where researchers may be required to include both female and male models. Some criticize these mandates for failing to address the gender inequities they aim to correct, and instead identify the mandates as *contributing* to data misinterpretation, reinforcing inaccurate perceptions of human differences and undermining recognition of the role of people's experiences in health disparities (Maney 2016; Richardson et al. 2015).

4.8 Attend to Power

Social sciences, writ large, are clear on the point that science, policy, and practice reflect, create but may also contest power inequalities. Many scientists understand gender to be a social category, but in the construction of policy, that understanding has to be augmented with knowledge about the use (and misuse) of gender/sex in policy. For example, in the Global North, gender/sex has historically been used to deny women the right to vote, to run for office, to enter professions, to own property, or to sign contracts. Through the institution of marriage and bans on same-sex marriage, gender/sex has been used to establish the inequality of women and lesbian, gay, bisexual, and gender-expansive people. In many contexts today, reductionist ideas about sex and gender are put to work in service of anti-transgender campaigns. Policy makers' reductive ideas of gender differences backed by the force of policy or law, can render—with astonishing speed—researchers' nuanced and complex accounts of the characteristics associated with gender/sex into hard and fast legal facts that harm many groups inequitably, including transgender people (see Chapter 15).

The term "science" is often used to anchor claims that a binary understanding of sex reflects "reality" and biology; that it rightfully aligns with common practice and "common sense." This suggests that arguments to the contrary are biased forms of "gender ideology" reflecting fictional or distorted interpretations of science and biology. These ideas and arguments leap over the complexities of gender/sex entanglement. Bridging these different interests and viewpoints is a challenge for politicians and policy makers, as evidenced by ongoing advocacy by a wide range of diverse interest groups, including gender critical feminists (e.g., Stock 2021; Sullivan and Todd 2024), conservative politicians, sport communities, education, and health-care providers.

The power inequalities that policy reflects and creates has greater consequences for more marginalized and vulnerable (e.g., minority) groups. This is because they stand to be disproportionately impacted by the popular uptake of knowledge claims about gender and sex that do not attend to entanglement and complex variations, notably racialized women as well as transgender and nonbinary communities and individuals. Consideration of these inequalities should be central to science in both practice and translation. Cisgender and transgender women continue to be harmed and limited by the popular uptake of claims about sex and gender differences, but harms are also pernicious and existential for all minorities and marginalized people who continue to live under patriarchy, face limited state protection, and experience high rates of discrimination and violence.

Some anti-trans lobby groups (as distinct from gender critical feminists) in the United States and United Kingdom regularly lean on the policy discourses connected to the SABV mandate to support their claims about sex as binary, fixed, and more foundational than gender to human experience. Scientific experts with opinions on sex differences—who may not be sex differences researchers—regularly participate in legal proceedings in support of plaintiffs seeking to impede the rights and inclusion of transgender and nonbinary people, particularly in the provision of health care (Montpetit and Gilchrist 2023). Adopting an entanglement framing may help scientific researchers and policy makers promote context-specific claims about gender/sex. It could also potentially help the wider public and policy makers understand the conflicting experiences of transgender and nonbinary people and other groups on their own terms, such as cisgender women desiring cisgender women-only spaces, or cisgender men who may similarly wish to retain cisgender men-only spaces. The realities of making policy and creating practice standards in a range of human endeavors is invariably a balancing act of competing rights, ideas of justice, universalism, and other political realities, all of which evolve and change over time in cultural contexts. As noted, there are many different theories from social science that provide lenses on this issue. For example, intersectionality is one that threads through this text that is particularly applicable to highly racialized societies, like the United States, with notable social and economic inequalities (e.g., Hull et al. 2023 for a relevant policy toolkit).

In community-based and community-engaged research, the power differentials and stakes can be especially high, and the capacity of research practices to do harm through scientific practice as well as its knowledge mobilization may result (e.g., Reid et al. 2017). New research models that are being developed and tested can center the need for more equitable and less harmful gender/sex science and knowledge mobilization while also attending to power dynamics (e.g., Poole 2012). General knowledge mobilization frameworks guide the work of international research (Graham and Tetroe 2007) as well as the work of knowledge mobilization for clinical practice (MacDermid and Graham 2009).

5 Final Thoughts

Currently, there is intense expansion of research and policy attention around gender and sex. This new focus is, in part, driven by institutional movements to consider sex and gender in research and the rising profile of clinical gender research and practice. Within this context, the meanings of the categories of sex and gender and their entanglement are rapidly changing, and consequently under discussion and challenge. Much more nuanced and complex approaches to researching gender/sex are clearly required (Ritz and Greaves 2024).

The ACL example discussed above and, more generally, sports policy illustrate how the mobilization of science to address "sex" and "gender" creates a challenging landscape for policy development, even though efforts are underway. An emergent Female Athlete Health paradigm is gaining institutional recognition, highlighting the gap in sports science and medicine regarding the specific needs of cisgender women. This has led to an endorsement by the International Olympic Committee (IOC) to proceed with defining what research is needed and where, and relatedly, what policy makers should be considering when issuing guidance, for example, about the prevention of injury in elite female athletes. Ideally, the IOC and advocates would agree that policy recommendations should be based on the most current science. In this case, the precedent for related prior efforts is to privilege treatments of "sex" as biology to reduce injury risks being faced by females in gendered sporting environments. A consensus then fails to situate the biological processes of individual bodies (e.g., ACL injury) in their gendered biosocial environment, and in doing so misses important points about where the most sustainable and effective policy could be focused.

Moreover, an entanglement framing could shift a Female Athlete Health paradigm toward one that makes room not only for cisgender women but also cisgender men and sex and gender minorities. It could create more space for policy and practice to advance human rights. However, as is the danger with inclusive policies, it could also lead to less precision in health practices, research, and treatment approaches, such as flattening the biology of women's bodies to a universal norm without attending to the biological diversity of bodies, the diverse factors that shape athletic performance, and their uneven distribution at a social and global level.

In the case of the IOC consensus report, embracing entanglement for better policy requires several changes. Scientific practice should begin with the assumption that sex and gender are entangled, that those advancing knowledge mobilization must recognize that sex and gender are entangled, and that resulting policies and practices must also reflect this. Addressing these challenges is the first step in minimally avoiding the pitfalls of reductionism around the notion of sex. It also means embracing gender/sex entanglement as a guiding principle to produce knowledge and in knowledge mobilization for policy and practice. The broader stakes are high, given discussion and policy making is often divisive and embedded in far more complex power dynamics that implicate human rights and gender equity.

To create better science and more equitable societal relations, gender/sex entanglement must be embedded in the material realities of funding, publishing, and educational policies as well as in notions of equitable participation in science. Much work remains to be done in scientific research, policy making, and the ways both interact. Social science is especially well equipped to test and refine necessary frameworks to move this forward.

References

Acker J (1990) Hierarchies, Jobs and Bodies: A Theory of Gendered Organizations. Gend Soc 4 (2):390–358. https://www.jstor.org/stable/189609

Adams M, and Danks B (2007) A Positive Approach to Addressing Indigenous Male Suicide in Australia. Aborig Isl Health Work J 31 (4):28–31. https://search.informit.org/doi/abs/10.3316/informit.956150326862037

Ainsworth C (2015) Sex Redefined. Nature 518 (7539):288. https://doi.org/10.1038/518288a

Al-Almaie SM, and Al-Baghli NA (2003) Evidence Based Medicine: An Overview. J Family Community Med 10 (2):17–24. https://pmc.ncbi.nlm.nih.gov/articles/PMC3425762/

Alabama State (2022) Senate Bill 184. 2022. https://legiscan.com/al/text/sb184/id/2527262 (accessed July 25, 2024)

Alcoff L, and Potter E (eds) (1993) Feminist Epistemologies. Routledge, New York

Andersen M, and Collins PH (eds) (2013) Race, Class and Gender: An Anthology. Thomson Wadsworth, Belmont

Baker KE, Wilson LM, Sharma R, et al. (2021) Hormone Therapy, Mental Health, and Quality of Life among Transgender People: A Systematic Review. J Endocr Soc 5 (4):bvab011. https://doi.org/10.1210/jendso/bvab011

Barad K (2007) Meeting the Universe Halfway: Quantum Physics and the Entanglement of Matter and Meaning. Duke Univ Press, Durham

Bennet A, Bennet D, Fafard K, et al. (2007) Knowledge Mobilization in the Social Sciences and Humanities. MQI Press, Frost, WV

Bluhm R (2012) Self-Fulfilling Prophecies: The Influence of Gender Stereotypes on Functional Neuroimaging Research on Emotion. Hypatia 28 (4):870–886. https://doi.org/10.1111/j.1527-2001.2012.01311.x

Brainard J (2023) In Some Scientific Papers, Words Expressing Uncertainty Have Decreased. Science. https://www.science.org/content/article/some-scientific-papers-words-expressing-uncertainty-have-decreased

Brown AD (2012) Addressing Cardiovascular Inequalities among Indigenous Australians. Glob Cardiol Sci Pract 1 (2):10.5339/gcsp.2012.5332. https://doi.org/10.5339/gcsp.2012.2

Brown AD, Mentha R, Rowley KG, et al. (2013) Depression in Aboriginal Men in Central Australia: Adaptation of the Patient Health Questionnaire 9. BMC Psychiatry 13 (1):1–10. https://doi.org/10.1186/1471-244x-13-271

CAS (2015) Dutee Chand v. Athletics Federation of India (AFI) & the International Association of Athletics Federations (IAAF). Lausanne: Court of Arbitration for Sport

Clayton JA (2016) Studying Both Sexes: A Guiding Principle for Biomedicine. FASEB J 30:519–524. https://doi.org/10.1096/fj.15-279554

Combahee River Collective (1977/1997) A Black Feminist Statement. In: Nicholson L (ed) The Second Wave: A Reader in Feminist Theory. Routledge, New York/London, pp 63–70

Connell R (1987) Gender and Power: Society, the Person and Sexual Politics. Allen & Unwin, Sydney

Crenshaw KW (1989) Demarginalizing the Intersection of Race and Sex: A Black Feminist Critique of Antidiscrimination Doctrine, Feminist Theory and Antiracist Politics. Univ Chic Leg Forum 1989 (1):8. https://chicagounbound.uchicago.edu/uclf/vol1989/iss1/8

Currah P (2022) Sex Is as Sex Does: Governing Transgender Identity. New York Univ Press, New York

Danielsen AC, Lee KM, Boulicault M, et al. (2022) Sex Disparities in COVID-19 Outcomes in the United States: Quantifying and Contextualizing Variation. Soc Sci Med 294:114716. https://doi.org/10.1016/j.socscimed.2022.114716

Davis AY (1983) Women, Race & Class, first edn. Vintage, New York

Devana SK, Solorzano C, Nwachukwu B, and Jones KJ (2022) Disparities in ACL Reconstruction: The Influence of Gender and Race on Incidence, Treatment, and Outcomes. Curr Rev Musculoskelet Med 15:1–9. https://doi.org/10.1007/s12178-021-09736-1

DuBois LZ, and Juster R-P (2022) Lived Experience and Allostatic Load among Transmasculine People Living in the United States. Psychoneuroendocrinology 143:105849. https://doi.org/10.1016/j.psyneuen.2022.105849

DuBois LZ, and Shattuck-Heidorn H (2021) Challenging the Binary: Gender/Sex and the Bio-Logics of Normalcy. Am J Hum Biol 33 (5):e23623. https://doi.org/10.1002/ajhb.23623

Epstein CF (2007) Great Divides: The Cultural, Cognitive, and Social Bases of the Global Subordination of Women. Am Sociol Rev 72 (1):1–22. https://psycnet.apa.org/doi/10.1177/000312240707200101

Fausto-Sterling A (1985) Myths of Gender: Biological Theories About Women and Men. Basic Books, New York

——— (2000) Sexing the Body. Basic Books, New York

——— (2012) Sex/Gender: Biology in a Social World. Routledge, New York/London

——— (2018) Why Sex Is Not Binary. https://www.nytimes.com/2018/10/25/opinion/sex-biology-binary.html

Fine C (2013) Is There Neurosexism in Functional Neuroimaging Investigations of Sex Differences? Neuroethics 6 (2):369–340. https://doi.org/10.1007/s12152-012-9169-1

Fox A, Bonacci J, Hoffmann S, Nimphius S, and Saunders N (2020) Anterior Cruciate Ligament Injuries in Australian Football: Should Women and Girls Be Playing? You're Asking the Wrong Question. BMJ Open Sport Exerc Med 6 (1):e000778. https://doi.org/10.1136/bmjsem-2020-000778

Frankenberg R (1993) White Women, Race Matters: The Social Construction of Whiteness. Univ Minnesota Press, Minneapolis

Gogovor A, Mollayeva T, Etherington C, Colantonio A, and Légaré F (2020) Sex and Gender Analysis in Knowledge Translation Interventions: Challenges and Solutions. Health Res Policy Syst 18:108. https://doi.org/10.1186/s12961-020-00625-6

Graham ID, and Tetroe J (2007) Whither Knowledge Translation: An International Research Agenda. Nurs Res 56 (Suppl 4):S86–88. https://doi.org/10.1097/01.NNR.0000280638.01773.84

Greaves L, Brabete AC, Maximos M, et al. (2023) Sex, Gender, and the Regulation of Prescription Drugs: Omissions and Opportunities. Int J Environ Res Public Health 20 (4):2962. https://doi.org/10.3390/ijerph20042962

Greaves L, Pederson A, and Poole N (eds) (2014) Making It Better: Gender Transformative Health Promotion. Canadian Scholars' Press, Toronto

Greaves L, and Ritz SA (2022) Sex, Gender and Health: Mapping the Landscape of Research and Policy. Int J Environ Res Public Health 19 (5):2563. https://doi.org/10.3390/ijerph19052563

Green EL, Benner K, and Pear R (2018) Transgender Could Be Defined out of Existence under Trump Administration. New York Times, Oct 21, 2018, https://www.nytimes.com/2018/10/21/us/politics/transgender-trump-administration-sex-definition.html

Greenblatt DJ, Harmatz JS, and Roth T (2019) Zolpidem and Gender: Are Women Really at Risk? J Clin Psychopharmacol 39 (3):189. https://doi.org/10.1097/JCP.0000000000001026

Hankivsky O (2012) Women's Health, Men's Health, and Gender and Health: Implications of Intersectionality. Soc Sci Med 74 (11):1712–1720. https://doi.org/10.1016/j.socscimed.2011.11.029

Haraway DJ (1989) Primate Visions. Gender, Race, and Nature in the World of Modern Science. Routledge, London/New York

——— (1991) "Gender" for a Marxist Dictionary: The Sexual Politics of a Word. In: Haraway D (ed) Simians, Cyborgs, and Women: The Reinvention of Nature. Routledge, London/New York, pp 127–148

Hernandez TK (2000) Sexual Harassment and Racial Disparity: The Mutual Construction of Gender and Race. Gend Race Just 4:183–224. https://ir.lawnet.fordham.edu/faculty_scholarship/12

Homan P (2019) Structural Sexism and Health in the United States: A New Perspective on Health Inequality and the Gender System. Am Sociol Rev 84 (3):486–516. https://doi.org/10.1177/0003122419848723

Hoyert DL (2022) Maternal Mortality Rates in the United States, 2020. https://www.cdc.gov/nchs/data/hestat/maternal-mortality/2020/maternal-mortality-rates-2020.htm

Hull SJ, Massie JS, Holt SL, and Bowleg L (2023) Intersectionality Policymaking Toolkit: Key Principles for an Intersectionality-Informed Policymaking Process to Serve Diverse Women, Children, and Families. Health Promot Pract 24 (4):623–635. https://doi.org/10.1177/15248399231160447

Johnson JL, Greaves L, and Repta R (2007) Better Science with Sex and Gender: A Primer for Health Research. Women's Health Research Network, Vancouver

——— (2009) Better Science with Sex and Gender: Facilitating the Use of a Sex and Gender-Based Analysis in Health Research. Int J Equity Health 8 (14):1–11. https://doi.org/10.1186/1475-9276-8-14

Kaiser A (2012) Re-Conceptualizing "Sex" and "Gender" in the Human Brain. Z Psychol 220 (2):130–136. https://doi.org/10.1027/2151-2604/a000104

Karkazis K (2008) Fixing Sex: Intersex Medical Authority and Lived Experience. Duke Univ Press, Durham

——— (2019) The Misuses of "Biological Sex." Lancet 394 (10212):1898–1899. https://doi.org/10.1016/s0140-6736(19)32764-3

Keith L, Herlihy W, Holmes H, and Pin P (2017) Breast Implant-Associated Anaplastic Large Cell Lymphoma. Baylor Univ Medical Center Proc 30(4), 441–442. https://doi.org/10.1080/08998280.2017.11930221

Krieger N (2003) Genders, Sexes, and Health: What Are the Connections—And Why Does It Matter? Int J Epidemiol 32 (4):652–657. https://doi.org/10.1093/ije/dyg156

——— (2005) Embodiment: A Conceptual Glossary for Epidemiology. J Epidemiol Community Health 59 (5):350–355. https://doi.org/10.1136/jech.2004.024562

——— (2012) Methods for the Scientific Study of Discrimination and Health: An Ecosocial Approach. Am J Public Health 102 (5):936–944. https://doi.org/10.2105/AJPH.2011.300544

Lampert FM, Schwarz M, Grabin S, and Stark GB (2012) The "PIP Scandal": Complications in Breast Implants of Inferior Quality. State of Knowledge, Official Recommendations and Case Report. Geburtshilfe Frauenheilkd 72 (3):243–246. https://doi.org/10.1055/s-0031-1298323

Laqueur T (1990) Making Sex: Body and Gender from the Greeks to Freud. Harvard Univ Press, Cambridge, MA

Latour B (1993) The Pasteurization of France. Harvard Univ Press, Cambridge, MA

Lister RL, Drake W, Scott BH, and Graves C (2019) Black Maternal Mortality-the Elephant in the Room. World J Gynecol Womens Health 3 (1) https://doi.org/10.33552/wjgwh.2019.03.000555

López I (2006) White by Law: The Legal Construction of Race. New York Univ Press, New York

Lugones M (2007) Heterosexualism and the Colonial/Modern Gender System. Hypatia 22 (1):186–219. https://doi.org/10.1111/j.1527-2001.2007.tb01156.x

——— (2010) Toward a Decolonial Feminism. Hypatia 25 (4):742–759. https://doi.org/10.1111/j.1527-2001.2010.01137.x

MacDermid JC, and Graham ID (2009) Knowledge Translation: Putting the "Practice" in Evidence-Based Practice. Hand Clin 25 (1):125–143. https://doi.org/10.1016/j.hcl.2008.10.003

Maney DL (2016) Perils and Pitfalls of Reporting Sex Differences. Philos Trans R Soc Lond B Biol Sci 371 (1688):20150119. https://doi.org/10.1098/rstb.2015.0119

Markowitz S (2001) Pelvic Politics: Sexual Dimorphism and Racial Difference. Signs 26 (2):389–414. https://doi.org/10.1515/9780295742595-006

Martin E (1991) The Egg and the Sperm: How Science Has Constructed a Romance Based on Stereotypical Male-Female Roles Signs. Signs 16 (3):485–501. https://doi.org/10.1086/494680

Mena E, Bolte G, et al (2019) Intersectionality-Based Quantitative Health Research and Sex/Gender Sensitivity: A Scoping Review. Int J Equity Health 18:199. https://doi.org/10.1186/s12939-019-1098-8

Mendoza B (2015) Coloniality of Gender and Power. In: Disch L, Hawkesworth M (eds) The Oxford Handbook of Feminist Theory. pp 100–121, https://doi.org/10.1093/oxfordhb/9780199328581.013.6

Montana State (2023) State Bill 0458, Section 1-1-201, https://archive.legmt.gov/bills/2023/billhtml/SB0458.htm (accessed Nov. 4, 2024)

Montpetit J, and Gilchrist S (2023) U.S. Conservatives Are Using Canadian Research to Justify Anti-Trans Laws. https://www.cbc.ca/news/investigates/james-cantor-gender-affirming-care-bans-1.6979356

Naidu T, Cartmill C, Swanepoel S, and Whitehead CR (2024) Shapeshifters: Global South Scholars and Their Tensions in Border-Crossing to Global North Journals. BMJ Glob Health 9:9:e014420. https://doi.org/10.1136/bmjgh-2023-014420

Nielsen MW, Stefanick ML, Peragine D, et al. (2021) Gender-Related Variables for Health Research. Biol Sex Differ 12 (1):23. https://doi.org/10.1186/s13293-021-00366-3

OECD (2020) Building Capacity for Evidence-Informed Policy-Making: Lessons from Country Experiences. OECD Public Governance Reviews. https://doi.org/10.1787/86331250-en

Oxman AD, Lavis J, Lewin S, and Fretheim A (2009) SUPPORT Tools for Evidence-Informed Health Policymaking (STP) 1: What Is Evidence-Informed Policymaking? Health Res Policy Syst 7 (Suppl 1):S1. https://doi.org/10.1186/1478-4505-7-S1-S1

Oyěwùmí O (1997) The Invention of Women: Making an African Sense of Western Gender Discourses. Univ Minnesota Press

Pape M, Miyagi M, Ritz SA, et al. (2024) Sex Contextualism in Laboratory Research: Enhancing Rigor and Precision in the Study of Sex-Related Variables. Cell 187 (6):1316–1326. https://doi.org/10.1016/j.cell.2024.02.008

Park BC, Das RK, and Drolet BC (2021) Increasing Criminalization of Gender-Affirming Care for Transgender Youths: A Politically Motivated Crisis. JAMA Pediatr 175 (12):1205–1206. https://doi.org/10.1001/jamapediatrics.2021.2969

Parsons JL, Coen SE, and Bekker S (2021) Anterior Cruciate Ligament Injury: Towards a Gendered Environmental Approach. Brit J Sports Med 55 (17):984–990. https://doi.org/10.1136/bjsports-2020-103173

Pederson A, Greaves L, and Poole N (2015) Gender-Transformative Health Promotion for Women: A Framework for Action. Health Promot Int 30 (1):140–150. https://doi.org/10.1093/heapro/dau083

Pederson A, and Tweed A (2004) Registering the Impact of Breast Implants. Centr Excell Womens Health Bull 3 (2):7–8, doi:10.1046/j.1365-2648.1994.20050844.x

Plumwood V (1993) Feminism and the Mastery of Nature. Routledge, New York

Poole N (2012) Boundary Spanning: Knowledge Translation as Feminist Action Research in Virtual Communities of Practice. In: Oliffe J, Greaves L (eds) Designing and Conducting Gender, Sex and Health Research. Sage, Thousand Oaks, pp 215–226

Reid C, Greaves L, and Kirby S (2017) Experience Research Social Change: Critical Methods, Third Edition. Univ Toronto Press, Toronto

Renstrom P, Ljungqvist A, Arendt E, et al. (2008) Non-Contact ACL Injuries in Female Athletes: An International Olympic Committee Current Concepts Statement. Brit J Sports Med 42 (6):394–412. https://doi.org/10.1136/bjsm.2008.048934

Richardson SS (2013) Sex Itself: The Search for Male and Female in the Human Genome. Univ Chicago Press, Chicago

———— (2022) Sex Contextualism. PTPBio 14:2. https://doi.org/10.3998/ptpbio.2096

Richardson SS, Reiches M, Shattuck-Heidorn H, LaBonte ML, and Consoli T (2015) Opinion: Focus on Preclinical Sex Differences Will Not Address Women's and Men's Health Disparities. PNAS 112 (44):13419–13420. https://doi.org/10.1073/pnas.1516958112

Ridgeway CL, and Correll SJ (2004) Motherhood as a Status Characteristic. J Soc Issues 60 (4):683–700. http://dx.doi.org/10.1111/j.0022-4537.2004.00380.x

Risman BJ (2004) Gender as a Social Structure: Theory Wrestling with Activism. Gend Soc 18 (4):429–450. https://doi.org/10.1177/0891243204265349

Ritz SA, and Greaves L (2022) Transcending the Male–Female Binary in Biomedical Research: Constellations, Heterogeneity, and Mechanism When Considering Sex and Gender. Int J Environ Res Public Health 19 (7):4083. https://doi.org/10.3390/ijerph19074083

———— (2024) We Need More-Nuanced Approaches to Exploring Sex and Gender in Research. Nature 629 (8010):34–36. https://doi.org/10.1038/d41586-024-01204-3

Rosenberg W, and Donald A (1995) Evidence Based Medicine: An Approach to Clinical Problem-Solving. Br Med J 310 (6987):1122–1126. https://doi.org/10.1136/bmj.310.6987.1122

Russett CE (1989) Sexual Science: The Victorian Construction of Womanhood. Harvard Univ Press, Cambridge, MA

Sanchis-Segura C, and Wilcox RR (2024) From Means to Meaning in the Study of Sex/Gender Differences and Similarities. Front Neuroendocrin 73:101133. https://doi.org/10.1016/j.yfrne.2024.101133

Schall TE, and Moses JD (2023) Gender-Affirming Care for Cisgender People. Hastings Cent Rep 53 (3):15–24. https://doi.org/10.1002/hast.1486

Schiebinger LL (2004) Nature's Body: Gender in the Making of Modern Science. Rutgers Univ Press, New Brunswick

Schuller K (2018) The Biopolitics of Feeling: Race, Sex, and Science in the Nineteenth Century. Duke Univ Press, Durham

Sievert LL, Huicochea-Gómez L, Cahuich-Campos D, Morrison L, and Brown DE (2021) When Does Fertility End? The Timing of Tubal Ligations and Hysterectomies, and the Meaning of Menopause. Anthropol Aging 42 (1):10–29. https://doi.org/10.5195/aa.2021.254

Snorton CR (2017) Black on Both Sides: A Racial History of Trans Identity. Univ Minnesota Press, Minneapolis

Springer KW, Stellman JM, and Jordan-Young RM (2012) Beyond a Catalogue of Differences: A Theoretical Frame and Good Practice Guidelines for Researching Sex/Gender in Human Health. Soc Sci Med 74 (11):1817–1824. https://doi.org/10.1016/j.socscimed.2011.05.033

Stock K (2021) Material Girls: Why Reality Matters for Feminism. Little Brown, Boston

Sudai M, Borsa A, Ichikawa K, et al. (2022) Law, Policy, Biology, and Sex: Critical Issues for Researchers. Science 376 (6595):802–804. https://doi.org/10.1126/science.abo1102

Sullivan A, and Todd S (eds) (2024) Sex and Gender: A Contemporary Reader. Routledge, London/New York

Tadiri CP, Raparelli V, Abrahamowicz M, et al. (2021) Methods for Prospectively Incorporating Gender into Health Sciences Research. J Clin Epidemiol 129:191–197. https://doi.org/10.1016/j.jclinepi.2020.08.018

Tannenbaum C, Greaves L, and Graham ID (2016) Why Sex and Gender Matter in Implementation Research. BMC Med Res Methodol 16:145. https://doi.org/10.1186/s12874-016-0247-7

Tuana N (1989) Feminism & Science. Indiana Univ Press, Bloomington/Indianapolis

Turban JL, Kamceva M, and Keuroghlian AS (2022) Psychopharmacologic Considerations for Transgender and Gender Diverse People. JAMA Psychiatry 79 (6):629–630. https://doi.org/10.1001/jamapsychiatry.2022.0662

UNFPA (2023) Maternal Health Analysis of Women and Girls of African Descent in the Americas. https://www.unfpa.org/publications/maternal-health-analysis-women-and-girls-african-descent-americas

van Anders SM (2015) Beyond Sexual Orientation: Integrating Gender/Sex and Diverse Sexualities via Sexual Configurations Theory. Arch Sex Behav 44 (5):1177–1213. https://doi.org/10.1007/s10508-015-0490-8

Villa P-I (2019) Sex-Gender: Ko-Konstitution Statt Entgegensetzung. In: Kortendiek B, Riegraf B, Sabisch K (eds) Handbuch Interdisziplinäre Geschlechterforschung. Geschlecht und Gesellschaft, 65 edn. Springer, Wiesbaden, pp 23–33

Wade L, and Marx-Ferre M (2015) Gender: Ideas, Interactions, Institutions. W. W. Norton, New York/London

West C, and Fenstermaker S (1995) Doing Difference. Gend Soc 9 (1):8–37. http://www.jstor.org/stable/189596?origin=JSTOR-pdf

WHO (2021) Evidence, Policy, Impact: WHO Guide for Evidence-Informed Decision-Making. https://iris.who.int/handle/10665/350994

Winny A, and Bervell R (2023) How Can We Solve the Black Maternal Health Crisis? https://publichealth.jhu.edu/2023/solving-the-black-maternal-health-crisis

Zhao FY, Xu P, Kennedy GA, et al. (2023) Identifying Complementary and Alternative Medicine Recommendations for Insomnia Treatment and Care: A Systematic Review and Critical Assessment of Comprehensive Clinical Practice Guidelines. Front Public Health 11. https://doi.org/10.3389/fpubh.2023.1157419

13

SABV Research Policies

From Distinctions to Entanglements

Madeleine Pape

Abstract   To understand how sex as a biological variable (SABV) research policies could better account for entanglement requires more than the input from biological sciences; it necessitates a joint scientific, political, and feminist response. This chapter analyzes the mandate put forth by the US National Institutes of Health (NIH) that requires consideration of SABV in animal research. Adopted in 2016, this policy has been a point of contention among feminist researchers. Advocates from women's health research policy argue that it corrects a long-standing reliance on male animals in many research fields, while skeptics and critics have expressed concern that the policy encourages overly binary, reductionist, and even inaccurate approaches to the study of sex-related variation. Although these two positions appear to be at odds, there is considerable common ground between them, pointing to the possibility of a productive exchange and a move toward mandates for research that emphasize *entanglements* of *sex* and *gender* as well as an embrace of more complex ways of describing variation within and across women and men and female- and male-classified animals. Research institutions, such as the NIH, as well as advocates of women's health research have the opportunity to move toward recognizing the value—for science, health, and equity—of approaching sex/gender variation in all bodies as irreducibly complex and dynamic.

Keywords   sex as a biological variable, gender, sex, entanglement, policy, public health

Madeleine Pape (✉)
Institute of Social Sciences, University of Lausanne, 1015 Lausanne, Switzerland
Email: madeleine.pape@unil.ch

© Frankfurt Institute for Advanced Studies (FIAS) 2025
L. Z. DuBois et al. (eds.), *Sex and Gender*, Strüngmann Forum Reports,
https://doi.org/10.1007/978-3-031-91371-6_13

1 Introduction

> Rather than casting the humanities and the sciences, and feminism and science, as binary oppositional practices, work and theories that stress the similarities, commonalities, and resonances may be a productive avenue for future collaborations.
> —Banu Subramaniam (2009:968)

> Why might it be useful, intellectually stimulating, and important to our personal and professional lives for feminists to know more about [biology] and for biologists to know how to develop a critical stance about their own lives and work? Let's consider, in other words, why feminists need science and why scientists need feminism.
> —Anne Fausto-Sterling (1992:339)

Feminist thought and mobilization have had important impacts on the production of biomedical knowledge. From holding institutions accountable for the lack of gender diversity in many fields of biomedical research and practice, to challenging scientific researchers themselves for how they produce claims about *sex* and sex-related variation, feminist interventions have supported critical reflection and actions to change the status quo (Bleier 1984; Fine et al. 2013; Keller 1985/1995; Richardson 2013). As in many other areas of feminist research and action, however, feminist engagement with the biomedical sciences has not been unified: rather, there exist varieties of feminism, which may well be unified by the ambition to overcome inequities in science, that are divided over how this should precisely be done. This is particularly clear when it comes to how sex and female–male difference are conceptualized, and how this relates to both women's health and the pursuit of rigorous, precise, and equitable science. Some advocates of women's health appear to embrace a binary, biological understanding of sex and its pursuit through research (Legato 2017; Mazure and Jones 2015; Woodruff 2014); others call for more attention to entanglements, such as between bodies and their environment as well as between and across bodies classified by sex and/or gender (Fine et al. 2013; Hankivsky et al. 2010; Richardson 2022; Ruzek et al. 1997; Springer et al. 2012). To what extent are these positions irreconcilable? Is there more common ground than appears at first glance?

In this context it is useful to examine recent policy mandates for biomedical research that shape how sex is enacted through biomedicine, specifically the *sex as a biological variable* (SABV) mandate put forth by the US National Institutes of Health (NIH). Announced in *Nature* in 2014 and implemented in 2016, this policy applies to preclinical research involving vertebrate animals, although researchers who use cell lines, primary cells, or tissue explants are also asked to account for sex (NIH 2018). It requires researchers to explain how sex will be factored into research designs and analyses, with the expectation that researchers will, at the very least, include both female and male animals (or offer strong justification for not doing so) and disaggregate their data by sex (NIH 2015). The initial policy announcement suggested that the mandate would also apply to cell lines (Clayton and Collins 2014). Later, NIH acknowledged that this is not practically possible, while still suggesting that it "is working to enhance strategies and techniques to address these challenges" (NIH 2018). The NIH is the first funding agency in the world to grant sex a dedicated inclusion policy, separate from the consideration of gender. Today, the Office of Research on Women's Health (ORWH)—the NIH office established in 1990 to advance research to improve women's health—considers the SABV policy to be one of its defining achievements. There is also evidence that the SABV policy paradigm has traveled to other contexts, including Canada and the European Union.

Importantly, the SABV mandate is not ahistorical: it emerged out of a specific history of mobilization by women's health research actors who targeted the NIH in an effort to ensure that the nation's preeminent funding body for biomedical research (and the world's largest) would account for women's experiences of health and illness (Epstein 2007; Pape 2021a). Extending back to at least the 1980s, this history was characterized by scholars as a fissure between feminists rooted in the grassroots women's health movement in the United States, which emphasized women's empowerment and intersectional justice in the face of the powerful institutions of medicine and science, and those who worked within those institutions and accepted the primacy of a biomedical approach to women's health (Ruzek and Becker 1999; Ruzek et al. 1997). While feminists who gained an institutional foothold working within institutions like the NIH have made progress on the inclusion of women in those spaces—as researchers, practitioners, and research subjects—more critical scholars have questioned whether in doing so they compromised the transformative potential of the women's health movement (Ruzek and Becker 1999).

As I discuss below, these differences among feminists have important implications for how sex, gender, and their entanglement are conceptualized by both policy makers and researchers. In the case of the SABV mandate in the United States, advocates have argued that the consideration of sex throughout the research spectrum—from basic to applied research—is a pillar of women's health research, and that studies should prioritize what these advocates consider to be the sex-related biological factors that underpin differences in women's and men's experiences of health and illness (Beery and Zucker 2011; Woodruff 2014; Woodruff et al. 2014). Others have argued that sex cannot so easily be divorced from the influences of gender and other contextual factors, and that human biology is always the sum of these parts, in ways that rarely resemble a simple female–male binary (Richardson 2022; Springer et al. 2012). The former position is currently institutionalized in US biomedical research policy. Here I argue that the SABV policy moment in the United States has actually brought to the surface the extent to research on sex is characterized by entanglement and complexity. Thus, I ask, under what (political) conditions might an entanglement approach become the dominant paradigm for the study of sex and gender within women's health research and biomedical research more broadly?

Before proceeding, a note on terminology: in this chapter, I define sex as a set of assigned categories (Massa et al. 2023). These categories are best treated as proxies in need of operationalization in the context of each individual research study to identify the variables of interest, which will often vary in ways that do not reflect a dimorphic female/male distribution (Pape et al. 2024; Richardson 2022). I define gender as a multilevel system of difference-making that organizes social life and is reproduced through major institutions, such as education, politics, the economy, and medicine (Connell 1987). Often, this system is built on assumptions about binary, biological differences between women and men and tends to reward individuals and practices that uphold these assumptions. Gender may also be part of the research setting, though the research environment could include a great number of factors beyond gender that influence studies of sex-related variation.

I consider entanglement to be an inescapable—and, for some, inconvenient— fact of sex and gender. Efforts to disentangle them are a form of epistemic politics, and here the playing field is not level: because of the extent to which it is accepted as "common sense," sex is often taken to exist prior to and independently of gender, to the detriment of our knowledge about the complexity of bodies, health, and disease. As part of this vision of entanglement, I also consider the binary categories of women/men and female/male to handicap our pursuit of knowledge intended to improve our knowledge of the body (Bauer 2023), and not only because such schemas exclude nonbinary people and other minority sex/gender groups. I find compelling the argument advanced by some researchers about the need to expand the methodological toolkit of researchers to get beyond simplistic categorical comparisons and toward the complexity of distributions across and within different sex/gender classified groups and individuals (Sanchis-Segura and Wilcox 2024).

Here I discuss how the SABV mandate emerged (Section 2.1) and what it has revealed about the challenges of defining sex as an object of scientific research (Section 2.2). The material I draw on comes primarily from my dissertation and postdoctoral research, which has analyzed the epistemic work of SABV advocates before as well as after the mandate was announced. Focus is on the efforts of policy makers from the ORWH and sex differences researchers to justify and operationalize the SABV mandate (for detailed description of data and methods, see Pape 2021a, b). I conclude by considering how researchers committed to a sex/gender entanglement might bridge feminist divides as part of strategically engaging with the political and institutional dimensions of policy making.

2 How Did We Get Here? Political and Institutional Contingencies

2.1 Creating the Conditions for an SABV Mandate

As I have written elsewhere, the reproducibility crisis in preclinical research created an opening for advocates of the SABV policy concept—and leaders at the ORWH, in particular—to make a case for the mandate (see Pape 2021a). Prior to announcing the SABV mandate, NIH leadership—including NIH Director Francis Collins—stated in *Nature* that "the complex system for ensuring the reproducibility of biomedical research is failing and is in need of restructuring" (Collins and Tabak 2014:612). ORWH Director, Janine Clayton, later reiterated that "a fundamental pillar of science—reproducibility—was buckling, threatening to collapse the entire edifice" (Clayton 2018:3). In a preclinical research context where many fields were excluding female models, SABV advocates were well positioned to convince NIH policy makers that mandating consideration of sex could be part of the solution.

Importantly, the groundwork for the SABV policy was laid well before 2014. There are at least two major historical "landmark" policy moments that created favorable conditions for the SABV mandate, both of which came about as a result of sustained organizing on the part of feminist actors committed to the study of sex in biomedical research (Epstein 2007). The first was the NIH Revitalization Act of 1993, which introduced race, gender, and age inclusion requirements for clinical trials. Second, in 2001, the Institute of Medicine (now known as the National Academy of Medicine) released a significant report—*Exploring the Biological Contributions to Human Health: Does Sex Matter?*—which lent legitimacy to the study of sex differences as a vital and neglected field of science. The report conceptualized sex in terms of biological female–male difference and stated that in contrast with clinical and social research on women's health, "scientists have paid much less attention to the direct and intentional study of these differences at the basic cellular and molecular levels" (Wizemann and Pardue 2001:1). The report can be credited with generating the oft cited (and contested) phrase that "every cell has a sex" (Wizemann and Pardue 2001:4).

As sociologist Steven Epstein (2007) has documented, both the 1993 NIH Revitalization Act and the 2001 IOM report relied on considerable mobilizing by actors both internal and external to the NIH (Epstein 2007). For example, Florence Haseltine, a physician and professor of obstetrics and gynecology, played a significant role in this, both internally through her director roles at the NIH (1985–2012) and externally as the founder of the Society for Women's Health Research (SWHR), a Washington-based lobby group. Haseltine leveraged connections to industry and the US Congress to position the SWHR as an influential voice in women's health, using this platform to frame sex differences research as essential. The SWHR played a leading role in bringing about the 2001 IOM report and funded early networks of sex differences research in the 1990s and 2000s (Epstein 2007). It supported the establishment of the Organization for the Study of Sex Differences in 2006, a professional society that continues to manage the flagship journal in the field, *Biology of Sex Differences*.

In analyzing the history and aftermath of the NIH Revitalization Act, Epstein described feminist actors as working within what he called the "inclusion and difference paradigm," which he argues has become the hegemonic way of conceptualizing equity in biomedical research in the United States. In the context of US identity politics, exclusion and inclusion is measured in terms of dominant identity categories, with the price of entry being an emphasis on difference between groups rather than similarity. That is, difference becomes the justification to include: by affirming (rather than challenging) presumed differences between women and men, women were able to gain entry to clinical trials. According to Epstein, this contributes to salient identity categories becoming imbued with biological meaning, at the risk of neglecting the social mechanisms that contribute to different life chances—and health outcomes—along categorical lines. While the answer to a one-size-fits-all approach (androcentrism) was not inevitably a two-sizes-fit-all approach (women as different from men), a number of feminist advocates of women's inclusion in clinical trials in the United States embraced the pursuit of difference—and particularly presumed biological differences—as in women's interests. This logic has today been extended to animal models via the SABV mandate.

2.2 Institutional Contingencies: Seeking Biomedical Solutions to Complex Health Challenges

An important precondition for the political work that built support for the SABV mandate is the strong tendency within the US context toward

biomedical-pharmaceutical-technological solutions to the nation's health challenges. These health challenges were well documented during the period leading
up to the adoption of the SABV mandate, most notably in a report by the National
Research Council (2013), which showed the recent decline of health outcomes in
the United States relative to high-income countries. This report also demonstrated
the decline of women's health in the United States, which is now significantly worse
than in peer countries, and prompted a follow-up meeting aimed at establishing the
contributing factors (NASEM 2016). According to the participating experts, the
countries where women (and men) experience better health outcomes "generally
have far more robust social service and related programs and policies than those
in the United States" (NASEM 2016:11), with more investigation needed to understand the precise causal mechanisms. Despite the relative decline in women's
health in the United States pointing to a need to better understand the contributing
social and structural factors, "sex differences research" was a leading priority of
the ORWH for many years (ORWH 2019).

This tendency aligns with what Epstein (2007) and others observed as a general
preference in the United States for biomedical solutions to health inequalities, in
part due to the pharmaceutical industry having a particularly pronounced influence
on the direction of research and policy (Epstein 2007). The women's health movement in the United States is characterized by a similar hierarchy between advocates
who focused primarily on the structural (and intersectional) underpinnings of health
inequalities and those who, through their embrace of biomedical and technological
solutions, found themselves more aligned with and accepted by dominant institutions like the NIH (Ruzek et al. 1997). Ruzek and Becker (1999) have documented
how women's health advocacy in the United States shifted over time from a predominantly grassroots movement to a movement in which professionalized groups
and individuals held the most sway over policy decisions and public understandings
of women's health, in part because of the institutional access and resources gained
through their embrace of biomedical solutions and corporate partnerships (see also
Bruch and Richardson 2023).

By imposing a hierarchical distinction between social and biomedical solutions
to health challenges, this institutional tendency can also be understood as curtailing
opportunities for the recognition and investigation of entanglement. Indeed, it is a
distinction that is reflected in the structure of the NIH, which has an office but no institute to study social and behavioral science—an arrangement with material consequences for the research funds and support dedicated to research on social determinants of health (Millstein et al. 2018). While the NIH Institute for Minority Health
and Health Disparities, which prioritizes structural health determinants related to
racialization, socioeconomic class, and rural communities, has existed since 2010,
gender-related health factors are not within its remit. Conceivably, this institutional
structural arrangement may undermine opportunities to pursue an entanglement approach to sex/gender and health. In the case of the ORWH—which is similarly an
office and not an institute and which arguably has to appeal to the dominant institutional logic of the NIH to ensure its influence on research funding and policy—an
entanglement approach to gender and sex could be a risk, given the lack of alignment with the wider NIH. Since 2015, the NIH also has a separate Office for Sexual
and Gender Minorities, which may further discourage the ORWH from pursuing a

more complex, entangled approach to sex and gender within its vision of women's health research (Epstein 2022).

2.3 SABV Mandate as Epistemic Politics and Frame Alignment

As Jasanoff (2004) holds, ideas that gain ascendancy and are translated into scientific research policy are those that represent a combined scientific and political achievement. The SABV mandate is no exception. In my work, I have argued that advocates of the SABV mandate were able to lean on the legacy of mobilization around gender inequality in biomedical research, succeeding in framing inattention to sex differences—and the exclusion of female animals from basic science—as continuations of androcentric tendencies and gender injustices of biomedicine (Pape 2021a). In a context where women have historically been excluded from scientific research[1] and remain under-represented in the senior ranks of many fields of medical practice, the exclusion of female models from many areas of animal research was experienced as yet another example of the gender biases of biomedicine, particularly once researchers debunked the idea that female animals were too hormonally variable to be reliable research subjects (Prendergast et al. 2014).

This was a message that resonated with the public. For example, prior to the mandate's announcement in *Nature* by the directors of ORWH and NIH (Clayton and Collins 2014), CBS ran an episode on 60 Minutes documenting the lack of female mice in NIH-funded preclinical research, which was equated to the historical exclusion of women from clinical trials and the centering of cisgender (white, middle-aged, middle class) men as the "universal" model (CBS 2014). This same theme was taken up on prime-time television by comedian Steven Colbert, who described the "long-standing tradition of testing on only male animals" as "based on the assumption that females are simply a variation on a theme. Folks, that's science. Male is default human" (Colbert Report 2014).

I have suggested that it was this sense of injustice and inequality, rather than concerns about reproducibility or scientific rigor per se, that gave the SABV mandate broader traction and buy-in from key political actors (Pape 2021a). The sentiment of injustice was at the heart of a January 2014 letter to the NIH from Congresswomen Nita M. Lowey and Rosa DeLauro, in which they demanded that the "gender bias in basic research" be rectified (Pape 2021a:347). Some SABV advocates argued that the mandate could contribute to improving women's place as scientists in the United States. As NIH policy advisor and SABV advocate, Teresa Woodruff et al. (2014:1183) argued:

> Sex-based research will not only improve health care into the future but will also send a message to rising young female scientists and the public in whose interest we work that from early discovery research to the pinnacles of science leadership, women and their cells have an equal place at the table.

---

[1] With regard to clinical trials, Epstein (2007) makes a compelling case that vulnerable women and minorities have very often been exploited for the production of scientific knowledge, rather than excluded from it.

While addressing gender equality in biomedical research is a worthy cause, such a framing risks overstating the potential for sex differences research to contribute meaningfully to these goals, particularly since such research risks reinforcing the stereotypical thinking that has undermined women's place in such institutions (Maney 2015). Overall, the approach of advocates to framing the benefits of the SABV mandate show that political mobilization and frame alignment are at least as much (if not more) important to the public legitimacy of a mandate than claims to "better science" and scientific rigor. In the US case, NIH policy changes do not happen without support from Congress, which in turn requires that a research agenda be articulated in ways that speak to the interests and priorities of a broad, nonscientific audience.

In sum, the SABV mandate, as it currently exists in the United States, was achieved through sustained political mobilization and is highly aligned with its institutional environment: the NIH (and Congress) proved receptive to a policy that prioritizes biomedical solutions to health challenges and appears to align with commitments to gender equality. That is, it is not scientific merit alone that decides which policies are adopted and when. This suggests that any attempt to embed a more complex, entangled approach to the study of sex-related variation in health and illness requires an articulation of how such an approach can align with the interests of diverse stakeholders.

3 Opportunities to Reimagine the SABV Mandate in the US

The SABV mandate has been criticized by some feminist scholars for reinforcing a biomedical model of health care, presuming binary, biological differences between women and men, and shutting down avenues for a more entanglement-focused approach to the study of how sex and gender shape experiences of health and illness (Maney and Rich-Edwards 2023; Richardson 2022). Yet perhaps counterintuitively, in my research on the SABV mandate, I found that this policy moment brought to the surface the very complexity of sex, both in terms of its underlying mechanisms as well as the diverse and complex ways that these vary across and within female- and male-classified groups. Following the policy announcement, the ORWH and other advocates coordinated various workshops, events, statements and other materials aimed at explaining what the mandate means in practice and how it can advance the study of sex-related variation. These activities brought many contradictions to the surface, including evidence that complexity and entanglement with gender and other environmental factors are precisely what researchers discover when researching sex, and pointed to the opportunity to move toward a different policy framework to study sex-related variation.

Has the SABV policy potentially led the research world closer to, rather than away from, an entanglement approach? And, if sex differences researchers were in fact already doing entanglement research, or at least revealing the seeds of an entanglement agenda, how might this be brought to the surface and institutionalized in both knowledge outputs and policy?

3.1 Variable or Category?

One of the stated goals of the SABV policy is to advance rigor in basic research. However, my research found that what is meant by sex as a "biological variable" is not entirely clear (Pape 2021b). Researchers and policy makers regularly referred to sex as simultaneously an assigned category, a biological mechanism in its own right, and an outcome (sex differences)—in other words, a looping cycle of sex as category, cause, and effect. According to the ORWH (2018) website:

> Sex makes us male or female. Every cell in your body has a sex—making up tissues and organs, like your skin, brain, heart, and stomach. Each cell is either male or female depending on whether you are a man or a woman.

That is, the distinction between sex as a set of assigned categories and what those categories are intended to represent remains unclear. As neuroscientist Daphna Joel (2016) has written, the idea that sex penetrates the entire organism—rather than simply being a classification, relying largely on a cursory examination of the genitalia—is pervasive in biomedical research. Evelyn Fox Keller (1995:33) has referred to this as the as a synecdochical error, in which "the same properties that have been ascribed to the whole [body] are then attributed to the subcategories of, or [biological] processes associated with, these bodies" as well as to behaviors and other characteristics. By moving toward a focus on operationalizing mechanisms, rather than relying on binary categories as proxies, researchers and policy makers can also avoid a related challenge: the tendency to represent sex-related variation as conforming to a neat female/male binary.

3.2 Beyond Simplistic Comparisons

A common tendency when describing sex (and gender) as complex is to assume that this is a reference to transgender and gender-diverse people as well as people with intersex characteristics. While such diversity is indeed part of the complexity of sex, this overlooks the fact that sex-related variation is complex in all bodies (Karkazis 2019): cisgender women and men, as well as female and male animals, have highly varied experiences of health and illness. While biomedical research brings this complexity to the surface, researchers often default to analyzing sex-related variation by comparing the means of female- and male-classified groups. Sanchis-Segura and Wilcox (2024) have persuasively argued that this statistical approach hampers the ability of researchers to describe and explain complex distributions, with implications in turn for the recommendations that are then carried into clinical settings.

Much of the discourse associated with the SABV mandate describes sex-related variation in terms of stark differences between women and men; namely, women and men have distinct biologies and experiences of illness that warrant, in turn, distinct treatments. For example, according to ORWH's Director (Clayton 2015): "At the most basic level, people—and animals—come in two forms: female and male." One of the most notable examples of this discourse concerns the case of Ambien (zolpidem), a popular sleep medication that attracted controversy for reported adverse drug effects in women. Numerous advocates of the SABV mandate have relied on a decision by the US Federal Drug Agency to recommend different dosages for women and men and claim that women have been harmed by an overreliance on male animal models (Clayton and Collins 2014). Yet Zhao et al. (2023) have shown that the claim of sex-specific dosage for Ambien has weak empirical support, with limited research to suggest that women's bodies process drugs like Ambien in an entirely different way to men's.

While the SABV mandate requires researchers to disaggregate their findings by sex category, which reinforces in turn the long-standing hyperfocus on establishing categorical female–male differences (DuBois and Shattuck-Heidorn 2021; Maney 2016; Ritz 2017), my research suggests that some sex differences researchers would support a more complex approach to the interpretation of sex-related variation. Consider the following excerpts taken from NIH-led meetings on the SABV policy mandate (Pape 2021a):

> [W]e have to be careful about sex dimorphism versus sex difference versus sex influence…There's way too much abuse of the term sex dimorphism, when most of what we're interested in is either sex influence or sex difference, with huge overlap.
> —Margaret McCarthy (neuroscientist)

> [W]e should get rid of this notion of either/or, binaries…we should talk about overlap and *pluralities* and be open to similarities.
> —Gillian Einstein (neuroscientist)

> If there is a statistically significant difference [between females and males] it can be biologically important. But there are people who think that because there's so much overlap, that sex differences aren't really important. And there are other people who simply want to think sex differences are bimodal: females are like this, and males are like that. What do we mean by sex differences? I think we need to think about sex differences as not being just one thing.
> —Jill Becker (neuroscientist)

This level of nuance is currently missing from the SABV policy which, with its emphasis on considering sex even when studies are not powered to detect a sex effect, can lead researchers to overextend in their pursuit of female–male comparisons and to conclude incorrectly the presence of a difference between the two groups (Maney and Rich-Edwards 2023). In some cases, researchers may miss an important main effect of a treatment because of relying instead on female–male comparisons. Can policy makers provide clearer guidance on how to present and interpret data, so as to comply with the SABV mandate while avoiding inaccurate conclusions or unsubstantiated sex differences claims? In addition to potentially improving scientific understanding of sex-related variation, moving beyond simple comparisons of sex-classified groups could have wider, positive impacts on public understanding of sex-related variation while also making room for sex and gender minorities within scientific research (Maney 2015).

3.3 Animal Models as Highly Variable Rather Than "Pure"

A further issue related to the variability of sex-linked factors concerns variation across models and their relevance to human experiences of illness and disease. In a policy context that favors comparisons of sex-classified groups in animal research, and discussion of implications for humans, it is common to overstate how findings of sex-related variation in animal models will translate to humans (Birke 2010; Eliot and Richardson 2016; Gungor et al. 2019; Pape et al. 2024). My research has shown that in practice, animal researchers often understand that their models present highly varied and even contradictory sex effects. As addiction researcher Jill Becker shared during an advisory committee meeting (ACRWH 2014):

> One of the things that people need to be aware of is that the social structure for other species is quite different. Rats are quite different from mice....the mom and the daughters stay with the natal group and the males disperse, [whereas] mice are harem breeders....so the males live with the females in a large group....A valid stress for a rat might not be the same thing as a valid stress for a mouse.

Indeed, in the meetings I analyzed, sex differences researchers discussed the diverse ways that their animal models varied with respect to sex-related findings. Geneticist Art Arnold, for example, shared how his bird models produced unexpected findings: "[My] birds weren't actually obeying the laws of the twentieth century with regard to sexual differentiation as developed predominately in research on mammals" (ACRWH 2015). In another meeting, Arnold discussed how, in the case of rodent research, "[y]ou have to look at more litters....it is really important because there are big litter effects, especially in some strains or some situations" (Pape 2021b).

SABV policy discussions have shown that at best, animal models should be treated as offering "clues" for sex-related illness or treatment pathways in humans, but cannot be taken as models of "pure" and universal biological mechanisms (Richardson 2022). As neuroscientist Emeran Mayer cautioned during an ORWH (2014) advisory committee meeting held around the time that the SABV mandate was announced:

> Sex differences are species dependent...The sex difference in some trait in a mouse and a rat does not necessarily translate to a sex difference in humans. We've found this in pain [research] many times, to the big surprise of everybody, that one group finds dramatic sex differences in terms of visceral pain in the animals, but we don't see this really in our human patients.

In sum, rather than show sex-related variation in animals to be more simple than in humans, the SABV policy moment brought to the surface the rich complexity of this research world. My point here is by no means to argue that researchers should abandon model species, but rather to embrace their complexity and specificity (Birke 2010; Haraway 2016).

3.4 Grappling with Gender/Sex Entanglement in Lab Research

While gender is not the only contextual factor to be considered when operationalizing sex in laboratory and clinical settings, it is often a focus of discussion among SABV advocates with regard to whether and to what extent it is relevant to basic research. Some SABV advocates have argued that while gender certainly shapes the health experiences of humans, only sex is relevant in the case of animal models (see Pape 2021b). Sometimes, this occurs when researchers define gender in terms of identity: for example, according to McCarthy (2015:1018), "we cannot know if animals have a perception of their sex, [therefore] the term gender cannot be applied." However, advocates of an entanglement approach argue for a more expansive approach to conceptualizing how gender—or more specifically, social environment/context—can be operationalized within a given experimental setting. To be clear, the purpose here is not necessarily to attempt to model human experiences of gender in animal models, but rather to recognize entanglement of sex with gender plus context as the inescapable condition of any research experiment.

The core of feminist research on the scientific laboratory is to show how gendered assumptions and biases enter the research process, shaping the construction of hypotheses, the interpretation of experimental findings, and even the questions deemed worthy of asking in the first place (Bleier 1984; Fausto-Sterling 2000; Fujimura 2006; Jordan-Young and Karkazis 2019; Richardson 2013). Building on the insight that gender in human experiences encompasses social relations, practices, and structures, an enlarged vision of gender in the case of animal models could also include attention to their social structure and its effects on the treatment being measured; in other words, an attention to the environment and diverse ways that relations between animals can vary (Gungor et al. 2019). Becker's comment above would fit nicely with this approach to "context": social structure is different for rats versus mice, and should inform how treatments are designed and interpreted. As Becker noted on a different occasion, "[a]s we think about rats and mice, we need to think about how they live in their world" (ACRWH 2016). Factors such as sex-specific housing, as well as grooming patterns and other dominance behaviors, could be important confounds that need to be considered when interpreting sex-related variation (Kalueff et al. 2006). Could reframing sex differences in terms of entanglement better reflect these realities?

Consider the following anecdote (cited in Pape 2021b), from Louise McCollough, a neuroscientist who studies the relationship between sex-linked factors and stroke:

> We knew that social isolation is as big of a risk for stroke as is hypertension, and it is as big of a risk for poor recovery from stroke…If you take two animals and separate them a week before you induce a stroke, their stroke is 50% bigger…and if you give them the exact same stroke, you leave them together and then separate them, the isolated one dies 30 days later.

While McCollough went on to describe this as a "very biological" phenomenon, here "biological" appears to encompass the effects of social relations on stroke severity and recovery; that is, an entanglement. As historian of science Londa Schiebinger emphasized to an audience of NIH policy makers and sex differences researchers in 2014, the effects of contextual/social factors "are precisely the ones that we might mistake for sex differences" (ORWH 2014). This points to the importance of conceptualizing biology as the interaction of bodies with their environment. The opportunity here is to consider how such an approach to sex—not as an isolatable biological variable, but as a biological entanglement—could be integrated into policy so as to better reflect what researchers are discovering in the laboratory.

4 Conclusion

In this chapter, I have sought to show that while sex differences researchers from the biomedical sciences might be depicted as being at odds with those feminist scholars and researchers who offer critical perspectives on entanglements of sex and gender, there are in fact many areas of common ground.

In the United States, the current policy environment encourages an understanding of sex that emphasizes a search for female–male differences, divorced from gender and other environmental factors; however, the actual insights that sex differences researchers bring to policy discussions point to opportunities to bridge these schools of thought and practice. This offers opportunities to elaborate a policy infrastructure that would support more complex, contextualist, and entangled approaches to the study of sex-related variation, although as described, there may be a number of institutional factors that block a body like the ORWH from pursuing such a change in approach.

My focus here has been on the SABV mandate in the United States, yet as Jasanoff (2004) demonstrates, each national context will approach research policy differently. Although the US experience of SABV will likely have consequences beyond US borders, there is much to learn from how feminist actors in other settings have approached the study of sex and gender in biomedical research, and to what extent they have succeeded in advancing an entanglement framework for doing so (see Chapter 14).

Policy has material consequences, since it creates the conditions of possibility for research and ultimately shapes what is possible to know about sex, gender, and their entanglement (Epstein 2007). As such, it is vital that feminist and queer scholars also understand the institutional and political factors that contribute to the emergence of research mandates in particular places and at particular moments in time. Put differently, neither good arguments nor evidence alone will lead to policies that support researchers to explore entanglement: political mobilization and institutional engagement must be part of the strategic vision. In the world of biomedicine in the United States, policy makers and researchers hold deeply held convictions about the injustices that women have endured and the need to continue to improve equity in science. They also have deeply held convictions that the women's health research agenda should embrace biomedicine, rather than shy away from it and remain on its margins (Ruzek and Becker 1999). As such, there is work to do to articulate how an entanglement approach is not anti-women, but rather can address inequities and lead to deeper and more relevant science.

While dialogue and difficult conversations are not easy, they are fundamental to feminist thought and mobilization (Ryan-Flood et al. 2023). ORWH staff and leadership have invested years in creating the conditions of possibility for the SABV policy and have genuinely sought to contribute positively to women's health. To tell such policy makers that their efforts and work have no merit and is bad for the people they claim to be serving is unlikely to garner their interest and support. One alternative is to approach the SABV mandate as a platform upon which one can build: as a wedge that has created the opportunity to expand the ways that researchers are engaging gender/sex—and entanglement more broadly—in their work. The purpose is not to return to a one-size-fits-all approach, where male models are privileged as universal, nor to settle for a "two-sizes-fit-all" approach. Rather, there is an opportunity to open up the universe of sex/gender research toward the language, frameworks, and methodologies that better reflect findings about the complex ways that bodies interact with their environment.

References

ACRWH (2014) 38th ACRWH Meeting (24 September) Videocast. https://videocast.nih.gov/summary.asp?Live=14687&bhcp=1 (accessed Nov. 4, 2024)
———— (2015) 40th ACRWH Meeting (20 October). Videocast. https://videocast.nih.gov/summary.asp?Live=15589&bhcp=1 (accessed Nov. 4, 2024)
———— (2016) 41st ACRWH Meeting (19 April). Videocast. https://videocast.nih.gov/summary.asp?Live=18490&bhcp=1 (accessed Nov. 4, 2024)
Bauer GR (2023) Sex and Gender Multidimensionality in Epidemiologic Research. Am J Epidemiol 192 (1):122–132. https://doi.org/10.1093/aje/kwac173
Beery AK, and Zucker I (2011) Sex Bias in Neuroscience and Biomedical Research. Neurosci Biobehav Rev 35 (3):565–572. https://doi.org/10.1016/j.neubiorev.2010.07.002
Birke L (2010) Structuring Relationships: On Science, Feminism and Non-Human Animals. Fem Psychol 20 (3):337–349. https://psycnet.apa.org/doi/10.1177/0959353510371324
Bleier R (1984) Science and Gender: A Critique of Biology and Its Theories on Women. Pergamon Press, Oxford
Bruch JD, and Richardson SS (2023) Women's Health, Inc. Lancet 401 (10384):1258–1259. https://doi.org/10.1016/s0140-6736(23)00736-5
CBS (2014) Sex Matters: Drugs Can Affect Sexes Differently. https://www.cbsnews.com/news/sex-matters-drugs-can-affect-sexes-differently/ (accessed Nov. 4, 2024)
Clayton JA (2015) Considering Sex as a Biological Variable in NIH-Funded Preclinical Research. Letter from the Director, National Institute of Arthritis and Musculoskeletal and Skin Diseases. 20 August 2015 https://www.niams.nih.gov/about/about-the-director/letter/considering-sex-biological-variable-nih-funded-preclinical-research (accessed Nov. 4, 2024)
———— (2018) Applying the New SABV (Sex as a Biological Variable) Policy to Research and Clinical Care. Physiol Behav 187:2–5. https://doi.org/10.1016/j.physbeh.2017.08.012
Clayton JA, and Collins FS (2014) Policy: NIH to Balance Sex in Cell and Animal Studies. Nature 509 (7500):282–283. https://doi.org/10.1038/509282a
Colbert Report (2014) Co-Ed Lab Rats. https://archive.org/details/COM_20140521_015700_The_Colbert_Report/start/27/end/87?q=the+caption (accessed Feb. 2, 2025)
Collins FS, and Tabak LA (2014) NIH Plans to Enhance Reproducibility. Nature 505 (7485):612–613. https://doi.org/10.1038/505612a
Connell R (1987) Gender and Power: Society, the Person and Sexual Politics. Allen & Unwin, Sydney
DuBois LZ, and Shattuck-Heidorn H (2021) Challenging the Binary: Gender/Sex and the Bio-Logics of Normalcy. Am J Hum Biol 33 (5):e23623. https://doi.org/10.1002/ajhb.23623
Eliot L, and Richardson SS (2016) Sex in Context: Limitations of Animal Studies for Addressing Human Sex/Gender Neurobehavioral Health Disparities. J Neurosci 36 (47):11823–11830. https://doi.org/10.1523/jneurosci.1391-16.2016
Epstein S (2007) Inclusion: The Politics of Difference in Medical Research. Univ Chicago Press, Chicago
———— (2022) The Question for Sexual Health. Univ Chicago Press, Chicago
Fausto-Sterling A (1992) Building Two-Way Streets: The Case of Feminism and Science. Natl Womens Stud Assoc J 4 (3):336–349. https://www.jstor.org/stable/4316219
———— (2000) Sexing the Body. Basic Books, New York

Fine C, Jordan-Young R, Kaiser A, and Rippon G (2013) Plasticity, Plasticity, Plasticity...And the Rigid Problem of Sex. Trends Cogn Sci 17 (11):550–551. https://doi.org/10.1016/j.tics.2013.08.010

Fujimura J (2006) Sex Genes: A Critical Sociomaterial Approach to the Politics and Molecular Genetics of Sex Determination. Signs 32 (1):49–82. https://doi.org/10.1086/505612

Gungor NZ, Duchesne A, and Bluhm R (2019) A Conversation around the Integration of Sex and Gender When Modeling Aspects of Fear, Anxiety, and PTSD in Animals. https://sfonline.barnard.edu/a-conversation-around-the-integration-of-sex-and-gender-when-modeling-aspects-of-fear-anxiety-and-ptsd-in-animals/

Hankivsky O, Reid C, Cormier R, et al. (2010) Exploring the Promises of Intersectionality for Advancing Women's Health Research. Int J Equity Health 9 (5) https://doi.org/10.1186/1475-9276-9-5

Haraway DJ (2016) Staying with the Trouble: Making Kin in the Chthulucene. Duke Univ Press, Durham

Jasanoff S (ed) (2004) States of Knowledge: The Coproduction of Science and the Social Order. Routledge, London

Joel D (2016) Captured in Terminology: Sex, Sex Categories, and Sex Differences. Fem Psychol 26 (3):335–345. http://doi.org/10.1177/0959353516645367

Jordan-Young RM, and Karkazis K (2019) Testosterone: An Unauthorized Biography. Harvard Univ Press, Cambridge, MA

Kalueff AB, Minasyan A, Keisala T, Shah ZH, and Tuohimaa P (2006) Hair Barbering in Mice: Implications for Neurobehavioural Research. Behav Processes 71 (1):8–15. https://doi.org/10.1016/j.beproc.2005.09.004

Karkazis K (2019) The Misuses of "Biological Sex." Lancet 394 (10212):1898–1899. https://doi.org/10.1016/s0140-6736(19)32764-3

Keller EF (1985/1995) Reflections on Gender and Science. Yale Univ Press, New Haven

———— (1995) Gender and Science: Origin, History, and Politics. Osiris 10 (1):26–38. https://doi.org/10.1086/368741

Legato MJ (2017) Gender-Specific Medicine: An Idea That Should Have Been Intuitive but Which Required the Efforts of an International Community to Establish. In: Legato MJ, Glezerman M (eds) The International Society for Gender Medicine. Academic Press, Cambridge, MA, pp 1–8

Maney DL (2015) Just Like a Circus: The Public Consumption of Sex Differences. Curr Top Behav Neurosci 19:279–296. https://doi.org/10.1007/7854_2014_339

———— (2016) Perils and Pitfalls of Reporting Sex Differences. Philos Trans R Soc Lond B Biol Sci 371 (1688):20150119. https://doi.org/10.1098/rstb.2015.0119

Maney DL, and Rich-Edwards JW (2023) Sex-Inclusive Biomedicine: Are New Policies Increasing Rigor and Reproducibility? Womens Health Issues 33 (5):461–464. https://doi.org/10.1016/j.whi.2023.03.004

Massa MG, Aghi K, and Hill MJ (2023) Deconstructing Sex: Strategies for Undoing Binary Thinking in Neuroendocrinology and Behavior. Horm Behav 156:105441. https://doi.org/10.1016/j.yhbeh.2023.105441

Mazure CM, and Jones DP (2015) Twenty Years and Still Counting: Including Women as Participants and Studying Sex and Gender in Biomedical Research. BMC Womens Health 15:94. https://doi.org/10.1186/s12905-015-0251-9

McCarthy MM (2015) Incorporating Sex as a Variable in Preclinical Neuropsychiatric Research. Schizophr Bull 41 (5):1016–1020. https://doi.org/10.1093/schbul/sbv077

Millstein RA, Quintiliani LM, and Sharpe AL (2018) Society of Behavioral Medicine (SBM) Position Statement: Increasing Funding for the NIH OBSSR to Promote Timely and Effective Behavioral Medicine Research. Transl Behav Med 8 (2):309–312. https://doi.org/10.1093/tbm/ibx022

NASEM (2016) Improving the Health of Women in the United States: Workshop Summary. National Academies Press, Washington, DC. https://doi.org/10.17226/23441

National Research Council (2013) US Health in International Perspective: Shorter Lives, Poorer Health. National Academies Press, Washington, DC. PMID: 24006554

NIH (2015) Enhancing Reproducibility through Rigor and Transparency. https://grants.nih.gov/grants/guide/notice-files/not-od-15-103.html (accessed Feb 2, 2025)

———— (2018) Frequently Asked Questions (FAQs): Rigor and Reproducibility. https://grants.nih.gov/faqs#/rigor-and-reproducibility.htm?anchor=question54422 (accessed Feb 2, 2025)

ORWH (2014) Methods and Techniques for Integrating the Biological Variable Sex into Preclinical Research, October 20. Videocast. https://videocast.nih.gov/watch=14501 (accessed Nov. 4, 2024)

———— (2018) Infographic: How Sex/Gender Influence Health & Disease. https://orwh.od.nih.gov/sites/orwh/files/docs/SexGenderInfographic_11x17_508.pdf (accessed Nov. 4, 2024)

———— (2019) The 2019–2023 Trans-NIH Strategic Plan for Women's Health Research. http://orwh.od.nih.gov/sites/orwh/files/docs/ORWH_Strategic_Plan_2019_508C_0.pdf (accessed Feb 2, 2025)

Pape M (2021a) Co-Production, Multiplied: Enactments of Sex as a Biological Variable in U.S. Biomedicine. Soc Stud Sci 51 (3):339–363. https://doi.org/10.1177/0306312720985939

———— (2021b) Lost in Translation? Beyond Sex as a Biological Variable in Animal Research. Health Sociol Rev 30 (3):275–291. https://doi.org/10.1080/14461242.2021.1969981

Pape M, Miyagi M, Ritz SA, et al. (2024) Sex Contextualism in Laboratory Research: Enhancing Rigor and Precision in the Study of Sex-Related Variables. Cell 187 (6):1316–1326. https://doi.org/10.1016/j.cell.2024.02.008

Prendergast BJ, Onishi KG, and Zucker I (2014) Female Mice Liberated for Inclusion in Neuroscience and Biomedical Research. Neurosci Biobehav Rev 40:1–5. https://doi.org/10.1016/j.neubiorev.2014.01.001

Richardson SS (2013) Sex Itself: The Search for Male and Female in the Human Genome. Univ Chicago Press, Chicago

———— (2022) Sex Contextualism. PTPBio 14:2. https://doi.org/10.3998/ptpbio.2096

Ritz SA (2017) Complexities of Addressing Sex in Cell Culture Research. Signs 42 (2):307–327. https://doi.org/10.1086/688181

Ruzek SB, and Becker J (1999) The Women's Health Movement in the United States: From Grassroots Activism to Professional Agendas. J Am Med Womens Assoc (1972) 54 (1):4–8.

Ruzek SB, Olesen VL, and Clarke AE (1997) Women's Health: Complexities and Differences. Ohio State Univ Press, Columbus

Ryan-Flood R, Crowhurst I, and James-Hawkins L (2023) Difficult Conversations: A Feminist Dialogue. Routledge, New York

Sanchis-Segura C, and Wilcox RR (2024) From Means to Meaning in the Study of Sex/Gender Differences and Similarities. Front Neuroendocrin 73:101133. https://doi.org/10.1016/j.yfrne.2024.101133

Springer KW, Stellman JM, and Jordan-Young RM (2012) Beyond a Catalogue of Differences: A Theoretical Frame and Good Practice Guidelines for Researching Sex/Gender in Human Health. Soc Sci Med 74 (11):1817–1824. https://doi.org/10.1016/j.socscimed.2011.05.033

Subramaniam B (2009) Moored Metamorphoses: A Retrospective Essay on Feminist Science Studies. Signs 34 (4):951–980. https://doi.org/10.1086/597147

Wizemann TM, and Pardue ML (2001) Exploring the Biological Contributions to Human Health: Does Sex Matter? National Academies Press, Washington, DC. doi:10.17226/10028

Woodruff T (2014) Sex, Equity and Science. PNAS 111 (14):5063–5064. https://doi.org/10.1073/pnas.1404203111

Woodruff TK, Kibbe MR, Paller AS, Turek FW, and Woolley CS (2014) Commentary: "Leaning in" to Support Sex Differences in Basic Science and Clinical Research. Endocrinology 155 (4):1181–1183. https://doi.org/10.1210/en.2014-1068

Zhao H, DiMarco M, Ichikawa K, et al. (2023) Making a 'Sex-Difference Fact': Ambien Dosing at the Interface of Policy, Regulation, Women's Health, and Biology. Soc Stud Sci 53 (4):475–494. https://doi.org/10.1177/03063127231168371

14

How Could a Gender Transformative Lens Foster the Integration of Sex/Gender into More Equitable Policy and Practice?

Lorraine Greaves

Abstract   Gender transformative policies and practices address underlying gender inequities and respond to specific social, health, or economic problems. Gender transformative approaches have historically focused on improving women's status by changing gendered power relations, redefining masculinities, and exhorting communities and institutions to address the drivers and root causes of problems. Sex and gender entanglement poses a challenge to making policies and practices that can better reflect gender transformative approaches in that such approaches need to be robust, precise, and based on evolving science. This chapter proposes expansions of theory and practice to progress gender transformative approaches that reflect both sex/gender entanglement and engage all gender groups in efforts to reduce gender inequity. Achieving these goals requires (re)committing to feminism, engaging as critically with femininities as masculinities, integrating corporeality, and recognizing individual and collective agency in responding to hegemonic gender. These actions need to recognize ongoing and evolving impacts of sex and gender and sex/gender entanglement. This approach will facilitate improvements in policy and support new areas of gender transformative practice that can be operationalized with more precision in a proportionate universalism framework that differentially attends to groups based on need and disadvantage. Examples of gender transformative policy approaches consider entanglements of sex/gender in regulatory, communication, and policy activities aimed at reducing gender inequity.

Keywords   gender transformative approaches, gender, sex, entanglement, policy, public health

Lorraine Greaves (✉)
Centre of Excellence for Women's Health, BC Women's Hospital + Health Centre, and School of Population and Public Health, University of British Columbia, Vancouver, BC V6H 3N1, Canada
Email: lgreaves@cw.bc.ca

© Frankfurt Institute for Advanced Studies (FIAS) 2025
L. Z. DuBois et al. (eds.), *Sex and Gender*, Strüngmann Forum Reports,
https://doi.org/10.1007/978-3-031-91371-6_14

1 Introduction

The ongoing intractability of gender inequity, most often depriving women and girls of their full lives, remains a wicked global problem. It clearly requires a more robust and precise assessment of sex/gender and its entangled nature to support more evolved gender transformative policy and practice. At base, gender transformative approaches to programs, policies, and research attend to reducing gender inequity *along with* the specific problem being addressed. While gender transformative approaches are supported in theory by progressive policy makers, gender transformative policy and practice is less easily operationalized as it requires critical thinking, identification of root causes and drivers, creativity, and sustained progressive political support.

To make a dent in the widespread inequity facing women and girls, and to increase social justice for all—men, women, trans, and gender-diverse individuals—gender transformative theory and practice must be expanded. This requires conscious consideration of reducing gender inequities as initiatives are created and theories or approaches are developed. This requires a sensitivity to sex/gender entanglement in research, best achieved by explicating sex- and gender-related factors in precise detail and naming the components and interactions as underlying evidence bases for policy and practice. Policy trends toward expansionary views of gender and inclusivity can, however, run the risk of eclipsing such detail, posing an added challenge (Greaves and Ritz 2022).

In addition, gender transformative thinking requires critical engagement with femininities as well as masculinities, attention to sex-based corporealities, and acknowledging and invoking individual and collective agency. Advancing such thinking will require the critical engagement of all genders and a fuller recognition of the contributions and interactions of both sex- and gender-related factors and their entanglement to overarching gender inequities. Collectively, these shifts could lead to more comprehensive, and perhaps effective and relevant, gender transformative policy and practice.

2 Designing Change: Gender Transformative Approaches

> Gender transformative approaches actively strive to examine, question, and change rigid gender norms and imbalance of power as a means of reaching health as well as gender equity objectives.
> —Elisabeth Rottach et al. (2009:8)

Gender transformative approaches directly aim to change gender norms to increase equity. In recognition of the complex dynamics upholding gender inequities, gender transformative approaches to program and policy have to improve a particular health, economic, legal, or social issue, and reduce gender inequality (formal equality) and/or gender inequity (substantive equality) *at the same time* (Greaves and Poole 2018). Typically, gender transformative approaches highlight the underlying and persistent gendered drivers that most often negatively affect girls, women, and females by honing the root causes and solutions to alleviate them. Specifically, they focus on identifying particular institutions or audiences that need engagement. Figure 14.1 illustrates these processes as they affect the field of women's health

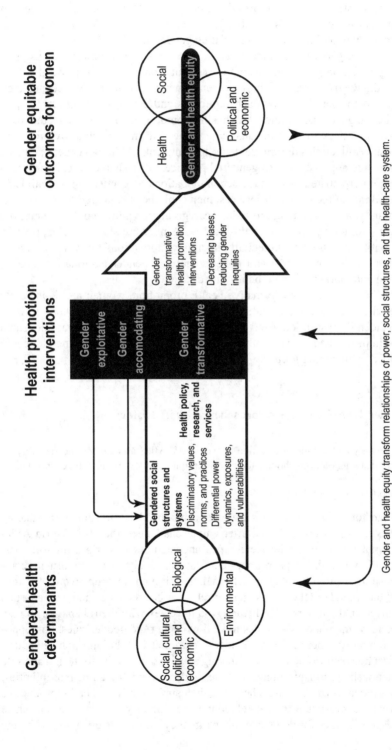

Figure 14.1 A framework for gender transformative health promotion which strives to improve the health and status of women. Reprinted with permission from Greaves et al. (2014).

promotion. In this example, biological, environmental, social, cultural, political, and economic structures act as health determinants. But standard interventions aimed at improving women's health often serve to reify discrimination and sex/gender roles by exploiting or accommodating gender norms.

For example, gender exploitative health promotion regarding tobacco use has, at times, pivoted on gendered assumptions about women's desire to be attractive to men and have used this to exhort women to not smoke as it might reduce their attractiveness to men. This approach reinforces and exploits gender stereotypes. Slightly better, gender accommodating health promotion recognizes gendered roles, norms, and practices such as unequal caregiving, and may respond by providing childcare at health clinics during women's appointments. Although beneficial in the immediate, this approach leaves gendered practices of unshared caregiving unaddressed and perpetuated. To reduce accommodations to gender norms, childcare could be offered at men's health services, men could be encouraged to revise ideas of masculinity, and a campaign to reduce caregiving inequities could be launched as a health-promoting solution. In this context, gender transformative approaches would identify self-directed, potentially liberating motivations for women's smoking reduction (Greaves 2014), and policies that ban gender exploitative tobacco advertising and marketing could be instituted.

Typically, in gender transformative health promotion, women are deliberately empowered to improve their health for their own reasons, exert agency in relationships, identify institutionalized gender influences, and engage in consciousness raising to improve their own and other women's health. All of these principles could be extended to all gender groups.

2.1 Gender Transformative Approaches: A Short History

To effectively address the intersection between HIV/AIDS and gender and sexuality requires that interventions should, at the very least, not reinforce damaging gender and sexual stereotypes.
—Geeta Rao Gupta (2000:8)

Gender transformative approaches were conceptualized in the course of reducing HIV/AIDS transmission in Africa after Gupta challenged the 2000 World AIDS Forum to not replicate gender stereotypes and inequities as it sought solutions. Although a low bar, this was an important first step. As gender transformative approaches spurred programming on HIV/AIDS with men, they uncovered the root causes of the spread of HIV/AIDS: gender relations between men and women, such as multiple sexual partners, MSM (men who have sex with men) behavior in men partnered to women, men's attitudes toward women, rigid gender roles, homophobia, and intimate partner violence. Gender transformative solutions included equalizing power between women and men, taking equal responsibility for health and safe sex in relationships, and questioned attitudes underpinning prevailing masculinities.

Gender transformative approaches were then applied in sexual and reproductive health where heterosexual gender relations were similarly renovated to introduce shared responsibilities for contraception, pregnancy, and childcare, along with new conceptualizations of fathering. Involving men in these issues required their engagement in previously unrecognized areas of men's concern, and the overt deconstruction of hegemonic masculine roles. This evolution of gender transformative approaches in sexual and reproductive health set a higher bar; that is, to engage men, shift gender relations, share responsibilities, and renovate masculinities. Gender transformative approaches were rapidly applied to a broader range of reproductive health issues such as obstetric fistula, early marriage, fetal alcohol spectrum disorder prevention, and ultimately to wider public health, education, or cultural issues, from sanitation and climate change to science, technology, engineering, mathematics, and medicine (STEMM), and female genital mutilation/cutting (FGM/C).

Gender transformative approaches evolved first in low-income countries, mostly in the Global South, to place power, knowledge, choice, and opportunity directly in women's hands, stressing empowerment and gender equity, and actively redrawing patriarchal assumptions and institutionally driven policies. More recently, gender transformative approaches have been taken up in middle- and high-income countries. For example, Our Watch, an Australian organization aimed at primary prevention of violence against women (VAW) identifies both root causes and drivers. These include the condoning of VAW, staying quiet when women are harassed, using gendered stereotyping with children and adults, and "mate ship" (OurWatch 2023). This example highlights the analyses, communication challenges, and change-making possibilities of gender transformative work. Our Watch has worked to develop targeted analyses by identifying primary prevention approaches with and for LGBTQ+ and Aboriginal and Torres Strait Islander communities.

In general, conceptualizations of gender transformative approaches are becoming more intersectional and encompassing more issues and approaches. Programs and policies have been developed to address masculinity and femininity, vulnerability to sexual assault and pregnancy, child marriage, sanitation, school access, gender roles and relations, gender stereotyping, and cultural and institutional practices that create and perpetuate sex/gender inequity. Indeed, gender transformative thinking can and should be applied to more issues and all gender groups and serve to keep sex/gender central while engaging with the multiplicities of factors embedded in diverse identities, situations, cultures, and developmental stages.

2.2 Engaging Men and Boys

Given the complexity of this multilayered task, and general resistance to it, a main focus of gender transformative approaches has historically been on improving the destiny of victims of gender inequality and inequity, while focusing on the attitudes and practices of the principal beneficiaries of gendered privilege and oppression. Hence, typical audiences for gender transformative programming have included engaging men and boys, doing community development and education, and sometimes engaging women and girls (Equimundo 2023; Hillenbrand et al. 2015).

Programs aimed at men and boys have not typically differentiated between groups of men and boys, one of the key criticisms of standard gender transformative programming. (Similarly, audiences of women and girls have typically not been disaggregated.) Hence it is not possible to assume audiences are cisgender, het-eronormative, in transition, experiencing gender oppressions or fluidities, or even relating to or upholding hegemonic masculinities and femininities, even when the programs are designed with that hegemony in mind. More recent programming for men appears to be better tailored and has addressed wider views of both sexual and gender identity in programming by identifying other ways of being distinct from heteronormative sexuality and hegemonic gender, which in turn leads to greater clarifications of audience and intent.

In addition, programming that addresses men and boys has often focused on individual or group behavioral change by developing a critique of prevailing, un-healthy masculinities and highlighting their limitations on men's experiences and emotional range. Initiatives often deconstruct and reconstruct hegemonic mascu-linities in a cultural context in an effort to undo unhealthy masculinities and replace them with broader versions. Programs that focus on fathering, the equalizing of caregiving and domestic work, healthy relationships, the prevention of domestic violence, shared decision making, and safe sexual practices enact a broader view of masculinity by breaking out of the restrictions of the "Man Box." While many of these initiatives began in low-income countries, they are now appearing outside those regions, facilitating research in assessing men's gender relations more broadly and with wider cultural inputs (Oliffe et al. 2023).

Community education to support and reinforce efforts to widen the understand-ing of masculinity and, more generally, to encourage broader attitudinal changes regarding the status of women, gender equality, and women's and girls' opportuni-ties and freedoms, is often coupled with engaging men and boys. Broad community campaigns can reinforce redefinitions of masculinities and focus on structural shifts in economic equality, prevention of VAW, and erasure of cultural practices such as menstrual isolation huts, FGM/C, dowry practices, or male initiation rituals, among other issues. Community education is an ongoing effort to dismantle the scaffolding that supports gender inequity.

2.3 Engaging Women and Girls

When gender transformative efforts focus on girls and women, they have generally sought to increase their empowerment. Less often, these initiatives have decon-structed and reconstructed power relations or developed critical thinking about fem-ininities, agency, and women's rights. Specifically, gender transformative programs and policies have not critically addressed unhealthy femininities, their roots, their perpetuation, or their impacts on multiple generations of women and girls. Beliefs about gender norms, femaleness, and womanhood have only recently been labeled a "gateway belief system" (TrueChild 2023) underpinning the structural barriers that confine girls and women, even if/when male norms shift, and even if/when women themselves become empowered.

Deconstructing femininities is tricky, as women (particularly grandmothers, aunts, mothers, sisters, peers, teachers) often act as enforcers and upholders of feminine norms, instrumentalizing the patriarchy. Much of this is carried out in the context of compulsory heterosexuality, where attractiveness to men and ultimately heterosexual marriage is seen as representing social success, status, security, and a future. Indeed, in the cases of FGM/C and cosmetic genital surgery, Jacobson et al. (2018) note that women are the "bearers of normal culture," highlighting these perpetuating roles. Despite this, blaming women for transmitting patriarchy is superficial, as they too are victimized by the same inequitable system reflecting intergenerational damage. While critically addressing femininities is essential, it is best done by generalizing its impact on all gender groups.

3 Expanding Gender Transformative Approaches

While some efforts to widen understandings of gender roles and stereotypes have been made, patriarchy will continue unabated if gender transformative programs and policies fail to include a society-wide, feminist approach to unpacking feminine norms as well as masculine norms. The focus on interventions with men and boys in gender transformative work has been critiqued on several fronts, including its reliance on hegemonic masculinity, a lack of intersectional views of men and masculinities, a focus on individual as opposed to structural change, and a lack of a concomitant focus on girls and women (Zielke et al. 2023). Girls and women continue to absorb, exhibit, and experience stereotypical and limiting feminine norms via internalized or intergenerational sexism. Similar to men and all gender groups, they can also act as bystanders involved in these processes, ultimately strengthening a society-wide scaffold for perpetuating acritical femininity.

To weaken the misogynistic drivers and the sexist, homophobic, transphobic, and femmephobic struts that maintain hegemonic masculinity, efforts must expand. Not surprisingly, homophobia and transphobia directed at lesbians, bisexual women, and trans men play directly into this state of affairs, as these identities and presentations creep across the heteronormative, male-oriented line. Hence, I suggest that gender transformative interventions and approaches must develop to engage fully, directly, and sensitively with femininities, encircle all gender identities, and encourage agency in critiquing and transcending hegemonic gender.

3.1 Acknowledging Realities of Sex/Gender

> The differential status of men and women in almost every society is perhaps the most pervasive and entrenched inequity.
> —Sir Michael Marmot (2007:1155)

To understand and disentangle the impacts of sex, gender, and gender inequity in a given context or for an issue, person, or community is complex. It requires scientists and policy makers to appreciate sex and gender science, to incorporate both biological and social evidence, and to address intersectional dynamics, cultural changes, temporality, politics, developmental stages, power, and agency. Translating these understandings into programs and policies that use a gender transformative approach is a challenging practical problem, but one urgently needed if society is to address the widespread and differential sex- and gender-based inequities evident today.

Sex- and gender-related characteristics and factors affect the opportunities, experiences, and access to resources of every individual and group in human society. Typically, sex- and gender-related factors interact to create societal responses to gendered categories of people in the form of attitudes, stereotypes, regulations, laws, treatments, and discrimination. Power relations in a patriarchy transmute through a sex/gender lens in everyday interactions, political decisions, and the granting or assumption of resources, rights, autonomy, or freedom of movement. Similarly, societal presumptions about sex and/or gender groupings evoke integration or marginalization, discrimination or freedom, violence or reverence, respect or dismissal.

Individuals experience the world via these sex/gender dynamics that operate constantly to create and maintain (substantive) gender inequity and/or (formal) gender inequality. As a result, people respond by exhibiting a wide array of degrees of agency and control, ranging from submission to resistance, and collectivities exert varying degrees of political and social agency to effect change. These interactions between the body and society form the heart of sex/gender entanglement and are often manifest in differential opportunities and responses.

Such inequities doggedly persist. A report from the World Economic Forum (2023) on the global gender gap indicates backsliding during the COVID pandemic, resulting in a forecast of 131 years to reach gender parity from 2023, compared to a forecast of 100 years in 2020. Similarly, the United Nations documented similar negative impacts on women's equality during COVID in areas such as housework, labor force participation, violence, poverty, health, and climate impact (UN Women 2023). And lest we think gender inequity and its effects are largely historical problems that do not require fresh analyses, Khan et al. (2023) remind us of global backlash against changing gender norms that includes women and LGBTQ+ groups in areas as diverse as educational curricula, reproductive health, abortion access, access to appropriate gendered health care, women's autonomy, girls' education, and freedom of movement.

These recent data amplify long-standing gender inequities that take place in countries with both weak and strong human rights protections. For example, in the United States, a country with strong human rights and criminal codes, increased numbers of immigrants from FGM/C-practicing countries between 1997 and 2012 resulted in concomitant increases of over 500,000 women/girls estimated to be at risk for FGM/C, reflecting the continuing persistence of this practice (Goldberg et al. 2016). Globally, entrenched gender inequality and sex discrimination, which enables fetal sex selection, selective abortion, and under nourishment of girls, has resulted in over 140 million "missing women" (i.e., females who have been quietly eliminated) and this number continues to rise (Bongaarts and Guilmoto 2015).

The sheer scale of these inequities is staggering, and response requires both increased scientific advocacy as well as more considered policy actions. Without an explicit commitment to the notion of sex/gender entanglement that recognizes such specific patterns, policy and practice decisions are easily made without due consideration of science and data regarding sex/gender inequities. Keeping the implications of inhabiting sexed bodies at the forefront and acknowledging the influence that gendered bodies and the mix of sex/gender has on life experiences must be a fundamental starting point for policy that aims to reflect and respond to sex/gender entanglement. Perhaps this is most critical and obvious in crimes of violence against females and women, female genocide and female apartheid, and reproductive health, where corporeal issues (sex-related realities) and gendered reflections (culture, social realities, politics) combine to determine vulnerability.

3.2 Understanding Dynamics

At root, sex/gender inequalities are based on interactions between sex-related factors, such as anatomy, hormones, physiological processes, and genetics, and gender-related factors, such as identity, norms, relations, and institutional practices (Canadian Institutes of Health Research 2023; Greaves and Hemsing 2020). This dynamic between corporeal and social reality is not fixed, neither is it culturally nor temporally stable. Rather, sex/gendered inequities are dynamic, fluid, and conjoined, perpetuated via institutional practices, expectations, rules, sanctions, laws, and policies. It stands, therefore, that such policies and practices (in media, religion, family, government, education, politics, and health) and evolving approaches (e.g., the use of artificial intelligence) can be key instruments, or repressors, of change.

For ease of understanding and resonance with the public, and for developing policy and programs, sex and gender are often separated and reified to consolidate this range of social and biological factors into ready approximated categories. Depending on the topic, cultural context, and political climate, sexed and gendered categories are used to exert, manifest, or upend sex/gender influences on everyday life and the distribution of power. While this separation can elide sex/gender entanglement and lead to oversimplification, it can also serve as a key entry point for understanding the importance of sex and gender and a first step in educating audiences, scientific or otherwise. The importance of this is not to be underestimated, and continues to form the basis of many trainings, resources, or guidelines in research and policy arenas.

Categorizations are reinforced by social institutions via formal policy and practices that are supported in turn by regulatory practices, laws, guidelines, and standards, communications messages, funding, survey techniques, benefit schemes, and affirmative actions. They are also supported by less formal practices of gender policing, socialization processes, and social and community dynamics, such as pressures to conform with gendered stereotypes, and stigmatization of those who do not. As a result, sex/gender categories are understood and experienced by the majority of the population as binary in an overriding heteronormative, patriarchal context. In addition, the dynamics surrounding these categories are also absorbed by individuals in *all* sex/gender categories, resulting in differing manifestations of identity formation and personal and collective agency. Unraveling the elements of such multiple sex/gender components, processes, and forces is the first task in forming gender transformative programs and policies that reflect the entanglements of sex/gender at their core.

3.3 The Body and Society

The body matters. All bodies, whatever their categorical labels (female, male, inter-sex) react to their genetics, reproductive and hormonal physiology and anatomies. These responses shift and change through developmental stages, modifications, and in response to a range of endogenous and exogenous factors and environments. Sex-related factors are shared, nuanced and do not manifest in a strictly binary pattern (Ritz and Greaves 2022). While some contemporary proponents of inclusive politi-cal and human rights for all genders prefer to erase or diminish the importance of sex, and/or infer that it is a questionable and elusive concept, some developmental and evolutionary biologists take a different position that emphasizes the realities of sexual reproduction as its base (Hilton and Wright 2024). More useful for in-tegrating sex/gender into science, Richardson (2022) suggests an approach called sex contextualism, which is focused on naming sex-related factors that matter in situation-dependent ways and are represented in more than binary essentialist ways. While this perspective recognizes that sex-related factors are networked into major bodily systems beyond reproduction, their very conceptualizations should rely on the purposes of the research into which they are being integrated.

Because sex-related factors shift, and sex and gender science evolves the com-plexities, overlaps, and anomalies of sex, its importance is not lessened, or its im-pact less real. The scientific challenges in categorically defining sex do not render sex as a concept soluble or indicate that concretizing it is reductionist or simplisti-cally supportive of a binary. Rather, evolving questions and queries about sex and gender indicate the need for more precise science and policy action and less confla-tion. Sex and gender impacts are anchored in both the bodily and social spheres, as it is largely societal responses to bodies (such as sex/gender assignment typically based on genital observation at birth), that create or deny opportunities to groups of people. Indeed, the concept of sex/gender entanglement recognizes this conjoined relationship, as should science and policy making.

Gender transformative approaches need to deliberately integrate the bio/social moving axis as a source of both drivers of, and solutions to, problems. Feminist, queer, and dis/ability studies have been at the forefront in emphasizing this dy-namic. For example, the lived experience of being dis/abled often involves transit-ing from abled to dis/abled in the course of a life. But disability is ultimately the re-sult of social and political impositions interacting with bodies, that produce or limit freedom or equity (Enke 2012). Similarly, gender identity is formed through varied impositions, assumptions, and experiences of social, cultural, and political life in-teracting with sexed bodies to create self. These include responses to hegemonic gendered norms, roles and expectations that are routinely assigned to apparently sexed bodies. While "felt identity" is often invoked by transgender individuals, the formation of body image and gender identity is universally experienced and itera-tive, with no guarantees of fit between sex/gender for self-described women or men, or *any* individual, for that matter. In light of this, Salamon argues that the lack of fit between felt sense of body and appearance of the body experienced by some trans people is, in fact, pervasive to all and inherent to the formation of normative gender identity, not a pathological process (Salamon 2010).

A reoccurring theme of centuries of feminist activism has been the questioning of the gendered assumptions and norms linked to female sexed bodies, many of which have hinged on the exploitation of female reproductive labor. Resisting such linkages has led to the loosening of the social equation of sex = gender and has underpinned continued political action to reframe gendered expectations, stereotypes, and categories, and the policies and practices that rest on them. This political action does not rest on erasing gender, but rather on loosening its social and political grip on the links to sex-related characteristics. This action requires detailed examination of both sex and gender as concepts in all of their manifestations, fluid and temporal, many of which are yet to be discovered.

The fluidity and context dependence of the relationship between the corporeal body and dynamic gender identity formation is highlighted in a study of Somali women in Toronto who had experienced FGM/C prior to migration (Jacobson et al. 2018). In discussing their bodies, pain, and everyday life, they sought to integrate their bodily experiences and structures within the sociocultural milieu in which they resided. The researchers reflect on this challenge, noting that "in the case of our participants, the juxtaposition of what was normal in Somalia with what is normal in Canada suggested a heartfelt struggle with their identity and trying to put the two pieces of their lives together" (Jacobson et al. 2018:18). Beyond the social-corporeal interaction, and possible sociosomatic pain, the authors invoke a link to wider bodily systems and neurological responses. Einstein (2008) notes that body, mind, and brain are affected by FGM/C, requiring further thinking about FGM/C-related rearrangements of neural networks, pain experiences, and chronic pain.

Schall and Moses (2023) argue that all genders can need or want (non-cosmetic) gender-affirming care using similar rationales, invoking examples of breast reconstruction or testicular implants. Their stance illustrates the generalizability of gender-affirming care and its established existence in the health-care system. Generalizing concepts such as gender identity formation and gender-affirming care is important to illustrate their applicability across populations and to broaden public understanding of the dynamics associated with these important aspects of gender. Such actions will also serve to avoid the frequent error made by many that *gender = gender identity*, eclipsing a host of other gendered processes such as norms, roles, institutional expressions of gender, discriminatory practices, stereotypes, and behaviors.

Addressing and integrating such sex/gender entanglements has a direct impact on policy and practice. Corporeal sex-specific genitals and their overlap with gender identity also feature in assessing the social and internal pressures that women face when choosing cosmetic labiaplasty as well as when gender-transitioning individuals make decisions about genital surgery (Walsh and Einstein 2020). Examples of FGM/C and patient-initiated genital surgery clearly illustrate how sex-based materiality interacts with social and cultural contexts to produce lived experiences. Such examples of sex/gender entanglement could be more consistently and widely addressed with an explicit view to adopting a gender transformative approach to policy and practice changes.

3.4 The Invisible Scaffold

More hidden, but as important, are the social processes that act as scaffolding for upholding sex/gendered inequities and inequalities. The perpetuation of gender inequity in popular culture reflects a deliberate adoption of the "right not to know," exercised by those with privilege who can ignore, trivialize, neutralize, deny, or ignore gender inequity (Feldthusen 1990). Among researchers these dynamics manifest in designing studies without using "we don't know what we don't know" as a starting point. Among policy makers, this "right" manifests as seeing critical thinking about equity as creating "lens fatigue" that hampers their work, thereby ditching the responsibility to deeply engage with inequities as outcomes of their work.

These forms of being willfully ignorant about blended sexed and gendered realities and impacts continue to delay progress, hide issues, and cost lives. Among the rest of society, the scaffold is strengthened by gender policing, intergenerational transmission of gendered stereotypes, and micro processes that centralize and hoard knowledge and power. Power issues are central to gender transformative approaches, especially when power is viewed as a binary (either won/lost), as opposed to a shared resource. Acknowledging power relations as underpinning gender inequity is a key step in developing sensitive gender transformative solutions (Ponic et al. 2014).

The invisible scaffold has directly facilitated morbidity and mortality of females and women due to centuries of ignoring the linked global epidemics of femicide, incest, rape in marriage, rape in war, intimate partner violence, sexual assault and harassment, acid attacks, stoning, dowry deaths, suttee, foot-binding, female genital mutilation, fetal sex selection, schoolgirl poisonings, and obstetrical violence (see Daly 1978). Indeed, these invisible processes affect all gender groups, especially those who reject and resist hegemonic gender norms. This is evident in the category of gender-based violence that encompasses transgender individuals and sexual minority and gender minority groups in almost any given cultural context. The roles of gender policing and willful ignorance are crucial in perpetuating sex/gender inequity. It stands, therefore, that its redress requires focused consciousness raising, (re) education and power sharing—key components of feminism, and ideally, of policy change. Hence, the attractiveness of seeking gender transformative approaches.

3.5 (Re)commitment to Feminism

The concepts of empowering women, challenging institutional sexism, and identifying structural causes of gender inequality are hallmarks of global feminist movements. The recognition of domestic labor and caregiving, valuation of reproductive labor, securing voting rights, abortion rights and reproductive freedom, ensuring property and land rights, resisting patriarchal and state control over women's bodies and movements, erasing VAW, preventing femicide, and arguing for affirmative action are but some of the ongoing global issues on the feminist menu. However, the persistence of women's inequality as a wicked problem illustrates that the resolution of such issues involves multiple, ongoing challenges that can be easily superseded in the political sphere by newer conversations about inequality.

Hence, (re)conceptualizing these issues requires an ongoing "consciousness raising" process that braids individual experiences into a collective story. While commonalities are shared, this process also can surface fault lines among groups of women with differing power (e.g., lesbians/heterosexual women, Black/White women, working class women/professional women, young/old). Globally, feminists have identified the augmentations of sex/gender oppression with racism, sexism, ageism, classism, and ableism among others, and a range of identity factors including race, gender identity, Indigeneity, ethnicity, and culture.

Despite wide differences among women, various contemporary equity-driven remedies have been achieved: affirmative access to STEMM, free menstrual products, restitution for historical gendered pay inequity, and reform of divorce laws. Reparation agreements are accelerating but so far include sexual slavery (i.e., Korean "comfort women"), impacts of residential schools (i.e., Canada's agreement with Indigenous survivors), and forced incarceration of mothers and child murder and relinquishment (i.e., Irish Mother and Baby Homes), as well as many agreements of financial payments to groups of sexual assault survivors. Such issues and campaigns identify and redress intersecting sex/gender discriminatory practices and create solutions aimed at power sharing and public accountability, fundamentally achieving wider social justice. Many of these reparation agreements depend on policy decisions that acknowledge sex/gender entanglements.

These understandings set the table for naming underlying drivers and developing gender transformative solutions. Addressing root causes is challenging and requires creative and courageous thinking, as well as support from a range of social justice movements. It requires scientists and policy makers to connect with all forms of evidence, wisdom, lived experience, and public opinion as social facts, not as either progressive or regressive. Such thinking has typically focused on changing gender norms affecting women and men but these processes and analyses offer expansionary room for all gender groups. Gender transformative thinking can and should be expanded to generate specific trans- and gender-diverse versions of gender transformative analyses and solutions, based on specific trans- and gender-diverse consciousness raising.

This expansive process will require broad and more open thinking and conceptualization of both sex and gender and their entanglements, starting with genuine efforts to rise above the clashes, tensions, and conflicts that have been glibly manufactured between trans and radical feminist activisms, or left- and right-wing approaches to sex/gender-related policies. This openness would not rely on the use of un-interrogated categories and terms such as "TERF" (Worthen 2022) and "woke" (Steel 2024) that has fed their pejoration in order to make points and distinguish arguments. Nor does it rely on under-interrogated divisional binary adjectives, such as trans and cis, to denote the relationships between gender and sex that ignore the fluid nature of gender identities.

In their multifaceted transfeminist critique of such terms, Enke (2012:76) states:

> Whatever else it may accomplish, cisgender forces transgender to "come out" over and over through an ever-narrower set of narrative and visual signifiers. This erases gender variance and diversity among everyone while dangerously extending the practical reach and power of normativity."

In short, despite the politicization of gender identity in modern discourse, it is important that more scientists and disciplines address sex/gender research and sex/gender entanglements with nuance, frequency, and clarity (Ritz and Greaves 2024).

Indeed, the powers and limitations of terms to enforce division and (often) binary understandings of sex/gender and their entanglements must be continuously interrogated in the interests of scientific and intellectual progress. This means engaging with science in terms of the intricacies of sex-related factors (e.g., brain, reproductive, or bodily functions) and sex as a legal foundation for discrimination or rights, along with gender as a concept that is focused on power and persisting and divisive norms and inequities as well as gender identity and fluidity of culture, time, and categories (Hilton and Wright 2024).

For science, it means critical and open engagement with the interactions between all of these concepts on an evolving tableau of human life and encouraging the flourishing of all inherent questions. Sadly, this remains an ideal, as sex and gender are still under interrogated in most science, and sex/gender entanglement is even less often conceptualized and discussed. Absent such openness, curiosity, and engagement among scientists with the important issues surrounding sex/gender entanglement, the scope and direction of gender transformative approaches and solutions is unlikely to expand.

4 Encouraging Individual and Collective Agency

To activate these changes, a full examination of the role of agency in responding to gender socialization and reacting to gender inequity is required (and long overdue). We can assume that sex/gender role socialization practices transmute via social institutions to recreate and perpetuate masculinities, femininities, and oppressive patriarchal practices. But that is far from the whole story. How are such attitudes, behaviors, and practices differentially received by both individuals and collectivities of all genders? How does intersectionality impact these teachings and receptivities? How are such teachings adopted or adapted or resisted, individually or collectively? Gender transformative theory and practice have not usually examined these questions in much detail and have typically assumed linear paths for binary-based, hegemonic audiences in their absorption of socialization.

These queries also impact the processes of identity formation. As Nordmarken (2023) explains, there are numerous routes to trans, nonbinary, and other gender identities that involve interactivity and numerous forms of agency. These diminish emphasis on a process of recognizing an innate or felt identity, and open up processes of gender development that apply to all genders. "Although the traditional 'doing gender' process and its epistemic foundations persist, the coming-into-identity process operates simultaneously, as an epistemic challenge that social actors of all genders can participate in" (Nordmarken 2023:613).

Agency needs more exploration as it directly relates to developing various program and policy positions that could be taken to achieve gender transformative change. Agency is dependent on the interaction between the body, personal power, and structural constraints or opportunities, and its manifestation is therefore entangled across all populations. Exploring agency could delineate a range of audiences living in different states of readiness and motivation for change in varied psycho-social-political and economic contexts that could be used to devise targeted gender transformative solutions. This will implicate all gender and sex groups, majority and minority, as the perpetuation of femininities and masculinities is enacted and reinforced across society by all genders, even implicitly, by nonbinary individuals. Agency involved in expressing gender or sexuality is, of course, dependent on context, political freedom, and social support, but collective program and policy decisions can reinforce, reify, critique, or erase femininities or masculinities.

For example, sex/gender expressions (e.g., butch and femme, transmasculine or transfeminine) and nonbinary gender expressions depend and rely upon widespread, central assumptions about femininities and masculinities, but these assumptions are, at least in the popular realm, rarely explicitly confronted. Addressing crosscutting issues, such as femmephobia among sexual and gender minorities, and its relationship to misogyny—"the law enforcement branch of a patriarchal order" (Manne 2017)—highlights these links. Femmephobia, referred to as the "devaluation and regulation of femininity," could be an intersectional axis linked to other forms of oppression, as femmephobia invades the gender binary, creating hierarchies within it (Hoskin 2020). Misogyny acts to police and govern the norms and expectations of the patriarchy, whereas sexism offers the ideology that rationalizes and justifies patriarchal relations (Manne 2017:79).

These interconnected processes affect all genders, albeit in differential ways. Hence, this type of thinking is essential to naming and working out collective and individual responsibilities and agency in changing gender inequities across all gender groups. Indeed, highlighting more critical thinking about femininities and femmephobia encircles a range of gendered inequities, such as those harming men and boys who are deemed inappropriately feminine, as well as gender policing practices that repress any resistance to hegemonic masculinist norms. The broadening of a discussion about agency across all gender groups is overdue and crucial to addressing sex/gender entanglement in policy and practice. Such a discussion will support linkages to shared drivers, common critiques and challenges in addressing the unhealthy aspects of femininities and masculinities that underpin all gender groups' experience.

4.1 Manifestations of Agency Regarding Hegemonic Gender

How does agency manifest with respect to gender? Table 14.1 begins to parse out examples of a range of responses to prevailing practices of men/boys and women/girls. I have labeled these responses from acceptance to rebellion, reflecting the mix of personal and structural power. These responses can uphold or erase gendered socialization practices, stereotyping, and the prolonging of sexist, homophobic, transphobic "humor" and media content. They can erase or challenge dress codes, bodily rites of passage, or sex- and gender-specific laws and policies. Ultimately these agency stances can influence stigmatization and marginalization as well as shunning, isolation, and segregation.

Table 14.1 Responses to hegemonic, culturally contextualized, gendered practices.

	Hegemonic Masculinities	Hegemonic Femininities
Acceptance	Act tough Exert authoritative, entitled, self-centered male privilege Oppress women and sexual and gender minorities	Nurturing, servile actions Internalize sexism Self-sacrificing, other centered, obedient Focus on presenting self to the male gaze
Adoption	Collect benefits of male privilege Sustain gender apartheid Operationalize sexism	Respond to male privilege Sustain gender apartheid Engage in practices that support male privilege and structural inequality Rationalize benefits of continued gender apartheid
Resignation	Quietly conform to observed masculinities Does not interfere when male oppression or violence is observed	Conform to gendered codes of expression, dress, occupation, and aspiration Silent on gender apartheid Avoid solidarity with women
Resistance	Identify as feminist men Call out male privilege Give up privilege to women Engage in equalizing nurturing and care work	Nonconform to gender hegemonies Support androgynous socialization of children Refuse stereotypical gender presentation of self Does not perpetuate rites of passage
Rebellion	Reject unhealthy masculinities and heteronormativity Engage in activism to address hegemonic male roles, structural oppression, and VAW and sexual and gender minorities	Engage in collective action to advance feminism and change structural oppression Connect oppressive systems Address global gender inequalities and micro-gendered assumptions

Although girls and women typically experience the most frequent and negative impacts of patriarchy-based power due to deep-rooted misogyny and femmephobia, this does not necessarily limit the range of responses among women and girls. Femininities are all too often understood and enacted in relation to hegemonic masculinity, not as qualitatively distinct. In Table 14.1, resistance and rebellion incorporate what Schippers (2007) termed "pariah femininities," thus setting the stage for carving out new ways of being women (and presumably feminist activists).

This examination of responses in the context of binarized sex/gendered normativity raises the question of what does agency around gender norms and inequity look like for sexual minority, gender minority, and gender-diverse individuals? It could be assumed that sexual and gender minorities, by definition, have entered into a form of resistance and/or rebellion with respect to standard binary sexual norms and gender identities. However, it cannot be assumed that members of sexual or gender minorities have analyzed gender inequity or embraced gender transformative thinking. It also cannot be assumed that they have called out misogyny or made connections to feminist goals.

Agency may be viewed from a different starting point, to see how stereotypical gender expression and norms are manifested within sexual and gender minorities, using a sex/gender lens. Among sexual minorities, for example, gendered behaviors, attitudes, and expressions could be viewed as resisting hegemonic gender expression as well as resisting heteronormativity.

Among gay men there can be a continuum of responses ranging from being closeted, engaging in MSM behavior while in heterosexual relationships, "passing" as heterosexual, to outwardly advocating for structural changes regarding heteronormativity, misogyny, and gender inequity. In between, expressions and behaviors may range from conscious adoption of masculine expressions or feminine expressions, differentiating between "masculine" and "feminine" sexual roles, developing a gay male esthete, resisting and confronting femmephobia, or developing an explicitly feminist activist stand.

Similarly gendered behaviors and attitudes of lesbians can be observed, ranging from remaining closeted within heterosexual marriages, to "passing" while living out feminine/masculine roles in private relationships, to the adoption of feminine or masculine expression, to developing a gender-neutral presentation or a new esthete. Approaches to inequities can include calling out sexism and misogyny or developing lesbian feminist activism to reduce structural inequities for lesbians, the erasure of misogyny in LGBT+ movements, and/or leadership in feminist actions to achieve a wider women-focused feminist political agenda benefiting all women.

Gender minority individuals can also exercise agency on different parts of a continuum: from conformity to stereotypical femininity or masculinity, to critiquing gendered assumptions, to engaging in feminist activism aimed at eroding misogyny and gender inequities. Trans individuals potentially have insights derived from experiencing the processes of rebuilding or repositioning their identities and varied social responses to their progress, all of which could inform a gender transformative understanding of inequity. Similarly, nonbinary individuals have insights arising from the experience of defying or selectively adopting binary hegemonic gender identities, and enacting resistance via gender expression or language, presumably also experiencing a range of societal responses. Harvesting such insights and experiences will be a key component in developing a more lateralized gender transformative programming and policy that reflects sex/gender entanglements.

However, if a lateralized gender transformative lens is not understood or applied, gender minorities could also perpetuate and affirm hegemonic and oppressive binary femininities and masculinities by conforming rigidly to stereotypical dress codes, aspirations, and expressions of their gender identity. If the intention is one of "passing" as a person who is not of trans experience, for example, or minimizing adverse societal responses, these adoptions may be personally and individually functional, but they do not necessarily operate to challenge hegemonic gender norms in ways that are gender transformative. Rather, gender transformative agency for all genders would mean actively working to dismantle hegemonic masculinities and femininities and diluting the strength of gender inequity by universally questioning and critiquing gendered assumptions, stereotypes, programs, and policies, *regardless of one's gender identity*. For example, trans men who adopt feminist analyses could ultimately dilute the strength of structural gender inequities by questioning, analyzing, and rejecting male privilege. Trans women could engage in the disassembly of structural gender apartheid and oppression by engaging in feminist work to neutralize the limits of gendered socialization, offering new perspectives and solutions to issues such as VAW and femicide, and actively supporting a feminist agenda for change. These expansionary approaches of understanding agency and hegemonic gender would deepen our understandings of gender and share responsibilities for reducing gender inequities more broadly.

4.2 Agency, Audience, and Purpose in a Feminist Context

If understandings of agency, audiences, and purposes of gender transformative programs and policies are explicitly widened, it becomes possible to develop crosscutting policy and programming to focus on a wider range of goals, factors, and conditions in their design. For example, if feminine norms were engaged with as critically as masculine norms have been in gender transformative programming, then solutions to dismantling the structures of female oppression could become more widely shared, understood, and applied across genders. If audiences other than men and boys were addressed in gender transformative approaches, a wider tableau emerges for reconstructing change.

Indeed, if all gender groups are engaged in critically understanding the drivers of gender inequity from their own context and positionality, then two things could happen. First, a more lateral understanding of sex/gender entanglements and their impacts on gender transformative programs and policies could emerge. Then, more targeted manifestations of gender transformative approaches could be built with the development of specific causes and solutions to gender inequities facing trans, gender-diverse, and nonbinary groups.

> Whereas misogyny upholds the social norms of patriarchies by patrolling and policing them, sexism serves to *justify* these norms, largely via an ideology of supposedly natural differences between men and women with respect to their talents, interests, proclivities, and appetites.
> —Kate Manne (2016)

One universal (age-old) driver is misogyny. As Loewen Walker argues, it is fundamental to understand its power to police not only women, but gay men, and trans women, by "enforcing intersecting systems of white supremacy, heteronormativity and cisnormativity" (Loewen Walker 2022:66). Expanding gender transformative solutions can be broadened by grappling with and naming fundamental historical drivers such as this. Were these expansions of gender transformative thinking adopted, the range of responses in policy and practice would be strengthened by a decidedly universal relevance. In short, gender transformative thinking would be progressed for all gender and sexual groups. In Table 14.2, I propose some programmatic and policy examples of this expansion and provide examples of men, women, sexual and gender minorities, and possible individual, program, and policy actions at micro, meso, and macro levels.

4.3 Generalizability of Gender Transformative Approaches

The notion of deriving root solutions to address root causes, though radical, is not new. It can be applied to numerous problems: The solution to poverty is money.

Table 14.2 Gender transformative programs and policies by audience and level.

Audience	Micro Level	Meso Level	Macro Level
Men and boys	Assess the limitations of masculinity and deconstructing masculinities (e.g., The Man Box; Equimundo 2023)	Shift sex/gendered experiences of masculinity within institutions to enable healthy relationship education, reproductive health planning, fathering and emotional expression, and shared housework and caregiving	Shift policies to change paternity leave and caregiving policies, provide affirmative action in education and access to occupations, ensure pay equity, and criminalize VAW
Women and girls	Assess the limitations of femininity (develop a "Woman Box"); deconstruct femininity and the "stereotypical Barbie" (Gerwig 2023)	Shift sex/gendered experiences of femininity within institutions: self-defense, occupational planning, body positivity, engage with women who perpetuate discriminatory practices, address internalized transgenerational sexism, critically analyze Western female body practices (e.g., cosmetic surgery)	Shift policies to ensure mandatory school attendance; eliminate child marriage; provide free menstruation products, WASH initiatives, paid maternity leave, STEMM education, free contraception and abortion; ensure access to leadership in religious institutions, media representation, sexist advertising; eradicate FGM/C, limits on movement, access to money and loans
Sexual minority groups	Assess the limitations and stereotyping of masculinity and femininity for selves and others	Rethink gender stereotyping of and by sexual minorities to critique and unlearn femmephobia, to address misogyny between and within sexual minorities, to promote the understanding of LGB issues in a feminist context, to link dynamics of bullying of gay males to victimization of women, and to introduce feminist approaches to gender inequity	Shift structural discrimination aimed at sexual minorities, tax, marriage and family formation policies. Address gendered discrimination by and among sexual minorities. Develop a feminist framework for policies addressing sexual minority issues (Friedman and Ayres 2013). Link victimization of women and girls to sexual minority boys and men (Dunn et al. 2017)
Gender minority and gender-diverse groups	Assess the limitations of masculinity and femininity as well as gender stereotypes in expression, behavior, and aspirations for selves and others	Rethink limitations of gender identity and its structures. Address processes that keep gender apartheid alive. Decouple gender identity from stereotypical gender expressions, femininities and masculinities, and gender relations from a trans-feminist perspective	Engage with feminisms and transfeminisms (Bettcher 2020). Shift policies and politics to support questioning and erasure of negative gender stereotypes. Develop new socialization approaches, health-care changes. Media representation; political representation

Solutions to dis/ability based-poverty is a guaranteed income. The solution to homelessness is housing (Ryan 1976). The solution to poor health is free, universal health care. In an analysis which focused on race and class (but unfortunately not gender, misogyny, or sexism) in the United States during the 1960s, Ryan (1976) found that domestic institutions and policies enunciated processes of ideological tautologies that kept low-income Black people underhoused, undereducated, and unhealthy. His recommendations were to move away from "blaming the victim" to universalism and structural change, an anathema to the politics of the time. While Ryan exhibited willful ignorance by ignoring sex, gender, and sexism, even when focusing on women-led Black families—a key US policy preoccupation of the era—his radical approach is linked to gender transformative thinking.

Sometimes the solutions are less linear. The solution to high rates of obstetric fistula in Sierra Leone turns out to be keeping girls in school and ending child marriage (Greaves and Poole 2018). To end FGM/C, the solution lies in challenging women's roles in informal intergenerational enforcement. The solution to racialized health inequity is to materially change the social and structural determinants of health and confront racism, colonialism, and discrimination. But these solutions need careful framing. Marmot and colleagues' approach to the distribution of resources provides an important guide for expanding gender transformative programs and policies:

> Proportionate universalism refers to the concept that people across the whole population gradient are entitled to social benefits proportionate to their needs. For policy, it encompasses both targeted and universal approaches to ensure the population as a whole is proportionately allocated benefits and services (Marmot and Bell 2012).

When designing gender transformative approaches in policy and programming, proportionate universalism that allocates energy and resources in proportion to the size of the group and level of disadvantage and inequity is critically important. It allows for the conceptualization of differentially targeted programs, policies, and messages to create specific solutions for various sex/gender (sub)groups who experience gender inequity in unique but shared ways, according to need, scope, numbers, and impact. It allows for remediation at scale for the impacts of historical and ongoing gender inequity. Blending sex/gender and feminist analyses provides direction for both targeted and universal gender transformative initiatives by identifying similarities and differences in drivers, scope, and need. This guiding principle assists in remembering the history and intractability of wicked problems as a component of current responses and facilitates linking these old problems to new emerging ones.

5 Conclusion

The purposes of gender transformative programs and policies are to recalibrate the distribution of health, social, and economic resources, to increase gender equity, and to rebalance individual and collective agency. As overwhelming data indicate, the persistence of global gender inequity facing women and girls reveals deep-rooted misogynistic, anti-women, anti-feminist, anti-female, and femmephobic dynamics which have, to date, resisted erasure. In the spirit of proportionate universalism, gender transformative practice, policy, and theorizing must engage in a targeted manner with a focus on reducing the most egregious, widespread, and frequent manifestations of patriarchy and gender apartheid, while applying intersectional feminist principles to assess the situations and agency of all gender groups.

Historically, most current gender transformative work has engaged undifferentiated groups of men and boys as key players in reducing gender inequality and inequity. This has usually incorporated new ways of being men/boys as well as naming potential benefits from undoing hegemonic masculinities. This is an apt first-line response, but they are not the only group that needs to be considered. Turning a critical engagement lens to women, girls, and sexual and gender minority groups is clearly overdue. In a (still) patriarchal society, with entrenched gender norms and strong gender policing, all women and girls and trans women, trans men, masculine women, feminine men, gay, lesbian, bisexual, nonbinary, and gender-diverse individuals need to understand and address their specific roles in upholding gender apartheid and gender inequity. The behaviors, beliefs, and stereotypes they maintain as well as their roles in upholding the invisible scaffold of gender inequity are important in identifying all gendered drivers.

The continuum of degrees of agency that people and groups can adopt to survive, acquiesce, or resist will continuously shift and change, and may involve a mix of psychological, biological, social, economic, and political factors. Distinguishing agency as constantly interacting with structure, resulting in perceived or actual power, is essential. These distinctions assist in understanding, assessing, and fostering the capacity for action without reverting to tired "victim-blaming" impulses.

A hallmark of gender transformative thinking is understanding feminism, critical thinking, and root causes, and using creativity to launch programs and design policies. All health inequity and social justice frameworks must improve their understanding and integration of sex/gender entanglements to support a platform of both crosscutting and differential policy and programming in search of gender equity. While this is not yet standard practice, making clear distinctions between sex and gender science, sex and gender-based analysis+, and research and policy can only help in mapping future actions in gender transformative thinking (Greaves and Ritz 2022). Such thinking must be sex inclusive and dynamic to change subjective and objective experiences of the corporeal throughout the lifespan for all genders, and that iterate constantly to produce shifting lived experiences and social and health consequences. To think otherwise is to ignore the entanglement of sex/gender and its scientific challenges.

This chapter has evolved gender transformative approaches to programs and policies by elaborating the roles of sex/gender, feminism, the body, and agency in gender transformative work. I have argued for both an extension and lateralization of gender transformative thinking to encompass all genders. This will give rise to discussions about how women and men, in all sexual and gender majority and minority groups participate in upholding and dismantling gender norms or gender inequity. It offers some bases for thinking laterally, for attributing agency to all human beings in all groups, and for challenging all people to sort out gender inequity and its root causes. In the context of a widespread global backlash to feminist advances favoring women's autonomy, changing gender norms, gender diversity,

LGBT+ initiatives, reproductive freedom, barrier-free health care, sex- and gender-specific care, and abortion (Khan et al. 2023), it is crucial that program and policy initiatives be viewed through an explicit, sex- and gender-informed, intersectional gender transformative lens.

In closing, the following examples are provided of gender transformative initiatives needed at different policy levels: funding, regulations, guidelines, and communications. Ideally, for gender transformative policies to be effective, program supports that offer training, education, community development, public education, and linked campaigns are also essential.

5.1 Funding: Sport Policies

How could gender transformative thinking help lessen the impact of unequal funding policies on sex/gender issues in sport?

Soccer injuries to women's knees result from both biological anterior cruciate ligament (ACL) structures, hormones, and anatomy as well as gendered sport policies that result in inadequate funding for trainers, training, and proper playing surfaces (CBC News 2023). Program interventions have focused on training preadolescent female athletes in landing and improving core strength to address aspects of this problem (Taghizadeh Kerman et al. 2023). However, gendered institutional sports policies are responsible for inequitable and unequal funding, pay, and media support, and this translates into inadequate training time and team-based treatment. Sport policies represent a nexus of sex/gender issues that evolve over the lifespan and developmental stages as well as corporeal experiences that are sex specific and impacted by gender and changes in gender identity, among other factors. Sport policies can be scrutinized at local to international levels, from minor sports to professional to Olympic, and are fluid, complex, and contentious. Gender transformative approaches that focus on refinancing, equal media representation, playing time, and sponsorships would address these drivers.

5.2 Regulations: Knowledge Translation for Medical Devices

How could gender transformative policies help lessen the impact of sex/gender blind knowledge translation regarding medical devices?

Breast implants are medical devices used for cosmetic breast augmentation (80%) and reconstruction (20%) (Pederson and Tweed 2004). Their approval and licensing is regulated rigorously in high-income countries but has still fallen short and resulted in scandals, catastrophic illness, death, implant-related cancer, and class action suits after faults and breakdowns in using these devices (Keith et al. 2017; Lampert et al. 2012). This is due to sex/gender blind regulatory processes, lack of sex and gender science, lack of funding for women's health research, and lack of sex/gender segregated post-market vigilance data, tracking, and reporting. Consumer and clinician monographs do not reflect complete risk analyses and adequate warnings. Gender transformative regulatory efforts would include development of knowledge products that directly address motivations and desires across gender groups for augmentation and reconstruction. Information and remedies linked to body image and gender identity issues, feminine aspirations, pornography, media representations of female bodies, cosmetic surgery industries, and gender stereotyping of female beauty could lessen risk by reducing demand.

5.3 Guidelines: Alcohol Consumption

How could gender transformative thinking lessen the negative impact of sex/gender blind, low-risk alcohol drinking guidelines?

Alcohol affects female bodies more negatively than male bodies at lesser amounts and facilitates gendered, negative social impacts, such as sexual assault, intensification of intimate partner violence, aggression, fighting, and crime (Greaves et al. 2022). Low-risk drinking guidelines in many countries do not account for these impacts; rather, they focus on general risk profiles, without identifying sex/gender components (Greaves et al. 2022). Evidence suggests that high-risk drinking by men is shaped and fueled in part by adoption of hegemonic masculinities, sexism and homophobia, and the rejection of femininity, femaleness, and homosexuality. Gender transformative guidelines could reduce high-risk drinking with messages to all gender groups about biological propensities and impacts (male/female) of ingesting alcohol across developmental stages (aging, brain development, hormonal changes), and undermining negative masculinities (toughness, ability to hold liquor, aggression) that impact women and gay men.

5.4 Communications: Violence Prevention

How can sex/gender-informed gender transformative thinking be used in messages about the prevention of violence against women and girls (VAWG)?

VAWG is a global pandemic, resulting in female infanticide, sex-selected abortion, sexual assault, gang rape, intimate partner violence, and femicide. The UN estimates that one-third of women experience some form of violence in their lives. Messaging about the prevention of violence could focus on the drivers, such as gender stereotyping, gender role socialization of children, condoning of VAW, and sexist humor. Solutions include equal rights and pay as well as legislation against restrictions on women's movement, autonomy, and decision making (OurWatch 2021). Messages and programming must address these issues in ways that reflect sex-based vulnerabilities (e.g., smaller stature, less strength, reproductive potential) as well as changes to create male power sharing, reduction of negative peer influences, and respect for women. On a societal level, regulations to prevent sexist advertising, unequal pay, and lack of access to professions and institutions can help shift the overarching balance of power. VAWG is a subset of gender-based violence, which encompasses broader misogynistic violence against trans women, gay men, and those seen as inappropriately feminine or transgressing sex/gender categorization.

Acknowledgments The author acknowledges helpful comments from Nancy Poole and Stacey Ritz on an earlier draft and the technical assistance of Andreea Catalina Brabete.

References

Bettcher T (2020) Feminist Perspectives on Trans Issues. https://plato.stanford.edu/archives/fall2020/entries/feminism-trans/

Bongaarts J, and Guilmoto C (2015) How Many More Missing Women? Lancet 386 (9992):427. https://doi.org/10.1016/S0140-6736(15)61439-8

Canadian Institutes of Health Research (2023) What Is Gender? What Is Sex? https://cihr-irsc.gc.ca/e/48642.html (accessed Nov. 4, 2024)

CBC News (2023) Knee Injuries Plague Women Soccer Stars. https://www.cbc.ca/player/play/2248187971638 (accessed Nov. 4, 2024)

Daly M (1978) Gyn/Ecology: The Metaethics of Radical Feminism. Beacon Press, Boston

Dunn HK, Clark MA, and Pearlman DN (2017) The Relationship between Sexual History, Bullying Victimization, and Poor Mental Health Outcomes among Heterosexual and Sexual Minority High School Students: A Feminist Perspective. J Interpers Violence 32 (22):3497–3519. https://doi.org/10.1177/0886260515599658

Einstein G (2008) From Body to Brain: Considering the Neurobiological Effects of Female Genital Cutting. Perspect Biol Med 51 (1):84–97. https://doi.org/10.1353/pbm.2008.0012

Enke AF (2012) The Education of Little Cis: Cisgender and the Discipline of Opposing Bodies. In: Transfeminist Perspectives in and Beyond Transgender and Gender Studies. Temple Univ Press, Philadelphia, pp 60–77

Equimundo (2023) Promoting Nurturing, Equitable, Nonviolent Masculinity since 2011. https://www.equimundo.org/ (accessed Nov. 4, 2024)

Feldthusen B (1990) The Gender Wars: "Where the Boys Are." Can J Women Law 4 (1):66–95. https://doi.org/10.51644/9780889208605-012

Friedman CK, and Ayres M (2013) Predictors of Feminist Activism among Sexual-Minority and Heterosexual College Women. J Homosex 60 (12):1726–1744. https://doi.org/10.1080/00918369.2013.824335

Gerwig G, dir. (2023) Barbie. Warner Bros Pictures and Mattel., Burbank

Goldberg H, Stupp P, Okoroh E, et al. (2016) Female Genital Mutilation/Cutting in the United States: Updated Estimates of Women and Girls at Risk, 2012. Public Health Rep 131 (2):340–347. https://doi.org/10.1177/003335491613100218

Greaves L (2014) Can Tobacco Control Be Transformative? Reducing Gender Inequity and Tobacco Use among Vulnerable Populations. Int J Environ Res Public Health 11 (1):792–803. https://doi.org/10.3390/ijerph110100792

Greaves L, and Hemsing N (2020) Sex and Gender Interactions on the Use and Impact of Recreational Cannabis. Int J Environ Res Public Health 17 (2):509. https://doi.org/10.3390/ijerph17020509

Greaves L, Pederson A, Poole N (eds) (2014) Making It Better: Gender Transformative Health Promotion. Canadian Scholar's Press/ Women's Press, Toronto

Greaves L, and Poole N (2018) Gender Unchained: Notes from the Equity Frontier. FriesenPress, Victoria, BC

Greaves L, Poole N, and Brabete AC (2022) Sex, Gender, and Alcohol Use: Implications for Women and Low-Risk Drinking Guidelines. Int J Environ Res Public Health 19 (8):4523. https://doi.org/10.3390/ijerph19084523

Greaves L, and Ritz SA (2022) Sex, Gender and Health: Mapping the Landscape of Research and Policy. Int J Environ Res Public Health 19 (5):2563. https://doi.org/10.3390/ijerph19052563

Gupta GR (2000) Gender, Sexuality, and HIV/AIDS: The What, the Why, and the How. Can HIV AIDS Policy Law Rev 5 (4):86–93. https://pubmed.ncbi.nlm.nih.gov/11833180/

Hillenbrand E, Karim N, Mohanraj P, and Wu D (2015) Measuring Gender-Transformative Change: A Review of Literature and Promising Practices. Care Working Paper. https://hdl.handle.net/20.500.12348/248 (accessed Feb. 2, 2024)

Hilton E, and Wright C (2024) Two Sexes. In: Sullivan A, Todd S (eds) Sex and Gender: A Contemporary Reader. vol 1. Routledge, Milton, pp 16–35

Hoskin RA (2020) Femininity? It's the Aesthetic of Subordination: Examining Femmephobia, the Gender Binary, and Experiences of Oppression among Sexual and Gender Minorities. Arch Sex Behav 49 (7):2319–2339. https://doi.org/10.1007/s10508-020-01641-x

Jacobson D, Glazer E, Mason R, et al. (2018) The Lived Experience of Female Genital Cutting (FGC) in Somali-Canadian Women's Daily Lives. PLoS One 13 (11):e0206886. https://doi.org/10.1371/journal.pone.0206886

Keith L, Herlihy W, Holmes H, and Pin P (2017) Breast Implant-Associated Anaplastic Large Cell Lymphoma. Proc (Bayl Univ Med Cent) 30 (4):441–442. https://doi.org/10.1080/08998280.2017.11930221

Khan A, Tant E, and Harper C (2023) Facing the Backlash: What Is Fuelling Anti-Feminist and Anti-Democratic Forces? https://www.alignplatform.org/sites/default/files/2023-07/align-framingpaper-backlash-web.pdf (accessed 6.11.24)

Lampert FM, Schwarz M, Grabin S, and Stark GB (2012) The "PIP Scandal"–Complications in Breast Implants of Inferior Quality: State of Knowledge, Official Recommendations and Case Report. Geburtshilfe Frauenheilkd 72 (3):243–246. https://doi.org/10.1055/s-0031-1298323

Loewen Walker R (2022) Call It Misogyny. Fem Theor 25 (1):146470012211199. https://doi.org/10.1177/14647001221119995

Manne K (2016) The Logic of Misogyny. https://www.bostonreview.net/forum/kate-manne-logic-misogyny/

———— (2017) Down Girl: The Logic of Misogyny. Oxford Univ Press, Oxford

Marmot M (2007) Achieving Health Equity: From Root Causes to Fair Outcomes. Lancet 370 (9593):1153–1163. https://doi.org/10.1016/s0140-6736(07)61385-3

Marmot M, and Bell R (2012) Fair Society, Healthy Lives. Public Health 126 Suppl 1 (Suppl 1):S4–S10. https://doi.org/10.1016/j.puhe.2012.05.014

Nordmarken S (2023) Coming into Identity: How Gender Minorities Experience Identity Formation. Gend Soc 37 (4):584–613. https://doi.org/10.1177/08912432231172992

Oliffe JL, Kelly MT, Gao N, et al. (2023) Neo-Traditionalist, Egalitarian and Progressive Masculinities in Men's Heterosexual Intimate Partner Relationships. Soc Sci Med 333:116143. https://doi.org/10.1016/j.socscimed.2023.116143

OurWatch (2021) Change the Story: A Shared Framework for the Primary Prevention of Violence against Women in Australia. https://media-cdn.ourwatch.org.au/wp-content/uploads/sites/2/2021/11/18101814/Change-the-story-Our-Watch-AA.pdf (accessed Nov. 4, 2024)

———— (2023) Change the Story. https://www.ourwatch.org.au/change-the-story/

Pederson A, and Tweed A (2004) Registering the Impact of Breast Implants. Centr Excell Womens Health Bull 3 (2):7–8

Ponic P, Greaves L, Pederson A, and Young L (2014) Power and Empowerment in Health Promotion for Women. In: Greaves L, Pederson A, Poole N (eds) Making It Better: Gender Transformative Health Promotion Canadian Scholar's Press/ Women's Press, Toronto, pp 42–58

Richardson SS (2022) Sex Contextualism. PTPBio 4:2. https://doi.org/10.3998/ptpbio.2096

Ritz SA, and Greaves L (2022) Transcending the Male–Female Binary in Biomedical Research: Constellations, Heterogeneity, and Mechanism When Considering Sex and Gender. Int J Environ Res Public Health 19 (7):4083. https://doi.org/10.3390/ijerph19074083

——— (2024) We Need More-Nuanced Approaches to Exploring Sex and Gender in Research. Nature 629 (8010):34–36. https://doi.org/10.1038/d41586-024-01204-3

Rottach E, Schuler SR, and Hardee-Cleaveland K (2009) Gender Perspectives Improve Reproductive Health Outcomes: New Evidence. USAID, IGWG. https://www.prb.org/wp-content/uploads/2021/01/030520210-genderperspectives.pdf (accessed Nov. 4, 2024)

Ryan W (1976) Blaming the Victim. Vintage, New York

Salamon G (2010) Assuming a Body: Transgender and Rhetorics of Materiality. Columbia Univ Press, New York

Schall TE, and Moses JD (2023) Gender-Affirming Care for Cisgender People. Hastings Cent Rep 53 (3):15–24. https://doi.org/10.1002/hast.1486

Schippers M (2007) Recovering the Feminine Other: Masculinity, Femininity, and Gender Hegemony. Theory Soc 36:85–102. https://doi.org/10.1007/s11186-007-9022-4

Steel J (2024) Free Speech, Cancel Culture and the "War on Woke." In: Steel J, Petley J (eds) The Routledge Companion to Freedom of Expression and Censorship. Routledge, Abingdon/New York, pp 232–244

Taghizadeh Kerman M, Brunetti C, Yalfani A, Atri AE, and Sforza C (2023) The Effects of FIFA 11+ Kids Prevention Program on Kinematic Risk Factors for ACL Injury in Preadolescent Female Soccer Players: A Randomized Controlled Trial. Children 10 (7):1206. doi:10.3390/children10071206

TrueChild (2023) Feminine Norms: An Overlooked Key to Improving Adolescent Girls' Life Outcomes? https://static1.squarespace.com/static/599e3a20be659497eb249098/t/59df186a18b27ddf3bb14668/1507793016558/__TrueGirl+%5BWMM%5D.pdf (accessed Nov. 4, 2024)

UN Women (2023) COVID-19: Rebuilding for Resilience. https://www.unwomen.org/en/hq-complex-page/covid-19-rebuilding-for-resilience?gclid=CjwKCAjw5remBhBiEiwAxL2M9x1SLyTN9ANu7gzm6FLFROs0cVSWIlVO60cFP6wKQr-MXj00KmTyahoC_GwQAvD_BwE (accessed Nov. 4, 2024)

Walsh R, and Einstein G (2020) Transgender Embodiment: A Feminist, Situated Neuroscience Perspective. INSEP–J Int Netw Sex Ethics Polit 8 (SI):9–10. https://doi.org/10.3224/insep.si2020.04

World Economic Forum (2023) Global Gender Gap Report 2023. https://www3.weforum.org/docs/WEF_GGGR_2023.pdf (accessed Nov. 4, 2024)

Worthen MGF (2022) This Is My TERF! Lesbian Feminists and the Stigmatization of Trans Women. Sex Cult 26 (5):1782–1803. https://doi.org/10.1007/s12119-022-09970-w

Zielke J, Batram-Zantvoort S, Razum O, and Miani C (2023) Operationalising Masculinities in Theories and Practices of Gender-Transformative Health Interventions: A Scoping Review. Int J Equity Health 22 (1):139. https://doi.org/10.1186/s12939-023-01955-x

15

Sex as a State Effect

Paisley Currah

Abstract Decisions made by a government agency on an individual's sex classification can differ from an individual's understanding of themselves as female, male,
or nonbinary. This disjunction has often been understood as a conflict between sex
construed as a biological phenomenon and gender as a psychological/social phenomenon. This article presents another way to understand conflicts and contradictory outcomes in gender/sex classification. Instead of thinking about problems in the
state management of gender/sex as consequences of getting the definition "wrong"
or as disagreements about "what sex really is" and its relation to gender, this article
suggests that much can be learned about the use of gender/sex in policy by adopting
the methodological starting point that "sex" is not a thing in itself but only an effect
of a particular state action. It focuses on one particular issue: the reclassification
of an individual as female, male, or nonbinary by state actors in the United States.
Examining these policies in the United States, three distinct phases are identified
that reveal the different governing and political rationales at play.

1 Introduction

On the first day of LGBTQ pride month in 2022, the US conservative news website
The Daily Wire released a 95-minute video *What Is a Woman?* featuring the conservative political commentator Matt Walsh. Like so much of the anti-trans rhetoric and images, it generated immense engagement on social media—170 million
views by June of 2023 (Chaitin 2023). The video included interviews with women's
and gender studies scholars, health-care professionals who work with transgender
people, students, people on the street, and later conservative thought leaders. In the
film, when gender studies academics explain that the answer is complicated, when

Paisley Currah (✉)
Dept. of Political Science, Brooklyn College, City University of New York,
Brooklyn, NY 11210, USA
Email: pcurrah@brooklyn.cuny.edu

© Frankfurt Institute for Advanced Studies (FIAS) 2025
L. Z. DuBois et al. (eds.), *Sex and Gender*, Strüngmann Forum Reports,
https://doi.org/10.1007/978-3-031-91371-6_15

they talk about the unknown etiology of gender identity or the many ways gender can be inhabited, the host just nods along politely. The conceit of the documentary, of course, is that it is a ludicrous question to ask—the answer is obvious, and Walsh's wife provides the answer in the last scene: a woman is an adult human female. (Immediately after saying this, she asks her husband to open a jar she's apparently too weak to open.) Further, liberal college students, gender studies academics, and psychologists who work with trans clients are portrayed as some version of pointy-headed intellectuals unable to see the truth right in front of them.

It is apparently self-evident that justice for transgender and nonbinary people often hinges on the meaning of female and male. Arguments for the equality of transgender people rely on the conceptual frameworks of gender and the gender binary, which position humans along a gendered continuum. Arguments in favor of denying transgender people equal rights assert the primacy of "biological sex." I use quotation marks here to indicate that the assumption that biological sex is neatly binary—genitals, chromosomes, gonads, hormone levels, and secondary sex characteristics all perfectly dimorphic and always in alignment—has not been borne out by research; for further discussion, see Chapter 2 and Chapter 5. In New York City, where I live, it is considered an illegal form of harassment to misgender someone consistently. By contrast, in some school districts in Florida, the opposite is true: teachers and staff cannot ask students to refer to them using a pronoun that does not correspond to the sex they (the teacher or staff person) were assigned at birth, even if they have lived in a different gender for decades (Palmer 2023). Arguments about the meaning of gender/sex are most vividly limned in the contrasting positions of two antithetical political groups. Transgender rights advocates rely on explanations about the mutability of gender, the secondary status of sex in determining social roles, the centrality of gender identity, and the nonbinary nature of characteristics generally associated with sex difference. In contrast, the "anti-gender" movement posits that gender identity does not exist, that gender is a synonym for sex, and that sex is binary, given by God or nature, and unchangeable (Butler 2024; Case 2019).

My interest in the problem of sex classification was sparked by the chaotic situation of sex classification policy in the United States. Whether one is classified as female or male—or, more recently as nonbinary—depends on the particular rules within a particular government agency. The same person could be classified as M on a birth certificate and F on a driver's license, be housed with men at a correctional institution and with women at a homeless shelter, be listed as F in Social Security records and an M on a pilot's license (Currah 2022a). One way to explain these contradictions would be to say that some policies reflect a pro-transgender position and others an anti-transgender position. Another explanation, which overlaps with the first somewhat, would be to suggest that for some policy makers, one classification as M or F is fixed at birth for life and for others it is not. My research, however, suggests a third approach. Instead of thinking about problems in the state regulation of the categories of female, male, and nonbinary as consequences of getting the definition of sex "wrong" or as disagreements about "what sex really is" and its relation to gender, I begin with the assumption that "sex" is not a thing in itself but only an effect of a particular state action. It is often not what sex "is" that matters, but what it "does" (Currah 2022a). The scholarship on "what sex is" (or is not) and the predicament faced by people whose gender identity is not traditionally associated with their birth sex is necessary and important (DuBois and Shattuck-Heidorn 2021; Fausto-Sterling 2020; Grant et al. 2011; Hyde et al. 2019; Jordan-Young and Karkazis 2019; Karkazis 2008; Richardson 2013). When trying to understand state policy, a focus on only the rightness or wrongness of a particular definition of sex misses the point. Stating any particular definition is a political decision. As Thomas Hobbes, the great theorist of state power put it: "Authority, not truth, makes the law" (Hobbes 2012:431). Failure to consider all the reasons why a particular definition was put in place might get in the way of grasping the larger state projects within which sex is embedded. Discrepancies in sex reclassification policy—from one period to the next, and from one agency to another, and even now, from blue (progressive) political jurisdictions to red (conservative) ones—are not always best explained as either transphobic or as trans positive. Without understanding why sex is defined the way it is in a particular context, analysis of an issue gets reduced to an overly simple identity politics analytic—a policy is intended to either harm or help transgender people. This is not to suggest that policies do not have transphobic effects and that they do not harm individuals; certainly, some are indeed intended to harm trans people. Nor is it to suggest that such policies should be left in place. I am merely suggesting that digging deeper into the often unstated governing logics may reveal much about the work that sex classification is doing to distribute rights and resources (Currah 2022a:95–96).

In what follows, sex is understood as an effect of governing and politics, not outside or prior to it. For the purposes of this research, I define sex not in relation to gender identity or chromosomes or genitalia—or even gender—but *as the outcome of a decision backed by the force of law.* In contrast to the narrowness of the working definition of sex, gender is defined very broadly to describe norms, narratives, practices, and conventions that arrange bodies, identities, roles, and expressions in hierarchies of difference based on notions of male/female, man/woman, and masculinity/femininity. I identify three distinct states that reveal different governing logics and political rationales at play. The first phase describes the result when, early on, individuals labeled "transsexual" in the language of the day attempted to have their sex marker changed on identity documents and collided with the massive state project of gender discrimination. The second phase looks at the contradictory sex reclassification policies that emerged as states gradually got out of the business of discriminating on the basis of sex and instead used sex classification as a more micro instrument of governing, its definition changing depending on the particular state project at issue. The analysis of the first two stages largely comes from my book, *Sex Is as Sex Does: Governing Transgender Identity* (Currah 2022a). The third phase describes the geographically bifurcated approach to sex reclassification of the contemporary moment in the United States, characterized by liberal policies of recognizing transgender identity claims in progressive "blue" states and conservative policies in conservative "red" states that do not. As will become clear, examining differences between these approaches to sex classification might reveal more about the administration of people and territories and the politics of inequality than staying on the level of arguing over the meaning of sex and its relation to gender identity. I conclude by raising questions about the broader applicability of this approach to questions other than sex reclassification and to political regions outside the United States.

2 No Change of Sex Designation Allowed

When a woman requested that the New York City Bureau of Records and Statistics change the sex on her birth certificate from M to F in the mid-1960s, officials did not know how to respond. They had granted a few such changes in the past but worried that these requests were becoming a trend. They asked a local medical institution, the New York Academy of Medicine, to advise them. Consequently, an ad hoc committee of doctors was formed and met several times before issuing the following recommendation: "The desire of concealment of a change of sex by the transsexual is outweighed by the public interest for protection against fraud." The committee decided that transgender women (they gave little consideration to transgender men) were "still chromosomally males while ostensibly females" (New York Academy of Medicine Committee on Public Health 1966:724). The City adopted this recommendation, said "no" to Anonymous, as she was identified in court records. It should be noted that the New York City policy on sex reclassification on birth certificates has a long and convoluted history, changing five times in five decades (Currah 2022a:31–38; 2022b). For the most part, before the twentieth century, such legal recognition had been impossible to imagine. Certainly, when identity documents were not yet required of everyone, White people who lived in a gender not traditionally associated with their birth sex sometimes could simply move away to "shed their past and adopt a new identity" (Bayker 2019:26). Once identity documents such as birth certificates became mandatory, changing one's sex classification was largely impossible. Only with the gradual and partial accessibility of gender-affirming surgeries in the latter half of the twentieth century did even making such a request seem imaginable.

Although the ad hoc committee of the New York Academy of Medicine was composed only of doctors, the official minutes reveal that the committee spent a great deal of time discussing the *legal* implications of sex reclassification. They noted, for example, that men and women were treated differently by governing agencies (e.g., conscription rules for military service, Social Security benefits, marriage). The rationale for saying "no" to requests for sex reclassification was not simply that sex is a "biologic phenomenon." If a transsexual woman remains "chromosomally male," why did these doctors spend so much time and effort talking about the *legal* ramifications of a sex reclassification request? For example, during this period, when a New York City official asked a federal official about the wisdom of changing sex markers on birth certificates, the official responded by pointing out that "in certain agencies, benefits to women differ from benefits to men" (Council 1965). These discussions reveal sex reclassification policy is never just about the thing we call sex itself but concerns the use of these classifications to further other government ends.

By asking to change their legal sex classification, transgender people inadvertently undermined a key part of the centuries-old state apparatus that ensured the subordination of women. Sex classification—understood as a binary of female and male—had long been cemented into legal architectures that used sex to ensure the denial of rights and resources to women. Transgender people appear as a "residual category," an unanticipated remainder that Susan Leigh Star and Geoffrey C. Bowker define as "that which is left over after a classification is built" (Bowker and Star 1999:300–301; Star and Bowker 2007:274). The problem faced by the apparently unforeseen category of people who want to change their sex can be seen as merely an accident, one to be rectified, or not, by policy. Discussions over the legal implications of transsexuality that Anonymous's petition raised, however, went beyond questions of definition and identity and raised the matter of unequal benefits, rights, and obligations. And in doing so, it partially revealed the contingency of the entire edifice of this classificatory regime.

It is generally accurate to say that women were treated differently than men by state actors, just as it is correct to say that White women were treated differently than Black women. However, suggesting that, for example, women were distinguished *because* they were women exhibits what Karen E. Fields and Barbara Fields describe, in the context of race and racism, a "weird causality," one that generally goes unnoticed (Fields and Fields 2012:17). The notion that women were denied rights and resources (available to men) because of their sex makes the action (i.e., treating women differently) disappear and transforms a characteristic (assumed to be femaleness) into the cause of the action. All the attention goes to sex difference and little or none to the forces and interests insisting on it.

People now referred to as trans or transgender were accidental beneficiaries of the movement to make women equal before the law. The regime of "no" gradually came to an end largely because gender was decommissioned, slowly and in fits and starts, from its role in *formally* distributing rights and resources over the course of the twentieth century. Beginning with women's constitutional right to vote in 1920, continuing with cases throughout the twentieth century that challenged the constitutionality of laws that treated men and women differently, and culminating in the 2015 Supreme Court decision that found bans on same-sex marriage unconstitutional, the barriers to trans people having their sex reclassification requests rejected out of hand crumbled. (In the United States, registering for national defense is one of the very few remaining state programs in which men are treated differently than women.) Saying "no" to sex reclassification petitions had not necessarily been a result of what we now call transphobia: allowing people to change their sex classification would call into question a key instrument (the sex binary) for maintaining gender hierarchies through state policy. Over time, as states got out of the business of "establishing" and enforcing gender hierarchy through the law (e.g., through marriage, benefits, and "protective" legislation), the rationale for saying "no" faded from view.

3 Sex Reclassification Allowed Sometimes

As gender was being disestablished in US law, transgender people began to see some successes in sex reclassification requests. There has, however, never been a uniform policy across all political jurisdictions, agencies, and courts; in the United States, every government agency, from the federal to the municipal, has the authority to define sex and to set the criteria for recognizing changes of sex. As a result, contradictions in sex reclassification policies emerged. It is possible and even likely that a person whose gender identity is not traditionally associated with the sex they were assigned at birth will have more than one sex classification (Grant et al. 2011). Different state actors have different policies—one can never change one's sex classification, one can change it if one could prove that one had gender-affirming surgery, one can change it without body modifications but with an affidavit from a medical professional affirming one's gender identity, one can simply check a box on a form to affirm one's gender identity. This creates a web of contradictions, as revealed by an analysis of sex classification policies in the United States (Currah 2022a): individuals could be categorized as male at a local prison, a female at a federal correctional institution, a male on their birth certificate, and female on their driver's license. A fragmented federal system—where power is distributed between federal, state, and municipal governments—increases the potential for contradictory outcomes.

When I began this research, I assumed that the different rules resulted from different understandings about the definition of sex. After sifting through this chaos and working with officials to change policy, patterns emerged. What I discovered was that in many cases the different rules for changing sex, while ostensibly based on definitions of sex, could be seen to reflect the different projects of particular state actors. While advocates (including myself at different times) put forth memos and position papers on ideal definitions of sex and gender, what sex really *is,* bureaucrats were often more concerned with what sex *does* in a given context. Comparing those policy documents and court decisions, it became clear that sex was not fixed but rather operationalized as a mobile technology of governing. This was especially obvious in legal briefs defending policies and in court decisions because, unlike concise statutes and regulations, legal briefs and judges' decisions usually articulate a rationale for a challenged government policy. The definition of sex in any particular policy depended on the outcome it was needed to produce. For transgender people asking whether their sex markers could be changed, if there could be a collective answer, it would be: "It depends."

I would like to be able to say that this finding came to me from my deep engagement with scholarship on sex and gender, statecraft, bureaucracy, and the law, but that would not be true. I came to this conclusion after working on New York City's sex reclassification policy on birth certificates. In 2005, I was invited to serve on an ad hoc committee of the New York City Department of Mental Health and Hygiene that was considering revising the birth certificate policy. By 2005, the policy had changed somewhat from that 1960s policy—individuals born in New York City who could prove they had "full convertive surgery" would be issued a new birth certificate that had no sex marker on it. For trans people, this policy effectively outed them as trans; people who look at identity documents for their work (e.g., human resources officials and frontline government workers) knew that only transgender people carried a birth certificate with no box for M or F on it. This committee, a reprise in some way of the committee from the 1960s, was composed of transgender advocates, health professionals who worked with transgender people, and public health officials. Advocates wanted M and F designations to be based on gender identity. (A surgeon on the committee argued, unsurprisingly, that surgery was the only true metric of a change of sex.) We drew on contemporary medical knowledge to convince others on the committee that the correct definition of female and male was one based on gender identity. Our arguments were made in the register of truth and expertise, but officials on the committee spoke in the register of governing and politics and were concerned primarily with the practical consequences of any change in the rules (Currah 2022b).

We eventually learned that our proposal to make gender identity the basis of sex classification was shot down when the Department of Health and Mental Hygiene circulated the proposal to other City agencies. Some agencies were fine with it, others were not. Each response depended on how sex classifications were implicated in the agency's work. Eventually, the City's health code was revised such that trans people could be issued a birth certificate with a new sex marker on it, but only if they had "full convertive surgery." For a supposedly progressive political jurisdiction like New York City, this was a disappointment.

A few years later, a trans rights organization brought a lawsuit challenging the birth certificate policy, pointing out that different agencies had different sex reclassification policies, and those discrepancies were "arbitrary," "capricious," and "irrational" (Berkley et al. v. Farley 2011a). In response, the City's attorneys argued that "the existence of different approaches to similar problems does not render and agency's rule irrational" (Berkley et al. v. Farley 2011b). In other words, the rationality of each agency's approach to sex classification depended on its remit, not what sex is in itself. Only when reviewing this lawsuit did it become clear that the City bureaucrats were the real Foucauldians: they understood that sex was not a thing in itself but something instrumentalized differently by different agencies.

The New York City policy deliberations helped me make sense of contradictions in other sex classification policies. Around the turn of this century, there had been a number of court cases challenging the validity of marriages in which one spouse was a transgender person. In almost all of the published appellate cases, the marriage was declared invalid because courts ruled that it was the sex assigned at birth that mattered for the purposes of marriage, even though all these individuals had changed the sex marker on most of their identity documents (Currah 2022a:99–118). Conversely, until recently, almost all states in the United States made it possible for people to change the sex marker on their driver's license. This seems like a contradiction: How could one be F for the purposes of a driver's license but M for the purposes of marriage? These situations are only contradictory if one assumes (a) that sex classifications refer to something outside of the law and (b) that "the state" is singular and unified. "Sex," however, is just as messy and diffuse a concept as "the state." Even within a sovereign entity, "the state" is not singular and hierarchically organized with every agency working in perfect unison with every other agency (Mitchell 1999). Thus, state definitions of sex are also plural. The purpose of a driver's license is to establish a relationship between it and the person who carries it. Sex markers on the document that do not reflect the gender presentation of the person who carries it weaken that connection. An F on the driver's license of a balding, bearded man (like me) hinders the public and private protectors of the security state. The relatively more liberal policies on sex reclassification with regard to identity documents reflect spatial logics, specifically the state's function of watching over individuals and tracking their movements over its territory. Marriage, on the other hand, furthers a very different kind of state project, a distributive one, and is enunciated in the language of property and temporality. Marriage groups individuals into formations that operate over time for the purposes of social reproduction and inheritance—otherwise known as the "family." Transgender spouses and parents upset the fiction that constructs families as biological entities, rather than the legal and social institution.

4 Culture Wars and Identity Politics

Outside of trans communities, few are aware of the Kafkaesque web of contradictions that impact trans people in terms of sex classification. But now that policing of the binary has been transformed from an unremarkable aspect of bureaucratic policy making to a weapon in culture wars, everyone is paying attention. As Donald Trump observed in June 2023: "It's amazing how strongly people feel about that. I talk about cutting taxes, people go like that [mimicking polite applause]. I talk about transgender everybody goes crazy. Five years ago you didn't know what the hell it was" (Palmer 2023).

The third regime describes the bifurcated approach to trans people happening right now in the United States. The division is now geographical, no longer tethered to the rationale of particular agencies. This regime is characterized by progressive policies, which recognize transgender identity claims in "blue" states, and simultaneously conservative policies in "red" states that do not. It's "identity politics" versus "the culture wars." In this regime, the "It Depends" framework has now largely been displaced by the push-pull of forces organized as for or against transgender people. Instead of differences between different kinds of agencies and similarities among similar ones—almost all red state Departments of Motor Vehicles allowed people to change their sex marker at one time—the differences are based on the governing party of the political jurisdiction.

Returning to the birth certificate example, New York City was firmly defending its refusal to amend birth certificates in 2011. However, by 2014, the city not only permitted transgender individuals to change the sex markers on their birth certificates but also removed the requirement for proof of gender-affirming care. What led to this change between 2011 and 2014? This victory was not the result of an agreement between advocates and the City about the ontological foundation of sex. Instead, contingent events made it possible to override some of the particular governing rationalities of the different agencies with regard to reclassifying sex: the election of a progressive mayor in 2013, the growing visibility of the transgender rights movement, and, most importantly, the legalization of same-sex marriage in New York State. Indeed, the issue of same-sex marriage had been raised in discussions several times at the City level. In 1966, the possibility of a person who was assigned male at birth using a new birth certificate to marry a man—and hence "fraudulently" enter into an opposite sex marriage—was a constant worry. (There was no corresponding concern about someone assigned female at birth marrying a man.) These same worries were raised during the 2006 deliberations, but the question of "ersatz" heterosexual marriages was rendered moot when the ban on same-sex marriage ended in 2011. Now sex classifications could take on a new role in the political algorithms of governance. Table 15.1 provides an overview of the three regimes; however, caution should be exercised when viewing the table, as it is difficult (if not impossible) to condense very complicated issues into such a frameworks.

Table 15.1 The three regimes of sex reclassification.

Regime 1: No Way	Regime 2: It Depends	Regime 3: Culture Wars vs. Identity Politics	
Grounded in the patriarchy of modern states	Began ca. 1980/1990s	Began ca. 2015 with the first anti-trans bathroom law	
		Conservative jurisdictions	*Progressive jurisdictions*
Sex classification built into the legal architecture to ensure the subordination of women	Sex used as a mobile tool for governing	"Gender ideology" and the specter of transgender children invoked to foment culture wars	Sex classification, an issue affecting the "transgender community," is a form of identity politics
Sex classification is static	Sex classification reflects what an agency does	Sex is fixed at birth	Classification (M, F, or X) based on an individual's gender identity
Over the course of the 20th century, US states (federal and individual states) stopped discriminating against women	Even conservative jurisdictions had policy areas where sex reclassification was allowed	State governments begin to define sex uniformly across the state, regardless of agency function	Sex classification policies gradually become uniform across the state, regardless of agency function

Instead of being used to maintain gender subordination before the law (the first regime) or to support the work of distinct state projects (the second regime), sex reclassification policies have become a vehicle of partisan politics. In progressive, blue states, they are now used as a political tool to recognize the needs of the constituency that identify as transgender, whereas in conservative red states, sex reclassification policies are used to incite a full-blown culture war. Indeed, in New York City, the rhetoric in support of making gender identity the only necessary condition for changing one's birth certificate highlighted the needs of a particular group. For example, in 2018, Mayor de Blasio announced his support of the policy, which also added nonbinary, this way: "New Yorkers should be free to tell their government who they are, not the other way around" (City of New York 2018). This policy change, however, was not accompanied by one that stopped assigning M or F on birth certificates to infants born New York City.

In other parts of the United States, transgender identity politics has been operationalized by the right wing very differently. Transgender people have been targeted as frauds and potential sex offenders, and over the past several years, conservative legislatures have passed hundreds of "anti-trans" bills and changed policy through executive action to affect legislation in many other policy areas. Bills have been passed that ban gender-affirming care for trans youth; others mandate that participation in school sports be based on assigned sex at birth. This makes it difficult or impossible for students to socially transition in school and compels people to use public bathrooms associated with their birth-assigned sex. Policy makers in Florida and Kansas, for example, have even passed legislation ending the ability of transgender people to change the sex marker on their driver's license and birth certificate.

Trans people were once accidental beneficiaries of the women's rights revolution. Now, in the current political moment in the United States, the tail is wagging the dog: a very targeted attack on a very small element of the population will undermine feminist achievements in general. Consider a small but telling example: In April 2023, the Texas Department of Agriculture announced that its employees must dress in clothes consistent with their "biological gender" (Levine 2023), seen as part of a broader array of anti-transgender policies enacted by the Republican administration of Texas. The policy, however, does not just impact trans individuals; it also reasserts gender norms more broadly. Under this policy, trans masculine people are forced to wear clothes traditionally coded as for women, but so too are cisgender women.

Feminist scholarship has centered on questions of distribution: how one's status as a woman is used to define opportunities and deny resources. As legal categories, it has largely taken M and F as given (Federici 2003; Gordon 1990; Mettler 1998; Mink 1990), with some exceptions (Johnson et al. 2007). Trans activism and scholarship have centered on questions of definition and the politics of recognition: who gets classified as male, who is recognized as female. Putting the two together, however, by paying close attention to the contours of sex classification in particular contexts reveals how and when it is used to distribute inequality, and how state projects deploy it differently depending on their remit. My aim has been not just to locate these issues of classification at the edges of governing and politics but to put them at the center of state policy, to show how these classifications create and maintain inequality between genders, between cisgender and transgender people, between the incarcerated and the free, and between citizens and noncitizens of all classes and races and gender identities. The analytical approach taken here has been *not* to center the injustice of sex classification policies but to figure out why they exist in the first place.

While trans advocates across the United States engage in battles over sex definition, access to sex-segregated spaces, accurate identity documents, and gender-affirming care, our opponents see themselves as resisting much broader social forces. Conservative opposition to reforming sex classification policies reflects animus against transgender people, to be sure. But it also indexes a much larger anxiety about the changes feminism has wrought and the effects of what the right calls "gender ideology." Gender subordination remains one of the organizing principles of domestic life and the workplace. Debates about sex reclassification and access to gender-segregated spaces give conservatives the opportunity to prosecute, yet again, the gender wars in the legal arena. It also uses the institutions of government to put state actors' imprimaturs on traditional visions of normative gender, from deputizing individuals to disputing someone's gender in public spaces to using the bully pulpits of state legislatures to instruct the public on womanhood and manhood. Even as one's status as female or male no longer carries formal distributive consequences in the United States—in that states cannot deny a right or responsibility based on it—the police powers of the state can still be wielded to decide precisely who is female and who is male, and in so doing, enforce traditional gender *norms* that do carry distributive consequences in the economy.

5 Conclusion

For any particular state apparatus at any given moment, the apparently minor issue of rules regarding sex reclassification might matter more than it seems; calling for their reform might involve more changes than we had anticipated, and consequently engender more resistance than initially seems reasonable. That is why it is important to understand, in each particular context (no matter how mundane it may appear), what sex is doing and how that "doing" works in tandem with other systems of social stratification. It is essential that we focus not simply on the recognition of these diverse sex classifications—or on a reparative project of proper classification—but also on the distributional consequences that follow from these classifications, in a variety of social and political settings: reproduction, the family, property, employment, and citizenship.

To return to the question posed at the beginning of this article: What is a woman (or a man)? While individuals can and certainly do hold a number of different beliefs about the answer, definitions promulgated by the state have material effects on people's lives (Cover 1986). From the constellation of entangled traits, properties, and characteristics housed under the carapace of gender/sex, policy makers have the sovereign authority to pick *one* as the deciding factor in determining one's sex classification. In 2023, legislators in the State of Kansas defined male and female as follows: "a 'female' is an individual whose biological reproductive system is developed to produce ova, and a 'male' is an individual whose biological reproductive system is developed to fertilize the ova of a female" (Kobach 2023). If the legal challenges to the law fail, someone who changed the sex marker on a state identity document—right before the law was enacted or decades ago—will have that change reversed on their documents. Certainly, legislators and administrators might draw on scientific knowledge to attempt to justify decisions about which single factor will be dispositive in state determinations of M, F, or X. It is, however, incumbent on scientists to refrain from making the leap from the knowledge they have about a particular property they have studied in humans or animals (e.g., chromosomes, reproductive tissues, hormones) to entangled and contested concepts—sex, gender, female, male—that carry so much weight in the world of policy making.

Acknowledgments I am grateful to Alexandra Brewer, Zachary DuBois, Anelis Kaiser, and Julia Lupp for their insightful comments on an earlier draft.

References

Bayker J (2019) Before Transsexuality: Transgender Lives and Practices in Nineteenth-Century America. https://rucore.libraries.rutgers.edu/rutgers-lib/60594/

Berkley et al. v. Farley (2011a) Petitioners Opening Brief, Supreme Court of the State of New York, County of New York (March 17, 2011).

——— (2011b) Respondents' Memorandum of Law, Supreme Court of the State of New York, County of New York (July 15, 2011)

Bowker GC, and Star SL (1999) Sorting Things Out: Classification and Its Consequences. MIT Press, Cambridge, MA

Butler J (2024) Who's Afraid of Gender. MacMillan, New York

Case MA (2019) Trans Formations in the Vatican's War on "Gender Ideology." Signs 44 (31):639–663. https://doi.org/10.1086/701498

Chaitin D (2023) "What Is a Woman?" Blows Past 170 Million Views on Twitter. https://www.dailywire.com/news/what-is-a-woman-blows-past-170-million-views-on-twitter

City of New York (2018) Mayor de Blasio Signs Historic Legislation Adding Third Gender Category to Birth Certificates. http://www.nyc.gov/office-of-the-mayor/news/501-18/mayor-de-blasio-signs-historic-legislation-adding-third-gender-category-birth-certificates

Council CR (1965) Letter from Charles R. Council, Chief Registration Methods Branch, Division of Vital Statistics, Department of Health, Education, and Welfare, Public Health Service, Washington, DC to Director of Bureau of Records and Statistics, NYC Department of Health

Cover RM (1986) Violence and the Word. Yale Law J 95 (8):1601–1629. http://digitalcommons.law.yale.edu/fss_papers/2708

Currah P (2022a) Sex Is as Sex Does: Governing Transgender Identity. New York Univ Press, New York

——— (2022b) What Sex Does. New York Review of Books. https://www.nybooks.com/online/2022/05/27/what-sex-does/

DuBois LZ, and Shattuck-Heidorn H (2021) Challenging the Binary: Gender/Sex and the Bio-Logics of Normalcy. Am J Hum Biol 33 (5):e23623. https://doi.org/10.1002/ajhb.23623

Fausto-Sterling A (2020) Sexing the Body: Gender Politics and the Construction of Sexuality, Updated edition. Basic Books, New York

Federici S (2003) Caliban and the Witch: Women, the Body, and Primitive Accumulation. Autonomedia, New York

Fields KE, and Fields BJ (2012) Racecraft: The Soul of Inequality in American Life. Verso, Brooklyn

Gordon L (ed) (1990) Women, the State, and Welfare. Univ Wisconsin Press, Madison

Grant JM, Mottet LA, and Tanis J (2011) Injustice at Every Turn: A Report of the National Transgender Discrimination Survey. https://transequality.org/sites/default/files/docs/resources/NTDS_Report.pdf

Hobbes T (2012) The Clarendon Edition of the Works of Thomas Hobbes, vol 4: Leviathan: The English and Latin Texts. 1668 edn. Oxford Univ Press, Oxford

Hyde JS, Bigler RS, Joel D, Tate CC, and van Anders SM (2019) The Future of Sex and Gender in Psychology: Five Challenges to the Gender Binary. Am Psychol 74 (2):171. https://psycnet.apa.org/doi/10.1037/amp0000307

Johnson CM, Duerst-Lahti G, and Norton NH (2007) Creating Gender: The Sexual Politics of Welfare Policy. Lynne Rienner Publ, Boulder

Jordan-Young RM, and Karkazis K (2019) Testosterone: An Unauthorized Biography. Harvard Univ Press, Cambridge, MA

Karkazis K (2008) Fixing Sex: Intersex Medical Authority and Lived Experience. Duke Univ Press, Durham

Kobach KW (2023) Attorney General Opinion No. 2023-2. https://ksag.washburnlaw.edu/opinions/2023/2023-002.pdf

Levine S (2023) Texas State Agency Orders Workers to Dress "Consistent to Their Biological Gender." https://www.theguardian.com/us-news/2023/apr/25/texas-agriculture-department-dress-memo-sid-miller

Mettler S (1998) Dividing Citizens: Gender and Federalism in New Deal Public Policy. Cornell Univ Press, Ithaca

Mink GM (1990) The Lady and the Tramp: Gender, Race, and the Origins of the American Welfare State. In: Gordon L (ed) Women, the State, and Welfare. Univ Wisconsin Press, Madison, pp 92–122

Mitchell T (1999) Society, Economy, and the State Effect. In: Steinmetz G (ed) State/Culture: State-Formation after the Cultural Turn. Cornell Univ Press, Ithaca, pp 76–97

New York Academy of Medicine Committee on Public Health (1966) Change of Sex on Birth
    Certificates for Transsexuals. Bull NY Acad Med 42 (8):721–724. https://pmc.ncbi.nlm.
    nih.gov/articles/PMC1806494/
Palmer E (2023) Trump Says Supporters More Concerned About Transgender Issues Than
    Taxes. https://www.newsweek.com/trump-transgender-supporters-north-carolina-1805783
Richardson SS (2013) Sex Itself: The Search for Male and Female in the Human Genome.
    Univ Chicago Press, Chicago
Star SL, and Bowker GC (2007) Enacting Silence: Residual Categories as a Challenge for
    Ethics, Information Systems, and Communication. Ethics Inf Technol 9 (4):273–280.
    https://doi.org/10.1007/s10676-007-9141-7

Strüngmann Forum Reports

Series Editor: Julia R. Lupp
Editorial Assistance: Aimée Gessner, Catherine Stephen
Lektorat: BerlinScienceWorks

The Nature and Dynamics of Collaboration
Edited by Paul F. M. J. Verschure
DOI: https://doi.org/10.7551/mitpress/15533.001.0001
ISBN: 9780262548144

The Frontal Cortex: Organization, Networks, and Function
Edited by Marie T. Banich, Suzanne N. Haber and Trevor W. Robbins
DOI: https://doi.org/10.7551/mitpress/15679.001.0001
ISBN: 9780262549530

Digital Ethology: Human Behavior in Geospatial Context
Edited by Tomáš Paus and Hye-Chung Kum
DOI: https://doi.org/10.7551/mitpress/15532.001.0001
ISBN: 9780262548137

Migration Stigma: Understanding Prejudice, Discrimination, and Exclusion
Edited by Lawrence H. Yang, Maureen A. Eger and Bruce G. Link
DOI: https://doi.org/10.7551/mitpress/15529.001.0001
ISBN: 9780262548120

Exploring and Exploiting Genetic Risk for Psychiatric Disorders
Edited by Joshua A. Gordon and Elisabeth Binder
DOI: https://doi.org/10.7551/mitpress/15380.001.0001
ISBN electronic: 9780262377423

Intrusive Thinking: From Molecules to Free Will
Edited by Peter W. Kalivas and Martin P. Paulus
ISBN: 9780262542371

Deliberate Ignorance: Choosing Not to Know
Edited by Ralph Hertwig and Christoph Engel
ISBN 9780262045599

Youth Mental Health: A Paradigm for Prevention and Early Intervention
Edited by Peter J. Uhlhaas and Stephen J. Wood
ISBN: 9780262043977

The Neocortex
Edited by Wolf Singer, Terrence J. Sejnowski and Pasko Rakic
ISBN: 9780262043243

Interactive Task Learning: Humans, Robots, and Agents Acquiring New Tasks through Natural Interactions
Edited by Kevin A. Gluck and John E. Laird
ISBN: 9780262038829

Agrobiodiversity: Integrating Knowledge for a Sustainable Future
Edited by Karl S. Zimmerer and Stef de Haan
ISBN: 9780262038683

© Frankfurt Institute for Advanced Studies (FIAS) 2025
L. Z. DuBois et al. (eds.), *Sex and Gender*, Strüngmann Forum Reports,
https://doi.org/10.1007/978-3-031-91371-6

Rethinking Environmentalism: Linking Justice, Sustainability, and Diversity
Edited by Sharachchandra Lele, Eduardo S. Brondizio, John Byrne,
Georgina M. Mace and Joan Martinez-Alier
ISBN: 9780262038966

Emergent Brain Dynamics: Prebirth to Adolescence
Edited by April A. Benasich and Urs Ribary
ISBN: 9780262038638

The Cultural Nature of Attachment: Contextualizing Relationships and Development
Edited by Heidi Keller and Kim A. Bard
ISBN (Hardcover): 9780262036900   ISBN (ebook): 9780262342865
Winner of the Ursula Gielen Global Psychology Book Award

Investors and Exploiters in Ecology and Economics: Principles and Applications
Edited by Luc-Alain Giraldeau, Philipp Heeb and Michael Kosfeld
ISBN (Hardcover): 9780262036122   ISBN (eBook): 9780262339797

Computational Psychiatry: New Perspectives on Mental Illness
Edited by A. David Redish and Joshua A. Gordon
ISBN: 9780262035422

Complexity and Evolution: Toward a New Synthesis for Economics
Edited by David S. Wilson and Alan Kirman
ISBN: 9780262035385

The Pragmatic Turn: Toward Action-Oriented Views in Cognitive Science
Edited by Andreas K. Engel, Karl J. Friston and Danica Kragic
ISBN: 9780262034326

Translational Neuroscience: Toward New Therapies
Edited by Karoly Nikolich and Steven E. Hyman
ISBN: 9780262029865

Trace Metals and Infectious Diseases
Edited by Jerome O. Nriagu and Eric P. Skaar
ISBN 9780262029193

Pathways to Peace: The Transformative Power of Children and Families
Edited by James F. Leckman, Catherine Panter-Brick and Rima Salah,
ISBN 9780262027984

Rethinking Global Land Use in an Urban Era
Edited by Karen C. Seto and Anette Reenberg
ISBN 9780262026901

Schizophrenia: Evolution and Synthesis
Edited by Steven M. Silverstein, Bita Moghaddam and Til Wykes,
ISBN 9780262019620

Cultural Evolution: Society, Technology, Language, and Religion
Edited by Peter J. Richerson and Morten H. Christiansen,
ISBN 9780262019750

Language, Music, and the Brain: A Mysterious Relationship
Edited by Michael A. Arbib
ISBN 9780262019620

Evolution and the Mechanisms of Decision Making
Edited by Peter Hammerstein and Jeffrey R. Stevens
ISBN 9780262018081

Cognitive Search: Evolution, Algorithms, and the Brain
Edited by Peter M. Todd, Thomas T. Hills and Trevor W. Robbins,
ISBN 9780262018098

Animal Thinking: Contemporary Issues in Comparative Cognition
Edited by Randolf Menzel and Julia Fischer
ISBN 9780262016636

Disease Eradication in the 21st Century: Implications for Global Health
Edited by Stephen L. Cochi and Walter R. Dowdle
ISBN 9780262016735

Better Doctors, Better Patients, Better Decisions: Envisioning Health Care 2020
Edited by Gerd Gigerenzer and J. A. Muir Gray
ISBN 9780262016032

Dynamic Coordination in the Brain: From Neurons to Mind
Edited by Christoph von der Malsburg, William A. Phillips and Wolf Singer,
ISBN 9780262014717

Linkages of Sustainability
Edited by Thomas E. Graedel and Ester van der Voet
ISBN 9780262013581

Biological Foundations and Origin of Syntax
Edited by Derek Bickerton and Eörs Szathmáry
ISBN 9780262013567

Clouds in the Perturbed Climate System: Their Relationship to Energy Balance,
Atmospheric Dynamics, and Precipitation
Edited by Jost Heintzenberg and Robert J. Charlson
ISBN 9780262012874
Winner of the Atmospheric Science Librarians International Choice Award

Better Than Conscious? Decision Making, the Human Mind, and Implications For
Institutions
Edited by Christoph Engel and Wolf Singer
ISBN 978-0-262-19580-5